Technology Management in Organizational and Societal Contexts

Andrew Borchers
Lipscomb University, USA

A volume in the Advances in Educational Technologies and Instructional Design (AETID) Book Series

IGI Global
DISSEMINATOR OF KNOWLEDGE

Published in the United States of America by
IGI Global
Information Science Reference (an imprint of IGI Global)
701 E. Chocolate Avenue
Hershey PA, USA 17033
Tel: 717-533-8845
Fax: 717-533-8661
E-mail: cust@igi-global.com
Web site: http://www.igi-global.com

Library of Congress Cataloging-in-Publication Data

Names: Borchers, Andrew, 1957- editor.
Title: Technology management in organizational and societal contexts / Andrew Borchers, editor.
Description: Hershey, PA : Information Science Reference, [2018] | Includes bibliographical references.
Identifiers: LCCN 2017038320| ISBN 9781522552796 (h/c) | ISBN 9781522552802 (eISBN)
Subjects: LCSH: Information resources management--Case studies. | Information technology--Economic aspects--Case studies. | Education--Data processing--Case studies. | Management information systems--Case studies.
Classification: LCC T58.64 .T425 2018 | DDC 658.4/038011--dc23 LC record available at https://lccn.loc.gov/2017038320

This book is published in the IGI Global book series Advances in Educational Technologies and Instructional Design (AETID) (ISSN: 2326-8905; eISSN: 2326-8913)

British Cataloguing in Publication Data
A Cataloguing in Publication record for this book is available from the British Library.

For electronic access to this publication, please contact: eresources@igi-global.com.

Advances in Educational Technologies and Instructional Design (AETID) Book Series

ISSN:2326-8905
EISSN:2326-8913

Editor-in-Chief: Lawrence A. Tomei, Robert Morris University, USA

MISSION

Education has undergone, and continues to undergo, immense changes in the way it is enacted and distributed to both child and adult learners. From distance education, Massive-Open-Online-Courses (MOOCs), and electronic tablets in the classroom, technology is now an integral part of the educational experience and is also affecting the way educators communicate information to students.

The **Advances in Educational Technologies & Instructional Design (AETID) Book Series** is a resource where researchers, students, administrators, and educators alike can find the most updated research and theories regarding technology's integration within education and its effect on teaching as a practice.

COVERAGE

- Educational Telecommunications
- K-12 Educational Technologies
- Online Media in Classrooms
- Bring-Your-Own-Device
- Higher Education Technologies
- E-Learning
- Game-Based Learning
- Collaboration Tools
- Classroom Response Systems
- Virtual School Environments

The Advances in Educational Technologies and Instructional Design (AETID) Book Series (ISSN 2326-8905) is published by IGI Global, 701 E. Chocolate Avenue, Hershey, PA 17033-1240, USA, www.igi-global.com. This series is composed of titles available for purchase individually; each title is edited to be contextually exclusive from any other title within the series. For pricing and ordering information please visit http://www.igi-global.com/book-series/advances-educational-technologies-instructional-design/73678. Postmaster: Send all address changes to above address.

Titles in this Series

For a list of additional titles in this series, please visit:
https://www.igi-global.com/book-series/advances-educational-technologies-instructional-design/73678

Supporting Multiculturalism in Open and Distance Learning Spaces
Elif Toprak (Anadolu University, Turkey) and Evrim Genc Kumtepe (Anadolu University, Turkey)
Information Science Reference • ©2018 • 381pp • H/C (ISBN: 9781522530763) • US $195.00

Innovative Technology Training Frameworks in CALL Emerging Research and Opportunities
Sandra Morales (Newcastle University, UK)
Information Science Reference • ©2018 • 100pp • H/C (ISBN: 9781522537649) • US $125.00

Handbook of Research on Pedagogical Models for Next-Generation Teaching and Learning
Jared Keengwe (University of North Dakota, USA)
Information Science Reference • ©2018 • 454pp • H/C (ISBN: 9781522538738) • US $295.00

Handbook of Research on Mobile Technology, Constructivism, and Meaningful Learning
Jared Keengwe (University of North Dakota, USA)
Information Science Reference • ©2018 • 454pp • H/C (ISBN: 9781522539490) • US $265.00

Deviant Communication in Teacher-Student Interactions Emerging Research and Opportunities
Eletra Gilchrist-Petty (The University of Alabama in Huntsville, USA)
Information Science Reference • ©2018 • 133pp • H/C (ISBN: 9781522527794) • US $125.00

Culturally Engaging Service-Learning With Diverse Communities
Omobolade O. Delano-Oriaran (St. Norbert College, USA) Marguerite W. Penick-Parks (University of Wisconsin – Oshkosh, USA) and Suzanne Fondrie (University of Wisconsin – Oshkosh, USA)
Information Science Reference • ©2018 • 359pp • H/C (ISBN: 9781522529002) • US $225.00

Enhancing Education and Training Initiatives Through Serious Games
John Denholm (The University of Manchester, UK & University of Warwick, UK) and Linda Lee-Davies (The University of Manchester, UK & Wroxton College of Fairleigh Dickinson University, USA)
Information Science Reference • ©2018 • 311pp • H/C (ISBN: 9781522536895) • US $185.00

For an enitre list of titles in this series, please visit:
https://www.igi-global.com/book-series/advances-educational-technologies-instructional-design/73678

IGI Global
DISSEMINATOR OF KNOWLEDGE

701 East Chocolate Avenue, Hershey, PA 17033, USA
Tel: 717-533-8845 x100 • Fax: 717-533-8661
E-Mail: cust@igi-global.com • www.igi-global.com

Table of Contents

Detailed Table of Contents

The current dynamic business landscape has compelled all organizations worldwide to innovate and strategize continually to remain competitive and achieve an edge in the market. Information technology has often made lofty promises to assist such endeavors. This chapter explores the management of large, integrated, and packaged software suites collectively called enterprise systems (ES) to enable innovation and strategic decision making in organizations. In particular, this chapter explains how a globally renowned and highly ranked university manages its enterprise-systems-based IT platform to offer new services, develop new processes, and make astute strategic decisions – all in an attempt to retain and improve its position in the global higher education market.

This chapter examines the merger of two cellular companies (MOBILINK and WARID) for sustainability in the telecommunication sector of Pakistan. During 2011-2013, WARID faced the news of the possible sales of the company due to falling revenues, constant poor network quality, and lower network coverage in rural areas. In 2014, all telecom players participated in the auction for 3G/4G licenses, but WARID did not participate due to its technology neutral license (TNL). Important decisions were taken in 2014 like launching of 4G/LTE services in major cities, US$500 million investment, increase of tower sites, opening of new regional sales offices. These decisions along with presence of TNL made WARID an attractive target for merger with MOBILINK. In 2016, the formal merger was finalized for benefits like synergies in CAPEX/OPEX, fastest 4G network, network reach to rural areas, roll-out of new services like the internet of things (IoT) and mobile banking. The merger proved successful. During the first quarter of 2017, the company generated PKR 38.7 billion in revenues, up from the same period a year before.

Section 2
Innovation in Information Technology

In this section of Technology Management in Organizational and Societal Contexts, our focus turns to how organizations can foster innovation through information technology. In particular the story of software developer PlantFactory addresses two important topics: home bias and the dynamics of competition.

Innovation ecosystems enable small- and medium-sized enterprises to compete with innovations that include the knowledge of others. Integrating external knowledge in innovation is however a complex process, and there are many pitfalls to consider. This chapter analyses a small company that leverages its innovation ecosystem to compete in a dynamic global industry characterized by rapid technological change. The company faces the challenge of creating and commercializing a breakthrough innovation for digital education, with educational content for teachers and easy-to-use authoring tools for modifying this content. As the company attempts to benefit from lead-user experiences, home bias towards knowledge from local users appears as an unexpected hurdle. The analysis of the case encompasses the firm, the industry, and the ecosystem levels, and indicates key concepts for innovation in ecosystem settings.

Section 3
Security

This section of Technology Management in Organizational and Societal Contexts has a single chapter devoted to information security. The number of chapters, however, does not represent the importance of the topic. Indeed, the promise of the digital revolution currently stands in peril due to security challenges by hackers and rogue governments. For the Internet to be a highway of information and payments, security challenges have to be overcome. In this chapter, Daya Sagar Guptaand and G. P. Biswas present a technical presentation on securing cloud storage using an algorithm with a homomorphic frame.

In this chapter, a cloud security mechanism is described in which the computation (addition) of messages securely stored on the cloud is possible. Any user encrypts the secret message using the receiver's public key and stores it. Later on, whenever the stored message is required by an authentic user, he retrieves the encrypted message and decrypts it by using his secret key. However, he can also request the cloud for an addition of encrypted messages. The cloud system only computes the requested addition and sends it to the authentic user; it cannot decrypt the stored encrypted messages on its own. This addition of encrypted messages should be the same as the encryption of the addition of original messages. In this chapter, the authors propose a homomorphic encryption technique in which the above-discussed scenario is possible. The cloud securely computes the addition of the encrypted messages which is ultimately the encryption of the addition of the original messages. The security of the proposed encryption technique depends on the hardness of elliptic curve hard problems.

Section 4
On Higher Education Applications

This section of Technology Management in Organizational and Societal Contexts speaks to information technology applications in higher education. The scope of these six chapters speaks to tremendous potential for Information technology to transform and enhance higher education. Some of these chapters speak of the exploit of a particular technology while others speak to how education methods have been enhanced through information technology.

Using QR codes to access videos in engineering laboratory classes might be a successful way of building a bridge from concrete to digital content. With QR codes placed on an apparatus, students know exactly which video to watch, allowing them to view the videos while performing the experiment or at home when writing the report. Low-cost videos do not require expensive equipment and software, and keeping them short assures a minimum download time for use with smartphones and tablets. The aim of this chapter is to evaluate the importance undergraduate engineering students attribute to these videos and their reaction to the possibility of accessing them with QR codes scanned by a smartphone or a tablet, using access statistics and video viewings to support the findings. Results show students attributed some importance to the videos, and that the QR codes are very helpful as means to quickly and easily access the videos.

Recent research indicates that even though social media networking sites are commonly used in higher education, very little empirical evidence is available concerning the impact of social media use on student learning and engagement. In this chapter, the experience of using Edmodo is analysed as learning aid to support group work in comparison with the level of familiarity, engagement, and frequency of use of social media technologies among university-level computer science students in Finland. The specific focus of the chapter is to examine the reception of the students towards the Edmodo platform. The data was collected through a social media familiarity questionnaire, Edmodo experience questionnaire, and interviews. The main findings are that the cohort was not very familiar with social media at the beginning of the course. This chapter provides experimental evidence that microblogging social networking sites and, more specifically, Edmodo can be used as an educational tool to help engage students more in the use of social media networking sites.

A typical task in language schools is to assign a level of skill to the students with respect to the language they are going to study. For this, the students are evaluated through a set of level tests that measure the student's competence. Normally, the assessment consists of an oral competency test, a written competency test, and a listening comprehension test. A characteristic of this type of test is the possibility of evaluating them remotely (usually they are tests with questions where every question has associated a set of possible answers where only one of the answers is correct) without the necessity of the presence of any evaluator. For this reason, it is possible to create computer applications that automate the evaluation and management of test results. This chapter presents a computer application that has been created with the aim of automating the process of management and evaluation of level tests for students of Spanish as a foreign language. The application has been tested at the University of Zaragoza.

Changes and developments in technology are also reflected in user behavior. Therefore, companies that monitor and analyze user behavior to gain competitive advantage will be one step ahead of others. Google Analytics is one of the most popular web analytics tools that perform this function. This chapter observes changes in user habits between 2011 and 2017 by using the Google Analytics data of Enocta Academic Education Platform (EAEP) that offers course content and learning management system (LMS) services for organizations and universities in Turkey. Thus, it advises to LMS providers in terms of improving their software considering changes in user habits. In addition, different approaches of improving e-learning services are discussed. This case study shows that Google Analytics, one of the web analytics tools, can provide useful information for e-learning platform or LMS providers in order to understand behaviors and needs of the visitors as well as their own strengths and weaknesses.

Project-based learning (PBL) is a powerful tool for teaching that helps students to get the best in terms of ratio effort/learning outcomes, especially in studies with a very practical basis, such as university degree studies in engineering. A way of getting even more out of this is by means of the adaptation of methodologies from different knowledge areas, because this allows the launch of innovative ways of working with certain guarantees of success from the very first moment, and at the same time to integrate skills from different fields within a shared context. Furthermore, it helps to put into practice some transversal competences, which are very useful for future professionals. The chapter also includes some case studies on the successful adaptation of different methodologies coming from different fields such as graphic design, biology, and social sciences in the context of a university engineering degree in industrial design and product development.

The objective of this chapter is to determine participants' perceptions about the characteristics that students must have to be successful in a blended learning environment. The authors adopted a qualitative approach in which 91 learners participated. Data collection was carried out through guided interviews over three months. The results of the research emphasized the need to have a good level of English and new digital skills as essential requirements to study an English degree in a blended learning modality. Participants were strongly aware of the necessity to master synchronous and asynchronous technological tools to communicate online with their virtual classmates, and to do assessment and self-assessment e-activities. Learners also considered that it was essential to be mature enough to be able to reflect and evaluate their own learning. Results showed that students must be organised and tenacious, and must have the ability to innovate and collaborate inside the created e-learning community.

Preface

INTRODUCTION

In the years since Nicholas Carr's classic 2003 article, "IT Doesn't Matter", each of us has witnessed the growing use of information technology (IT) in a wide range of organizational and societal contexts. Although Carr's article was hotly contested at its writing, his observation of information technology ubiquity is hard to argue with. Indeed, literally billons of people today have smartphones with powerful apps running on cellular networks around the world. What is more easily argued is his assertion that the strategic importance of IT has declined. Clearly, firms continue to invest large sums of money on increasingly sophisticated IT projects. Indeed, this author would update Carr's statement of "IT Doesn't Matter" to a more nuanced "While IT may not matter in differentiating the firm for long, the absence of IT does matter and can hobble a firm." Certainly, most organizations have determined that at least some level of up-to-date IT is a cost of doing business if they hope to remain competitive.

Gilbert and Ward (2012) provide an interesting response to Carr. They wrote in a period where a sea of contrary responses argued that Carr's points were overblown, overbroad and that knowing how to use IT was indeed a rare skill. Gilbert and Ward's works suggests three disagreements with Carr, but interestingly their data analysis suggests that Carr may indeed be right.

Gilbert and Ward's (2012) disagreements are as follows. First, they argue that IT innovations are not necessarily available to all. Although the window of opportunity to differentiate one's self may be short, organizations are constantly seeking new technologies before their rivals can adopt it. Indeed, IT innovations may not have disappeared, but their lifespans may simply have been shortened. Second, Gilbert and Ward argue that while IT technology may be increasingly ubiquitous, IT execution is not. While even commercial off the shelf software (such as ERP packages) is available for all to purchase, the actual implementation of this software requires significant skill and success is not guaranteed. In this volume, Abdullah Alkhuraiji's chapter on the impact of top management characteristics on enterprise systems implementation is instructive on this point. Third, they argue that Carr's use of analogy does not constitute proof. Carr's work, for example, refers to electricity as compared to information technology. The analogy is limited, however, as electricity and IT have fundamentally different characteristics.

Most notable about Gilbert and Ward's work is their data analysis on the importance CIOs put on information technology in firms in various industries. Working with 104 participants, they compared responses between executives in high, medium and low innovation industries. They found significant differences in the importance of software development and managing application portfolios in high innovation industries. These results certainly seem to support Carr's contentions.

Irving Wladawsky-Berger, writing ten years after Carr, asserts that we live in a post-technology era. This is not to say, that technology is irrelevant. Rather, the components of IT, such as disk drives, CPUs, operating systems and the like, are no longer the focus for any but a handful of technologists. Indeed, as this author teaches business schools students in information systems classes, his focus has shifted from technology details to what business results one can achieve with technology. This includes concepts like lean business processes and data driven decision making.

Irving Wladawsky-Berger summarizes the ten years following Carr's article as a move to an open, integrated, heterogeneous IT infrastructure. He notes three key shifts. First, increased standardization has largely removed proprietary systems and architectures. Second, horizontal integration of business processes (via enterprise information systems) has extended across the entire supply chain. Third, utility computing has, as noted above, shifted the focus off of hardware and onto applications. A careful reading of this volume will certainly support each of these points. Yet Wladawsky-Berger does not believe the transition is complete. He predicts there are decades to go, leaving space for innovators to bridge technology and business goals for many years to come.

For managers with technology responsibilities (seemingly all managers today!) there is a critical need for an understanding of the potential and risks of information technology. As Evans and Wurster pointed out in 1997, technology has the potential to make venerable firms vulnerable and to undo long standing successful business models. At the same time, employed well information technology can create new enterprises with value that exceed that of well-known competitors. For example, at the time of this writing the market capitalization of long time automotive makers Ford and GM are trailing that of technology driven Tesla.

CONTEXT OF THIS VOLUME

This volume provides a collection of research work from a global group of scholars. Originally published as articles in the *Journal of Cases on Information Technology* or another IGI Global publication, each chapter has been revised, expanded and brought up-to-date with new references. The chapters have been organized into four sections with introductory words for each section.

CLASSIC THEORETICAL FRAMEWORKS

Carr's arguments notwithstanding, as one considers technology management in organizational and society contexts, it is hard to ignore the essential role that technology plays in organizational strategy and innovation. A major challenge for leaders comes in determining how to best employ technology to achieve organizational goals. Leaders are well advised to consider three useful frameworks for how new technology is accepted and integrated into organizations and society as a whole. The following sections speak to three classic frameworks: diffusion of innovation, disruptive innovation and the technology acceptance model. In pondering the application of new technology, these frameworks can provide useful guidance to leaders.

Diffusion of Innovation

The first framework relevant to this discussion is diffusion of innovation. Everett Rogers, a communications professor, popularized the idea beginning in 1962. His book, now in its 5th edition, has been cited over 88,000 times according to scholar.google.com. Pictured in Figure 1, the theory explains how new ideas and technology are accepted. Combined with the concept of the "Tipping Point" (popularized by Malcolm Gladwell, 2006), leaders can better understand how difficult it can be to launch new ideas until a critical mass is reached. Further, this framework reminds one that significant numbers of users may be slow to ever switch to a new technology. Note also, that Rogers identifies four key elements that influence adoption of new ideas: the innovation itself, communications channels, time and the social system.

Disruptive Innovation

A second framework that leaders need to consider in bringing new innovation to organizations and society is disruptive innovation. First introduced by Clayton Christensen in *The Innovator's Dilemma* in 1997 and updated most recently in 2016, the idea is pictured in Figure 2. In this figure, one can see how disruptive technology moves rapidly from poor performing to high performing compared to traditional technology. Classic examples include computer disk drives and digital cameras.

Disruptive innovation is perhaps better understood by comparison with sustaining innovation. The latter do not significantly alter markets. For example, the move from

Figure 1.
Source: Rogers Everett [Public domain], via Wikimedia Commons

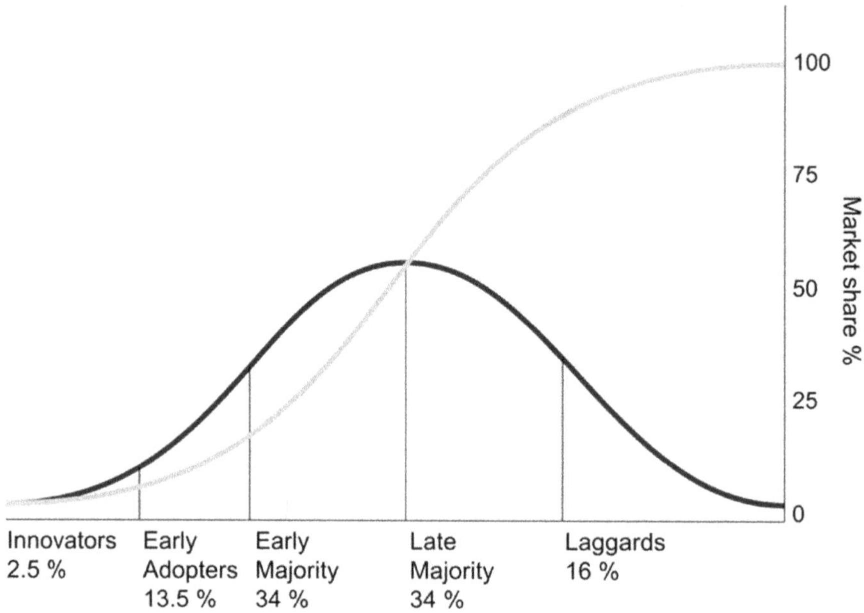

horse driven carriages to early automobiles was not a significant market change given the relatively high cost of vehicles and the wealthy customers involved. Henry Ford's Model T, is a different story, as his product dramatically dropped the cost of automobiles to a level that middle class customers could afford one. This change, along with installment purchasing, made for a disruptive innovation in personal transportation.

Technology Acceptance Model

A third and final theoretical framework useful to this discussion is the technology acceptance model (TAM). Developed by Davis, et al (1989), this framework is shown in Figure 3. The model is straightforward, but powerful. It describes how users come to accept and utilize new technology. User decisions on new technology are driven by two factors: perceived usefulness of the new technology and perceived ease of use. One can use TAM to consider successful innovations (such as smartphones) and unsuccessful innovations (such as VCR recording). Further research has proposed a TAM 2 and TAM 3.

Taken as a whole, these three theoretical foundations are quite useful in understanding the acceptance of innovation in organizational and societal contexts.

Figure 2.
Source: Megapixie [Public domain], via Wikimedia Commons

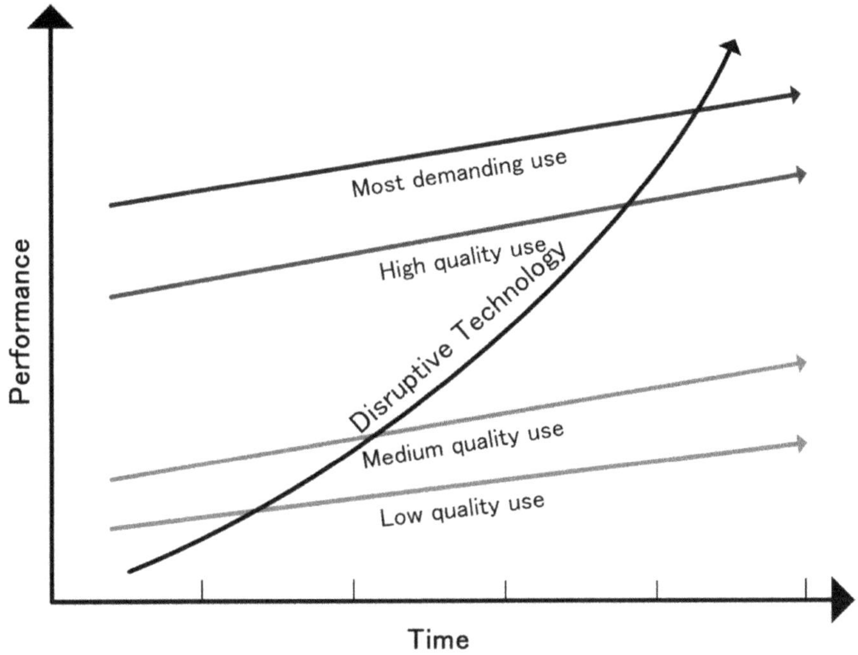

Figure 3.
Source: Nippie (Own work) [CC BY 3.0 (http://creativecommons.org/licenses/by/3.0)], via Wikimedia Commons

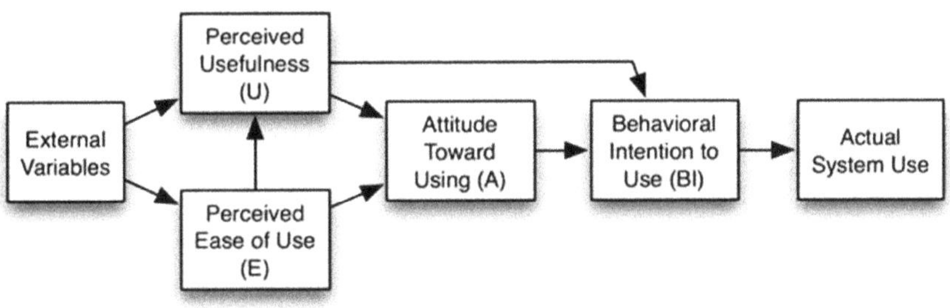

ORGANIZATION

This volume is organized into four sections. Section 1 addresses the strategic impact of information technology with a set of three case studies from disparate parts of the world, Australia, Saudi Arabia, and Pakistan. Section 2 presents a single case study, PlanetFactory, that speaks to the power of information technology in the development of a new enterprise in Spain. Section 3 presents a single chapter on the critical subject of security. Section 4 contains six chapters on innovative applications in higher education.

REFERENCES

Carr, N. (2003). IT Doesn't Matter. *Harvard Business Review*. PMID:12747161

Christensen, C. M. (2016). *The innovator's dilemma: when new technologies cause great firms to fail*. Harvard Business Review Press.

Davis, F. D., Bagozzi, R. P., & Warshaw, P. R. (1989). User acceptance of computer technology: A comparison of two theoretical models. *Management Science*, *35*(8), 982–1003. doi:10.1287/mnsc.35.8.982

Evans, P. B., & Wurster, T. S. (1997, September). Strategy and the new economics of information. *Harvard Business Review*, 65–82. PMID:10170332

Gilbert, A. H., & Ward, S. G. (2012). Does" IT Doesn't Matter" Matter?: A Study Of Innovation And Information Systems Issues. *The Review of Business Information Systems (Online)*, *16*(4), 177. doi:10.19030/rbis.v16i4.7437

Gladwell, M. (2006). *The tipping point: How little things can make a big difference*. Little, Brown.

Rogers, E. M. (2003). *Diffusion of innovations*. Free Press.

Wladawsky-Berger, I. (2013, May). *'IT Doesn't Matter:' 10 Years Later*. Retrieved from: https://blogs.wsj.com/cio/2013/05/24/it-doesnt-matter-10-years-later/

Section 1

Strategic Impact of Information Technology:

Introduction

The opening section of *Technology Management in Organizational and Societal Contexts* addresses the strategic impact of information technology in organizations. The first two articles share a common focus on enterprise wide systems, identified as enterprise systems (ES), enterprise information systems (EIS) and enterprise resource planning systems (ERP). The third article speaks to organizational strategy for firms in the Pakistan telecommunications industry.

In "The Influence of Top Management Characteristics on the Success of Enterprise Information Systems," Abdullah Alkraiji employs a qualitative case analysis method to study two government agencies in the Kingdom of Saudi Arabia (KSA) with contradictory approaches to enterprise systems. The authors identify top management characteristics in the two organizations with respect to leadership styles and leader's views of technology.

In "Managing IT to Innovate and Strategize in the Higher Education Sector: The Role of Enterprise Systems," Prithvi Bhattacharya also utilizes a case study approach to see how enterprise systems enable an organization to compete. The organization studied is a major Australian university with diverse faculties. Prithvi chronicles the way that an enterprise system enables leaders to make critical decisions in the competitive marketplace of higher education.

In "Synergy for Sustainability in the Upcoming Telecommunications Revolution: The Case of a Developing Economy," Abdul Rafay and Arsala Khan chronicle the development of two cellular companies as they navigate the changing technological landscape of 4G, 5G and the Internet of Things (IOT). Beyond technological change, strategic realities brought the two firms to successfully merge their operations in an effort to better compete.

Chapter 1
The Influence of the Top Management Characteristics on the Success of the Enterprise Information System

Abdullah Ibrahim Alkhuraiji
King Fahd Security College, Saudi Arabia

ABSTRACT

The purpose of this chapter is to investigate the effect of top management characteristics on the overall success of enterprise information systems. It aims to offer an in-depth understanding of the necessary characteristics required for the implementation of an enterprise resources planning system. It applies a case study methodology of two contradictory government organizations in the Kingdom of Saudi Arabia: one with successful implementation and the other with a complete failure. The results expose various characteristics such as those related to business leadership and abilities and technological cognition and awareness. The results can be developed into a strategy to enhance awareness as well as top management participation.

INTRODUCTION

Overcapacity, reengineering, globalization, and dealing with constant change are some of the prime reasons why many organizations implement Enterprise information systems (EISs) (Loonam & McDonagh, 2005). EISs can replace multiple independent systems in an organization to allow sufficient interaction between the delivery of high-quality products and services to stakeholders within the optimal

DOI: 10.4018/978-1-5225-5279-6.ch001

time (Loonam & McDonagh, 2005). Through the streamlining of workflow, EISs can help organizations to achieve profitable growth, such as ensuring competency and competitiveness, increasing productivity, reducing expenses, and improving decision-making quality and resource control (Howcroft et al., 2004). In addition, EIS software packages are built based on a suite of best practices afforded to implementing organizations; this would help organizations to efficiently improve their overall performance (Liang et al., 2007).

Despite such attractive strategic importance for EISs, the literature exposes the fact that obtaining EIS benefits is not as straightforward as organizations would like to believe (Buonanno et al., 2005; Howcroftet al., 2004; Yen Teoh &Pan, 2008). One common challenge is that EISs limit the overall flexibility of the organization's operations owing to the technical rigidity of the EIS by nature (Yen Teoh &Pan, 2008). This requires the implementer organizations to align their business practices—including their strategies and cultures—to the EIS practices with limited customizations. In some cases, the standardization of practices—according to predefined standards—cannot provide a competitive advantage (Mayère et al., 2008; Yen Teoh &Pan, 2008); therefore, cultural factors should be taken into account during EIS selection (Mayère et al., 2008), as the organizational culture becomes strategic capital and thus a source of competitive advantage (Yen Teoh &Pan, 2008).

Scholars have highlighted multiple and multi-dimensional influences in EISs project success, for example, top management (TM) support has been repeatedly emphasized as a direct predictor of project success (Liu et al., 2015). The role of TM support is not only allocating the required resources, but most importantly their participation, involvement and commitment throughout the project implementation stages (Dong et al., 2009). Although top management (TM) is consistently identified as the key social player in reaping the potential success of EISs and should not be a cause of disagreement between the researchers, nonetheless, there is a need to understand the top management characteristics (TMCs) that promote TM involvement as well as strong commitment to and support for successful implementation. According to Armstrong and Sambamurthy (1999) and Hambrick and Mason (1984), TM is defined as the person(s) with overall responsibility for the organization (Armstrong & Sambamurthy, 1999; Hambrick & Mason, 1984). Existing literature on EIS project implementation tends to focus primarily on the different required supports by TM, leaving a gap in knowledge concerning the TMCs and their effects on successful EIS projects.

For example, researchers should ask why there is strong participation, involvement, and commitment from TM during EIS project implementation in some organizations, whereas there are fewer supportive actions in other organizations. This means gaining an understanding that the necessary supports and dissimilar needs from TM might be less important than understanding the TMCs and their influence on the success of EIS projects in order to stimulate these characteristics for more involvement and stronger commitment. In so doing, there is a need to investigate TMCs that likely can promote the involvement and strong commitment, support, and interaction of TM during the implementation of EIS projects.

The purpose of this study is to develop a comprehensive empirically set of TMCs influencing the successful implementation of Enterprise Resources Planning (ERP) system. The remainder of the paper is organized as follows. First, the influences of the socio-technical on the success of EIS project is explained followed by an introduction to the ERP system. Then, the paper describes the research method adopted for this study. Next, the paper presents the data analysis, results, and discussion. The paper concludes by providing some implications on theory and practice, recommendations, and a direction for future research.

The Influences of Socio-Technical Aspects on the Success of EIS Projects

Improving the successful implementation rate of EISs and the management practices required are key focal areas for organizations today. EISs continue to be a phenomenon of considerable research interest owing to their significant and profound effects on contemporary global, distributed and networked societies and organizations (Van Wart et al., 2017; McLeod & Doolin, 2012). EISs have become more and more sophisticated and integrated and, as a result, the possibility of consequences that are unpredictable or unintended has increased. Thus, the nature of the success of EISs that has previously been assumed is now continually challenged by aspects that suggest EIS project development needs to be more contingent and tailored to specific situations (Gasson, 2006). Scholars (e.g. Sawyer & Jarrahi, 2014; McLeod & Doolin, 2012) have argued that misunderstandings concerning the complexity and multi-dimensional nature of the development and implementation of EISs, as well as the 'cause-and-effect relationship' between interrelated and interdependent social and technological determinants, make the success of EIS projects an elusive goal. Liu et al. (2015) explained that only a third of EIS projects succeed in organizations while the majority of organizations enjoying such success fail to recoup a return on their investment.

Gasson (2006) similarly confirmed the need to consider more contingent and situational dimensions of EIS projects. Other studies, such as those by Gregor and Jones (2007) and Khan et al. (2010), for example, have also argued that the implementation of EISs should involve, not only prescriptions for designing technological tools, methods and applications, but also an understanding of organizations' socio-technical systems so that methodologies and interventions can be designed accordingly. McLeod and Doolin (2012) stressed that the boundaries between technological and social aspects are less clear-cut in the implementation of an EIS; therefore, the implementation should be conceptualized as a complex socio-technical process (Orlikowski & Iacono, 2001). According to Mayère et al. (2008), identifying the various dimensions of the technical and social aspects of organizations and their impact on EIS implementation is a major and ongoing open problem. The underlying premise of the studies concerning socio-technical factors involves achieving a better understanding the complex scenarios regarding relationships among the social environment of an organization, the social enablers enacting business processes, and the system that supports such processes (McLeod & Doolin, 2012).

Greasley and Wang (2016) asserted that the poor level of adoption of EISs (i.e. ERP systems) is often considered to be linked to a loss of interaction between the system itself and the enterprise social of the systems. Appelbaum (1997) repeatedly emphasized that integrating an EIS intervention with organizational development into a total complex system is a complicated and delicate process that requires expertise and judgement in social, technological and systems in terms of both theory and practice. Appelbaum (1997, p. 452) described this phenomenon thus: "It is complicated because of the many areas and systems involved and it is delicate because of the dynamic relationship among these systems within the environment." Accordingly, the adoption of an EIS necessitates the creation of a flexible organizational change model designed from a socio-technical basis in order to fit the complex social networks of the organization (Appelbaum, 1997).

One key and essential stakeholder during the implementation of an ERP system is the TM (Liu et al., 2015). The TM is responsible for setting the organization's overall direction while ensuring that major organizational objectives are achieved. In this regard, one significant duty is to decide what achievements it seeks to accomplish and how the organization will grow while going forward (Tihanyi et al., 2000). The TM also controls decision making, which affects everyone and every entity in the organization; therefore, understanding the effects of any decisions helps the TM to adjust and carry out changes as and when necessary (Liu et al., 2015). In addition, the TM acts as an interface between the organization and the external environment to establish goals and make strategic decisions that encourage the organization and employees to grow. Accordingly, the TM is considered a key person in an organization in providing strategic context and guidance (Liu et al., 2015). Recently, the role of the TM has become more complex, owing to today's fast-paced, dynamic, and changing business environment (Bansal & Agarwal, 2015; Loonam & McDonagh, 2005; Momoh et al., 2010) and this new environment requires smart and innovative practices centered on effectively optimizing the exploitation of all available tools (Bansal & Agarwal, 2015).

For EIS projects, the adoption decisions made by the TM become exceptionally important because they must be timely; they also require major startup investment, the settings of precedents and the commitment of resources (Van Wart et al., 2017; Liu et al., 2015). Poor adoption decisions and initialization choices have enormous down-stream effects, particularly in those government organizations where failure has been widely reported (Anthopoulous et al., 2016). EISs have had a huge impact on government organizations in in terms of increasing their efficiency, effectiveness, inclusiveness and transparency (Anthopoulous et al., 2016). According to Elbanna (2010), one essential task for the TM during the implementation of an EIS project is to act as an interface between the project team and routine functional operations in order to increase the acceptance and support of the project stakeholders. Scholars of socio-technical research in EIS projects have agreed that TM support is critical for the success of EIS project and that this can therefore be considered as a direct predictor of success (Liu et al., 2015). One main reason for this is that EIS projects are highly task-interdependent and cross-functional and so the role of the TM is critical in facilitating decision-making across stakeholders, groups and sponsors, and also in enhancing user participation and interdepartmental communication (Liu et al., 2015). Another reason is that EIS projects involve a lengthy implementation process with complicated architectures and modules (Aloini et al., 2007). As a result, continuous TM support is crucial for successful implementation (Liu et al., 2015).

Dong et al. (2009) categorized TM support into three critical components: resource provision, participation and involvement. EIS projects in particular require the establishment of dynamic methods and contexts for resource allocation to facilitate the flow of resources. The TM's presence and visibility throughout the implementation stages and processes are crucial in terms of repairing dysfunctional organizational structures or processes, and in solving management problems. The involvement of the TM during the implementation affects, among other things, the motivation felt by employees, and so the TM needs to provide sincere support and make strenuous efforts to connect with and show an interest in the staff while establishing the project as the top priority. Other ERP studies have stressed different supportive actions that should be adjusted by the TM. These include change management (i.e., actions related to fostering the organization's receptivity of a new system (McAdam & Galloway, 2005)) and building strong and strategic partnerships with functional areas (Liang et al., 2007; Momoh et al., 2010). In spite of the substantial amount of literature indicating the importance of TM support in the introduction of ERP systems into an organization, thus far, there is little insight into how EIS project teams and members can obtain TM support if it is not forthcoming. In addition, certain scholars (e.g. Liu et al., 2015; Schmidt et al., 2001) confirmed that acquiring the support, participation and involvement of the TM is often considered outside the control of the EIS project team. For example, a TM with a poor IT background, skills and vision might have little real interest in IT projects (Van Wart et al., 2017; Liu et al., 2015). Furthermore, contemporary organizations are multiple and project environments may compete so that the attention of the TM may be focused on other areas or projects (Elbanna, 2010).

Van Wart et al. (2017) reported that, according to the estimation of experts, the number of EIS projects that are underperforming in government sectors worldwide are even higher than the estimated percentage that is reported. One reason for this is that there is a lack of TMs with skills in technology across government areas (Van Wart et al., 2017). Avolio et al. (2014) asserted that the relationship between the TM and IT is recursive and therefore scholars should not only consider how advanced IT systems mediate and influence TM processes but, most importantly, should also describe how TMs influence the adoption and implementation of IT systems. Avolio et al. (2014, p.105) stressed that the interaction between TMs and advanced IT systems 'remains at the very nascent stages of development'. Similarly, Van Wart et al. (2017, p.528) confirmed the need to study 'antecedent conditions for TMs in facilitating the adoption of EISs. Van Wart et al. (2017) asserted that the literature lacks an exploration of the traits, skills, styles and behaviors that lead some TMs to do a better job at EIS adoption and implementation than others do.

ERP System

ERP system (i.e., one of the most common examples of EISs) continues to be a source of great concern and frequent frustration for both practitioners and researchers. ERP systems are a complex piece of software; therefore, many implementations reported have been found to be difficult, lengthy, and over budget, with some terminated before completion (Buonanno et al., 2005; Mayère et al., 2008; Momoh et al., 2010) or otherwise noted as failing to achieve their business objectives—even a year after implementation (Mayèreet al., 2008). According to Mayère et al. (2008), the paradoxical abilities of ERP systems and the interaction with the majority of the functions in the organizations mean that the ERP system is a risky project. Mayère et al. (2008) illustrated the paradoxical abilities of ERP systems, stating that they are, on the one hand, designed based on best practices to improve organizations' daily activity functions; on the other hand, they are intended to be configured and customized according to the specific requirements, dedicated tools, and methods of each organization.

Organizational fit and adaptation are major factors because the ERP system is built with re-determined business processes (Mayère et al., 2008). The complexity of ERP implementation also comes from the fact that its influence on processes, structures, cultures, and politics is much broader and more profound than less complicated technologies (Liang et al., 2007). The early success of one ERP function or module might later become a failure when moving targets (Liang et al., 2007; Momohet al. 2008); owing to the fact that the total business value of the ERP system in an organization cannot be fully realized until ERP implementation is complete (Liang et al., 2007). Previous research on EPRs (e.g., Buonanno et al., 2005; Dong et al., 2009; Elbanna, 2013; Liang et al., 2007; Lyytinen et al., 2009; Sarker & Lee, 2003; Somers & Nelson, 2004) advocated that strong managerial and strategic competence are required in order to achieve fit between the business peculiarities and the system itself and to avoid undesirable organizational impact induced throughout the implementation.

Existing literature on ERP systems stresses different supportive actions that should be adjusted by TM, such as resources provision (Dong, Neufeld, & Higgins, 2009; Loonam & McDonagh 2005), change management (Donget al., 2009; McAdam &Galloway, 2005), and building strong and strategic partnerships with functional areas (Liang et al., 2007; Momohet al., 2010). In spite of substantial literature indicating the importance of TM support to the introduction of ERP systems in an organization, thus far, little insight into the TMCs needed to maintain the required support during the implementation stages—which in turn means revealing a weak understanding of this area—has been recognized.

RESEARCH METHOD

An exploratory interpretive approach was conducted to engage the supported TMCs during the implementation of ERPs. A multiple-case study methodology was selected to carry out this research. One potential advantage of conducting a case study is the opportunity to garner better insight into and understanding of the issues related to the critical managerial actions and then to interpret how such actions might influence employees 'attitudes and behaviors (Eisenhardt, 1989; Irani et al., 1999). The researcher adopted the empirical inquiry of qualitative studies by Bryman (2015). The starting point was through the puzzling assumptions in the literature, where different studies (e.g., Donget al., 2009; Liu et al., 2010; Sarker & Lee, 2003) suggested a straightforward causal relationship between TM support actions and ERP success. Elbanna (2013) stressed the importance of constant and consistent TM-level support of ERP implementation in the organizational context, which often is recognized as being outside the IT project teams' control (Elbanna, 2013). It is not clear how we can guarantee constant and consistent TM support. One method might concern understanding TMCs in relation to EIS implementation. Accordingly, the author conducted two case studies in the Kingdom of Saudi

Arabia (KSA): one with successful implementation of an ERP system and another demonstrating a complete failure. Through the two contradictory cases, the author was able to develop a near-to-practice understanding of the aforementioned issue.

Site Selection

The researcher looked for two contradictory government cases in the KSA: one with a successful story and another with a complete ERP failure. Government organizations in the KSA have similar regulations and legislations across all levels. The researcher exploited his close relationships with some chief information officers in some government organizations in Riyadh (the capital of the KSA) to establish the most appropriate case sites. The search process led the researcher to several suitable case sites for conducting the study. The researcher contacted the IT department in each organization in an effort to garner the necessary consent for conducting the case site study. The IT departments were given enough information concerning the research as well as further assurance that their information would be treated with high confidentiality and anonymity. The researcher contacted the first two contradictory case sites; they agreed to volunteer for the study.

Data Source

The researcher utilized multiple data sources, including focus groups, semi-structured interviews, and unstructured interviews. As the researcher comes to the case sites free of hypotheses or preconceptions with the objective to investigate and describe the research phenomena, the phenomenological approach was conducted (Sanders, 1982). A phenomenological approach is concerned with the study of experience from the perspective of the individual and gaining insights into his or her motivations and actions through intensive information probing based on subjective interpretation (Sanders, 1982). Phenomenology does not place strong emphasis on the number of participants; rather, attention is directed toward the quality of participants in enriching the study with in-depth information (Donget al., 2009); hence, careful selection of participants is required (Sanders, 1982). For reasons of confidentiality, the participants are kept anonymity throughout the paper. The researcher selected the key social players in the ERP project across the two case sites; the total number of informants to complete the focus group or interviews was 13 (9 in Organization A and 4 in Organization B; see Table 1). The researcher had a meeting with four persons only in Organization B because the TM and ERP vendor had left the organization. The data collection journey took place from September 2015 and was ongoing until December 2015.

Table 1. General information about the informants

Organization	No.	Informants' General Information			
		Coded Name	Position	Experience by Year	Qualification
A	1	Informant 1	General manager	23	Master in CS
	2	Informant 2	Business developing consultant	22	PhD in IS
	3	Informant 3	IT applications consultant	17	Bachelor in CE
	4	Informant 4	IT affairs manager	18	Bachelor in IS
	5	Informant 5	Data center manager	15	Bachelor in IS
	6	Informant 6	ERP project manager	18	Bachelor in CS
	7	Informant 7	Executive manager of vendor	21	MBA
	8	Informant 8	Project manager of vendor	19	Master in CS
	9	Informant 9	ERP project manager of vendor	17	Bachelor in CS
B	1	Informant 10	IT affairs manager	30	Master in IS
	2	Informant 11	Data center manager	28	PhD in MIS
	3	Informant 12	IT applications manager	25	Bachelor in CS
	4	Informant 13	ERP project manager	23	Master in CS

The author began the data collection by inviting IT affairs managers from both case sites to a focus group meeting with some senior representatives concerning the ERP project. The researcher held the focus group meeting in Organization A with informants 4, 6, and 9; in Organization B, informants 11 and 12 were involved. Both focus group meetings were recorded and took place for approximately one hour. The focus groups led the researcher to gather detailed information surrounding the main question of this study: Why does some TM push the ERP system toward successful implementation while others are less keen in this regard? Based on focus group outcomes, the semi-structured interview questions were formalized and finalized (see Table 2). The researcher conducted in-depth, tape-recorded interviews for approximately 40 minutes. In addition, the researcher carried out several rounds of follow-up unstructured interviews with informants to clarify various issues and further extend understanding.

Case Organizations Description

The following sections present background on each case organization and the ERP implementation context.

Table 2. The semi-structured interview questions

Questions
1. Why did TM decide to adopt the ERP project?
2. Was TM enthusiastic about and keen on the success of the ERP project?
3. Why did the ERP system succeed or fail?
4. How did TM contribute toward ERP success or failure?
5. What was the role of TM concerning the IT-business strategic alignment?
6. How did TM consider the ERP project in the organization?
7. Can you describe the relationships between TM and IT staff?
8. Can you describe the TMCs required to achieve success on the ERP project?

Organization A

Organization A is a government sports entity, established in 1965, with the objective to undertake the organization and management of horse sports. The board of directors and the headquarters of Organization A are located in Riyadh. Organization A has two main international equestrian fields located in Riyadh and the west region of the KSA. It oversees 11 equestrian fields, all scattered throughout regions of the KSA. The overall total number of employees in the headquarters is approximately 800, with tens of part-time consultants related to horse race regulations, arbitration, and oversight.

The majority of the legacy systems in Organization A originally were developed between 2001 and 2003. In 2012, the board of directors appointed a new General Manager (GM) to Organization A. In 2012, the GM in Organization A issued a strategic IT project supported by the board of directors to provide all services through the organization's portal. In so doing, two plans were formulated. Plan 1 started in actual process in 2013 and was ongoing until 2016, with an approximate budget of 68,000,00SR (Saudi Riyal) (US$18,000,000). The main projects in this plan included the development of an integrated horse information system and the ERP system. Plan 2 began in 2015 and is expected to be completed in 2017, with an approximate budget of 4,000,000 SR (US$10,666,000), which includes the development of the data warehouse, business intelligence system, main portal, and mobile and smart device support.

The integrated IT infrastructure is expected to serve the drafted projects in plan 2, such as the e-services introduced in the main portal and the applications of the business intelligence system. The methodology for selecting the ERP system went through rigorous evaluation procedures of vendors: for example, one condition confirmed that the candidates' outsourcing companies had to have at least five references (i.e., five government organizations) in the KSA, all with successful stories. In addition, the GM obligated the dedicated ERP team to formulate an evaluation form to be completed by the team members during the reference site visits; based on this, the final decision then would be made. After evaluating a variety of ERP companies, a national vendor was chosen. The vendor had 26 references in the KSA, 19 of which were government organizations, encompassing 23 success stories and 3 partial success stories. The ERP implementation process began in mid-2013, comprising different iterative phases. The informants confirmed that the most critical stage is the post-implementation of the ERP system, which took six months before successfully closing the project.

Organization B

Organization B is a government educational entity established in 1954 as a training center. In1985, a royal order was issued approving the transformation of the organization from a training center to an academic college and a research institution of security sciences in the KSA, servicing different academic levels, including undergraduate and higher education studies. Organization B is located in Riyadh, with just one main campus. Organization B adopts the same national university academic and curriculum system, as issued by the Ministry of Higher Education. There are approximately 2,000 students enrolled in the academic college program each year, and approximately 3,000 enrolled in the diploma programs of the higher institution of security and training on an annual basis. The security research center issues an academic scientific journal on a quarterly basis, which is peer reviewed, publishing original research papers concerning security studies. The overall total number of employees is approximately 2,000, 360 of whom are faculty members, and 420 are assistant instructors.

In 2006, Organization B made the decision to remove the entire legacy systems developed in-house with different integration drawbacks. In 2007, Organization B signed a five-year contract with one of the leading software development companies in the region. Organization B dedicated a limited ERP budget; therefore, IT affairs had to identify a solution that addressed the special needs and requirements of the organization and that ensured affordability. The contract stated that Organization B undertook the salary of the vendor developer teams and their expenditures, and the vendor undertook the remaining finical expenditures. The vendor owned the system; therefore, it had the right to resell the system. The vendor presumed that Organization B could be a useful environment in which to develop its ERP system, which might have a good return on investment in the long term.

The stages of the ERP system were to build human resources, finance, and inventory modules first, then to build the education modules, and at the last stage to build the supported systems, such as accommodation, medical, monitoring, security, and the like. In late2011, the vendor and project team announced the suspension of the contract and the failure of the ERP project. The key factors of the failure were the weakness of the vendor's expertise, the resistance to change, and the complexity of the bureaucratic government regulations. The project's total budget was commissioned at approximately 12,000,000 SR (US$3,200,000), with almost US$610,000 paid by Organization B and the rest paid by the vendor.

Data Analysis

Data analysis was achieved using a hybrid approach, including within-case and across-case comparisons. The within-case approach adopted the six-step guidelines by Braun and Clarke (2006) in consideration of thematically analyzing the qualitative data, as qualitative analysis methods are grounded in general in thematic analysis approaches (i.e., thematizing meanings [Holloway &Todres, 2003] and thematic coding [Ryan &Bernard, 2000]). The first step is formalizing the collected data; this involves transcription and scanning the transcripts for the purpose of prior knowledge, initial analytic interests, and thoughts. This step is also important to provide additional concepts to guide the later investigation of the coding process. For example, the interviewees related the identified concepts of characteristics reflectively to TMCs related to business leadership and abilities and to technological cognition and awareness. These themes are consistent with previous studies. For example, Watson and Schneider (1999) declared that the main benefits and strengths of the ERP system derive from two sides: business-wise, as ERP systems provide comprehensive business functionality, and technology-wise, as ERP systems provide fully integrated solutions. These initial constructs were then used as high-level coding categories for data analysis where the subthemes were grouped.

The second, third, fourth, and fifth stages involve iterative procedures, including, respectively, generating initial codes and searching for themes, reviewing themes, and defining and naming themes. The aforementioned steps were iterated several times to ensure that the researcher completely understood the concepts behind the empirical data. In between these steps, the themes were compared with the literature for more refined constructs and then were incorporated into the coding process. In addition, the researcher used the literature to increase the overall validity of the findings (Yin, 2003). One key feature throughout the qualitative data analysis is the flexibility and freedom for the researcher to make adjustments relating to the relevance of the codes and themes to the overall research purpose (Dong et al., 2009). However, ambiguous codes and themes were clarified by the informants through the adoption of follow-up unstructured interviews. Once the within-case analysis was completed for the two case sites, the cross-case analysis of pattern matching between the two case sites was conducted to prepare for a list of other concepts based on the similarities and differences between the cases (Yin, 2003). Although the cross-case analysis was based on two contradictory cases—meaning only the differences can be demonstrated—the pattern matching between the informants across the two sites can be used to support the chain of evidence of such characteristics (Donget al., 2009)

RESULTS

This section presents the analysis of the potential TMCs themes that are classified based on business leadership and abilities and technological cognition and awareness.

TMCs Related to Business Leadership and Abilities

The empirical data exposed different TMCs related to business leadership and abilities that supported successful ERP in Organization A whilst in the other side some negative TMCs resulting in a complete failure of ERP implementation in Organization B. The informants in Organization B agreed that the EISs such as ERP requires different TM with contemporary charismatic leadership in order to success. According to the informants in Organization B, ERP would affect every single entity in the organizations necessitating long term implementation time and budget and therefore it requires champions, strong influences and commitments, power, team trust, fast interventions and decision-making, adjustments, authority, patience, incentives, inspirations and motivations … etc. The participants advocated that the traditional bureaucratic TM practice management in ways that do not allow for changes, improvements, developments, proper use of authorities and decision-making systems, decentralization and innovations … etc. which are essential aspects during the implementation process of ERP.

In contrast, the informants in Organization A explained different charismatic leadership traits that pushed up the ERP project to success. For example, informant 9 when asked about business–ERP alignment, he stated:

We presumed challenges during the documentary sessions and cycles but the GM addressed these issues through regular meetings, listening and listening, and taking advice from stakeholders, urgent decision-making, power and rigorous follow-up.

In addition, several other informants emphasized the need of TM to have full authority during the ERP project to ensure that the speed-up of the implementation could be handled and to tackle the different resistance challenges. Informant 4 mentioned:

The GM drives his authority from the board of directors . . . this helps him to be outside the legal accountability in changing some policies and procedures.

Other two valuable TMCs related to business leadership and abilities are the organization vision and understanding organization core business. According to informant 12, one reason as to why Organization B could not sustain successful progresses in ERP implementation was due to the organization's lack of a clear strategic vision for the present and future. Informant 12 agreed:

The GM in the government sector in general lacks a proper . . . understanding of the core business and that is why they are unable to develop strategic vision.

In addition, informant 13 explained similar issue in this regard:

The majority of the government organizations cannot succeed because the ERP is not only based on automation and integration but also most important are monitoring, planning, transparency, accountability and cost cut down . . . these are outside the accounts of the government organizations.

TM's lack of clear vision and understanding core business practices are thought to have contributed to the ERP failure in Organization B. Contrary to Organization B, the TM in Organization A developed a clear organization vision that serves as a roadmap of future development. Informant 3 reported this statement made by the GM:

Our core business is Speed Horse Racing . . . we must focus based on this vision . . . we must define our resources and planning accordingly.

According to informant 1, both the board of directors of Organization A and the council of ministers confirmed and approved this direction of vision. Other valuable TMCs related to business leadership and abilities that had a positive impact of the ERP success in Organization A are passion in transparency and KPIs. Informant 2 justified the need for transparency:

We have three major issues in the Speed Horse Racing including registration, polling and results . . . the owners and trainers are sometimes sceptical about the credibility of the organizational committees.

Assessment of the performance indicators were and still are obsessed with the GM. Informant 1, when asked about how TM considers the ERP system, stated:

We are specialized in the Speed Horse Racing . . . but how [do] we know if we are doing well . . . we have now unbiased comprehensive reports based on our systems concerning the races statistics and indicators and then we can compare us with others through international KPIs [key performance indicators].

The empirical data also exposed other positive TMCs related to business leadership and abilities contributing to the successful implementation of ERP system in Organization A, including open to the external environment and passion in innovation. These can be understood from different informants, such as informant 5, who reported:

The GM has several consultants from USA and UK regarding the horse affairs and horse races and they helped us to visit some worldwide places to get to know about the newest concerning speed horse races and to accommodate changes based on [this information].

When the informants in Organizations B asked about the aforementioned issues as to confirm, they explained that those issues are behind the agenda of the traditional bureaucratic TM. However, the informants in Organizations B emphasized on a major issue that is related to TM stability for a reasonable period. Informant 13 explained that, during the last 10 years, they have had three GMs, with every GM holding a different vision and strategic outlook:

When the GM was replaced by another GM before the implementation, we lost the support and moreover the new GM did not produce any interest at all since he had no credit in so on.

With frequent reshuffles, TM is thought to have a negative effect on the ERP system. Table 3 summarizes the TMCs related to business leadership and abilities showed how these characteristics are being addressed in the two contradictory sites.

TMCs Related to Technological Cognition and Awareness

The empirical data showed various TMCs related to technological cognition and awareness, which positively contributed to the success of the ERP system in Organization A while their absence contributing directly to the failure of ERP implementation in Organization B. The empirical data exposed that TM in Organization B did not direct more attention to IT in general; therefore, the IT affairs department staff were frustrated and thus faced challenges on all fronts in every single system implementation. In contrast, Different informants in Organization A

Table 3. TMCs Related to Business Leadership and Abilities and how are being addressed in the two contradictory sites

No.	Subthemes	Organization A	Organization B
1	Leadership style	Contemporary charismatic	Traditional bureaucratic
2	Organization vision	Clear	Unclear
3	Understanding organization core business	Yes	No
4	Passion in transparency	Contemporary leadership	Traditional leadership
5	Passion in KPIs	Passion	Not interested
6	Relationship with external environment	Open	Not interested
7	Passion in innovation	Passion	Traditional leadership
8	TM reshuffles	Acceptable constant	Frequent

reported that the TM believes and trust in IT as a very important component towards the overall development of the organization. Informant 3 explained another valuable TMC in Organization A when answering question 6 (see Table 2):

The GM looks at the IT as a strategic partnership and investment that can lead to a major development in the organization.

The former positive characteristic, namely strategic outlook on IT affairs and investment, led the researcher to identify two other characteristics: cohesion and strong relationship with the IT staff and passion in chairing and steering IT project committees. For instance, when asked about question 7 (see Table 2), the informants in Organization A declared that the GM is considered almost a part of the IT team, chairing the key IT committees and sometimes participating in technical meetings. Data analysis revealed that TM that has an acceptable level of understanding information systems infrastructure in the organization could lead to successful implementation. According to informants 6, 7, 8, and 9, the GM in Organization A was keen to attend the awareness session for every launched ERP systems module or tested various ERP functions.

Contrary to Organization A, the TM in Organization B believes in IT with no return of media momentum and looking at the IT department as a service-oriented department. These characteristics were found to have a negative impact on the ERP project in Organization B. Informant 11 illustrated that the GMs in the KSA's government organizations generally look at innovation as the balance between gaining media momentum and implementation cost and time:

In order to understand the relationship of the GM with the IT systems you should investigate first the relationships between the government organizations and the GM . . . the government organizations are not profit or being monitored and therefore the equation and the priorities are different . . . the priorities are the GM priorities.

According to informant13, the former issue results in a traditional outlook on the IT departments in the government sectors as service-oriented departments. Informant 11 agreed with informant 13:

One of the problems that confront us and I think the majority of government organizations face is the outlook to the department of IT affairs as a service-oriented department . . . this traditional outlook by the GM contributes to less form of support towards particularly the strategic IT projects such as ERP.

Another positive TMC found in Organization A is the understanding of the organization's information systems' infrastructure. Accordingly, this led to another valuable characteristic: understanding the ongoing challenges of IT affairs. Different challenges were sorted out directly by the GM in Organization A during the ERP implementation. The informants explained that the data center required different amendments and changes during implementation, and the project team appreciated a fast response by the GM. The informants expressed that the GM looked at the ERP system as one component focused on the development of the organization's information infrastructure. The GM needs the data to support the board of directors and to exchange data with national and international bodies relating to horse races. Hence, the characteristic acknowledge the role and importance of the information is considered especially important. Table 4 summarizes the TMCs related to technological cognition and awareness exposed during the analysis of the empirical data and how these characteristics are being addressed in the two contradictory sites.

Table 4. TMCs Related to Technological Cognition and Awareness and how are being addressed in the two contradictory sites

No.	Subthemes	Organization A	Organization B
1	IT relationship	Believe and trust	Caution and fear
2	Outlook on IT affairs	Strategic investment	Service oriented
3	Relationship with IT staff	Cohesive and strong	Formal and superficial
4	Chairing and steering IT projects	Passion	Not interested
5	IT infrastructure	Acceptable understanding	IT illiterate
6	IT challenges	Acceptable understanding	IT illiterate
7	Information role and importance	Acknowledge	Not interested

DISCUSSION

The cross-case analysis approach based on two contradictory government organizations in the KSA—one with successful implementation and the other with a failed ERP revealed several TMCs recognized as having a direct effect on ERP projects implementation. These characteristics varied between business leadership and abilities and technological cognition and awareness (see Table 3 and 4). Concerning TMCs related to business leadership and abilities, the first and essential of these characteristics is contemporary charismatic leadership. The former characteristic were strongly emphasized by the informants in Organization A, whereas the bureaucratic leadership in Organization B significantly contributed to the immobilization of the important leadership traits essential to the successful implementation of ERP system. This result is in consistent with some related studied such as Alkraiji (2016) and Neufeld et al. (2007) which revealed that projects champion charisma has a positive impact on different IT projects aspects such as the increased performance expectancy, effort expectancy, social influence and facilitating condition perceptions of users.

In addition, the empirical data showed that the behavior of the bureaucratic leadership is not consistent; therefore, they sometimes produce acceptable levels towards pushing up the ERP implementation process, whereas in some similar occasions, they do not show any reaction; Streatfield (2001) refers to this as the paradox of control. This cannot work for ERP projects, where constant and consistent TM support is essential throughout the ERP project (Elbanna, 2013). According to Streatfield (2001), the collective emergent outcomes in the organizations with bureaucratic leadership cannot be traced back to specific actions; these represent the reality of managerial contexts not being in control. Within such managerial contexts, the employers struggle to create meaning and to maintain a sense of self in the face of such uncertainty (Uhl-Bien & Marion, 2009).

Nevertheless, some informants explained some reasons of the lack of contemporary charismatic leadership in the KSA. The majority of the government organizations in the KSA are not for-profit organizations, and therefore the government organizations lack an effective system of overseeing and accountability. Other informants showed that the culture and its different dimensions, rather than competency, plays a major role in selecting the TM of government organizations in the KSA. This issue (along with a lack of national systems of accountability) has a negative impact on the overall development, including the IT projects (Alkraiji, 2017). Therefore, it is expected that the most important indicators capable of measuring productivity, transparency and innovation are missing from the TM agenda in the government organizations in general (Alkraiji, 2017).

In the relevant literature, measuring performance with KPIs was found to directly contribute to ERP usage and value (Ruivo & Neto, 2011). In addition, the GM in Organization A looked to transparency as a power tool, particularly in regard to enhancing the credibility of the horse races; this may explain the significant relationship between GM transparency and attitude toward ERP value (Al-Jabri & Roztocki, 2015; Zhu et al., 2010). In contrast, the GM in Organization B, as noted by the informants, was concerned about whether important or confidential information would be leaked, thus placing the organization in a vulnerable position. The feature of transparency within the ERP system could be counterproductive because the relationship between transparency and system usefulness is reciprocal (Al-Jabri & Roztocki, 2015).

An additional valuable characteristic intertwined with the contemporary charismatic leadership is the open to the external environment. The cross-case analysis revealed that the GM in Organization A was open to the external environment and produced a positive effect on the innovation concerning international speed horse races and accommodated changes based on that, unlike the GM in Organization B. The literature shows that the external environment might have an influence on TM, which in turn affects behaviors and actions (Yigitbasioglu, 2015).

The empirical data exposed another TMCs related to business leadership and abilities that is the organization vision. While the GM in Organization A had a clear vision of the current state and future growth of the organization, the GM in Organization B, in contrast, lacked a clear vision, which is recognized as crucial in determining the resources and planning necessary for ERP implementation. According to Donget al. (2009), when TM fails to clearly declare the organization's vision, they fail to convey the clear vision required concerning ERP implementation in an organization and to emphasize the vision behind the system implementation.

The informants in Organization B explained that the GM could not formulate a clear vision owing to the fact that he did have the required understanding of the organization's core business. According to the informants, the GM could provide a theoretical description as to what the organization and growth future are about; however, when it came to practical reality, overlap and conflict among departments and developed programs were recognized; this led to wasted resources, duplications in funding, and a lack of coordination in terms of managerial control. According to the informants, different academic programs and initiatives were launched and then terminated and then re-launched within a short period. In Organization B, the TM's lack of understanding core business characteristics might be one of the reasons behind the short tenure of TM, unlike in the case of Organization A, where TM had less turnover. It is held that the short tenure affects the required heterogeneity between the TM and the organization, thus causing an imbalance in managerial contexts (Tihanyiet al., 2000). TM reshuffling is seen as a major characteristic in the success of ERP projects.

The cross-case analysis of the TMCs related to technological cognition and awareness exposed that the GM in Organization A was strongly involved in IT affairs and in the implementation of the ERP system, unlike the GM in Organization B, who held the view that IT systems do not give enough media momentum compared to the costs and time of implementation. This construct would be consistent with the literature: for example, Uhl-Bien and Marion (2009) noted that, if the innovation in the bureaucratic organizations does not gain momentum, then the innovation would die or disappear, as TM would not constitute the required effective leadership. In addition, the GM in Organization B viewed the IT department as a service-oriented department; therefore, the GM strived to make it fully operational while mitigating as much of its expenses as possible. According to the informants, this approach meant that the IT department was developed based only on service needs rather than on a strategic investment. This would limit the professional development of the IT department (Loonam & McDonagh, 2005). The GM in Organization A, in contrast, trusted in IT and further believed that the organization could not evolve or move forward without robust IT infrastructure.

Research in organizational innovation and strategic management distinguishes between two concepts of TM involvement: belief and participation (Chatterjee et al., 2002; Walsh, 1988). According to Jarvenpaa and Ives (1991), belief is a psychological state with the potential of a certain initiative, whereas participation refers to the behaviors and actions performed to accomplish an initiative. According to Yigitbasioglu (2015), TM behaviors and actions toward EISs are the mental image of a desired future state for the organization, which in turn might increase the level of support. Yigitbasioglu (2015) asserted that TM participation in EIS projects mainly is driven by belief rather than by objective reality. In so doing, TM participation is much more than merely giving blessings to the ERP system (Sarker & Lee, 2003): the participation must not be limited to the ERP conception but should also imply a willingness to serve on steering committees or to oversee the ERP implementation, including addressing various challenges through personal interventions (Chen, 2001). The results of the TMCs related to technological cognition and awareness can be grounded into the former discussion concerning the belief and participation of TM in acquiring innovation.

CONCLUSION

This study provides an in-depth understanding of the effect of TMCs on EIS projects through the completion of a cross-case analysis approach based on two contradictory government organizations in the KSA, one with successful ERP implementation and the other with failure. The cross-case analysis revealed various TMCs (see Table 3 and

4) that are grouped into two main themes: TMCs related to business leadership and abilities and TMCs related to technological cognition and awareness. The implication of this study fills the knowledge gap in the EIS literature. In addition, the literature concerning government organizations—particularly those in developing countries—is much in need, as TM has more power in the decision-making process in terms of control and mechanisms. The identified characteristics can be also incorporated into the development of such a strategy to enhance overall awareness and TM participation in EISs in order to maintain a constant and consistent support level.

However, it is worth noting that the results of this qualitative study are bound within the context of the two-government case sites in the KSA and therefore it may be difficult to suggest generalizing to a broader context (Eisenhardt, 1989). Moreover, the researcher did not directly examine the influence of TM demographic characteristics such as tenure, age, educational level, and functional background (Herrmann & Datta, 2005). It is suggested that future research should transform the TMCs in the study and the demographic characteristics into a large-scale survey questionnaire to offer the opportunity to establish generic significance to the research results.

REFERENCES

Al-Jabri, I. M., & Roztocki, N. (2015). Adoption of ERP systems: Does information transparency matter? *Telematics and Informatics*, *32*(2), 300–310. doi:10.1016/j.tele.2014.09.005

Alkraiji, A. I. (2016). Cross-Case Analysis of Top Management Characteristics and Enterprise Information System Success. *Journal of Cases on Information Technology*, *18*(3), 40–58. doi:10.4018/JCIT.2016070104

Alkraiji, A. I. (2017). The Efficiency of the Top Management and the Reality of the Decision Support Systems in Saudi Government Organizations. In New Contribution in Information Systems and Technologies, Advances in Intelligent Systems and Computing (vol. 569). Springer International Publishing.

Aloini, D., Dulmin, R., & Mininno, V. (2007). Risk management in ERP project introduction: Review of the literature. *Information & Management*, *44*(6), 547–567. doi:10.1016/j.im.2007.05.004

Anthopoulos, L., Reddick, C. G., Giannakidou, I., & Mavridis, N. (2016). Why e-government projects fail? An analysis of the Healthcare.gov website. *Government Information Quarterly*, *33*(1), 161–173. doi:10.1016/j.giq.2015.07.003

Appelbaum, S. H. (1997). Socio-technical systems theory: An intervention strategy for organizational development. *Management Decision*, *35*(6), 452–463. doi:10.1108/00251749710173823

Armstrong, C. P., & Sambamurthy, V. (1999). Information technology assimilation in firms: The influence of senior leadership and IT infrastructures. *Information Systems Research*, *10*(4), 304–327. doi:10.1287/isre.10.4.304

Avolio, B. J., Sosik, J. J., Kahai, S. S., & Baker, B. (2014). E-leadership: Re-examining transformations in leadership source and transmission. *The Leadership Quarterly*, *25*(1), 105–131. doi:10.1016/j.leaqua.2013.11.003

Bansal, V., & Agarwal, A. (2015). Enterprise resource planning: Identifying relationships among critical success factors. *Business Process Management Journal*, *21*(6), 1337–1352. doi:10.1108/BPMJ-12-2014-0128

Braun, V., & Clarke, V. (2006). Using thematic analysis in psychology. *Qualitative Research in Psychology*, *3*(2), 77–101. doi:10.1191/1478088706qp063oa

Bryman, A. (2015). *Social research methods* (2nd ed.). New York, NY: Oxford University Press.

Buonanno, G., Faverio, P., Pigni, F., Ravarini, A., Sciuto, D., & Tagliavini, M. (2005). Factors affecting ERP systems adoption: A comparative analysis between SMEs and large companies. *Journal of Enterprise Information Management*, *18*(4), 384–426. doi:10.1108/17410390510609572

Carlsson, S. A., Henningsson, S., Hrastinski, S., & Keller, C. (2011). Socio-technical IS design science research: Developing design theory for IS integration management. *Information Systems and e-Business Management*, *9*(1), 109–131. doi:10.1007/s10257-010-0140-6

Chatterjee, D., Grewal, R., & Sambamurthy, V. (2002). Shaping up for e-commerce: Institutional enablers of the organizational assimilation of web technologies. *Management Information Systems Quarterly*, *26*(2), 65–89. doi:10.2307/4132321

Chen, I. J. (2001). Planning for ERP systems: Analysis and future trends. *Business Process Management Journal*, *7*(5), 374–386. doi:10.1108/14637150110406768

Dong, L., Neufeld, D., & Higgins, C. (2009). Top management support of enterprise systems implementations. *Journal of Information Technology*, *24*(1), 55–80. doi:10.1057/jit.2008.21

Eisenhardt, K. M. (1989). Building theories from case study research. *Academy of Management Review*, *4*(4), 532–550.

Elbanna, A. (2010). Rethinking IS project boundaries in practice: A multiple-projects perspective. *The Journal of Strategic Information Systems*, *19*(1), 39–51. doi:10.1016/j.jsis.2010.02.005

Elbanna, A. (2013). Top management support in multiple-project environments: An in-practice view. *European Journal of Information Systems*, *22*(3), 278–294. doi:10.1057/ejis.2012.16

Gasson, S. (2006). A genealogical study of boundary-spanning IS design. *European Journal of Information Systems*, *15*(1), 26–41. doi:10.1057/palgrave.ejis.3000594

Greasley, A., & Wang, Y. C. (2016). Building the hybrid organization through ERP and enterprise social software. *Computers in Industry*, *82*, 69–81. doi:10.1016/j.compind.2016.05.007

Gregor, S., & Jones, D. (2007). The anatomy of a design theory. *Journal of AIS*, *8*(5), 312–335.

Hambrick, D. C., & Mason, P. A. (1984). Upper echelons: The organization as a reflection of its top managers. *Academy of Management Review*, *9*(2), 193–206.

Herrmann, P., & Datta, D. K. (2005). Relationships between top management team characteristics and international diversification: An empirical investigation. *British Journal of Management*, *16*(1), 69–78. doi:10.1111/j.1467-8551.2005.00429.x

Holloway, I., & Todres, L. (2003). The status of method: Flexibility, consistency and coherence. *Qualitative Research*, *3*(3), 345–357. doi:10.1177/1468794103033004

Howcroft, D., Newell, S., & Wagner, E. (2004). Understanding the contextual influences on enterprise system design, implementation, use and evaluation. *The Journal of Strategic Information Systems*, *13*(4), 271–277. doi:10.1016/j.jsis.2004.11.010

Irani, Z., Ezingeard, J., Grieve, R., & Race, P. (1999). Case study approach to carrying out information systems research: A critique. *International Journal of Computer Applications in Technology*, *12*(2), 190–198. doi:10.1504/IJCAT.1999.000203

Jarvenpaa, S. L., & Ives, B. (1991). Executive involvement and participation in the management of information technology. *Management Information Systems Quarterly*, *15*(2), 205–227. doi:10.2307/249382

Khan, G., Moon, J., Park, H., Swar, B., & Rho, J. (2010). A socio-technical perspective on e-government issues in developing countries: A scientometrics approach. *Scientometrics*, *78*(2), 267–286.

Liang, H., Saraf, N., Hu, Q., & Xue, Y. (2007). Assimilation of enterprise systems: The effect of institutional pressures and the mediating role of top management. *Management Information Systems Quarterly*, *31*(1), 59–87.

Liu, G. H. W., Wang, E. T. G., & Chua, C. E. H. (2015). Leveraging social capital to obtain top management support in IT projects. *Journal of the Association for Information Systems*, *16*(8), 707–737.

Liu, S., Zhang, J., Keil, M., & Chen, T. (2010). Comparing senior executive and project manager perceptions of IT project risk: A Chinese Delphi stud. *Information Systems Journal*, *20*(4), 319–355. doi:10.1111/j.1365-2575.2009.00333.x

Loonam, J. A., & McDonagh, J. (2005). Exploring top management support for the introduction of enterprise information systems: A literature review. *Irish Journal of Management*, *26*(1), 163–178.

Lyytinen, K., Newman, M., & Al-Muharfi, A. (2009). Institutionalizing enterprise resource planning in the Saudi steel industry: A punctuated socio-technical analysis. *Journal of Information Technology*, *24*(4), 286–304. doi:10.1057/jit.2009.14

Mayère, A., Grabot, B., & Bazet, I. (2008).The mutual influence of the tool and the organization. In B. Grabot, A, Mayère, & I Bazet (Eds.), ERP systems and organizational change (pp. 1–11). London: Springer-Verlag.

McAdam, R., & Galloway, A. (2005). Enterprise resource planning and organizational innovation: A management perspective. *Industrial Management & Data Systems*, *105*(3), 280–290. doi:10.1108/02635570510590110

McLeod, L., & Doolin, B. (2012). Information systems development as situated socio-technical change: A process approach. *European Journal of Information Systems*, *21*(2), 176–191. doi:10.1057/ejis.2011.43

Momoh, A., Roy, R., & Shehab, E. (2010). Challenges in enterprise resource planning implementation: State-of-the-art. *Business Process Management Journal*, *16*(4), 537–565. doi:10.1108/14637151011065919

Neufeld, D. J., Dong, L., & Higgins, C. (2007). Charismatic leadership and user acceptance of information technology. *European Journal of Information Systems*, *16*(4), 494–510. doi:10.1057/palgrave.ejis.3000682

Orlikowski, W. J., & Iacono, C. S. (2001). Research commentary: Desperately seeking the "IT" in IT research—A call to theorizing the IT artifact. *Information Systems Research*, *12*(2), 121–134. doi:10.1287/isre.12.2.121.9700

Ruivo, P., & Neto, M. (2011). Sustainable enterprise KPIs and ERP post adoption. In *Proceeding of the 6th Iberian Conference on Information Systems and Technologies* (pp. 1–7). Piscataway, NJ: IEEE.

Ryan, G., & Bernard, H. (2000). Data management and analysis methods. In N. K. Denzin & Y. S. Lincoln (Eds.), *Handbook of qualitative research* (2nd ed.; pp. 769–802). Thousand Oaks, CA: Sage.

Sanders, P. (1982). Phenomenology: A new way of viewing organizational research. *Academy of Management Review*, *7*(3), 353–360.

Sarker, S., & Lee, A. (2003). Using a case study to test the role of three key social enablers in ERP implementation. *Information & Management*, *40*(8), 813–829. doi:10.1016/S0378-7206(02)00103-9

Sawyer, S., & Jarrahi, M. H. (2014). Sociotechnical approaches to the study of Information Systems. In A. Tucker & H. Topi (Eds.), Computing Handbook (3rd ed.; pp. 5.1-5.27). New York: Chapman and Hall. doi:10.1201/b16768-7

Schmidt, R., Lyytinen, K., Keil, M., & Cule, P. (2001). Identifying software project risks: An international Delphi study. *Journal of Management Information Systems*, *17*(4), 5–36. doi:10.1080/07421222.2001.11045662

Somers, T., & Nelson, K. (2004). A taxonomy of players and activities across the ERP project life cycle. *Information & Management*, *41*(3), 257–278. doi:10.1016/S0378-7206(03)00023-5

Streatfield, P. (2001). *The paradox of control in organizations*. London: Routledge. doi:10.4324/9780203361504

Tihanyi, L., Ellstrand, A., Daily, C., & Dalton, D. (2000). Composition of the top management team and firm international diversification. *Journal of Management*, *26*(6), 1157–1177. doi:10.1177/014920630002600605

Uhl-Bien, M., & Marion, R. (2009). Complexity leadership in bureaucratic forms of organizing: A meso model. *The Leadership Quarterly*, *20*(4), 631–650. doi:10.1016/j.leaqua.2009.04.007

Van Wart, M. (2015). Evaluating transformational leaders: The challenging case of Eric Shinseki and the US Department of Veterans Affairs. *Public Administration Review*, *75*(5), 760–769. doi:10.1111/puar.12393

Van Wart, M., Roman, A., Wang, X., & Liu, C. (2017). Integrating ICT adoption issues into (e-) leadership theory. *Telematics and Informatics*, *34*(5), 527–537. doi:10.1016/j.tele.2016.11.003

Walsh, J. (1988). Selectivity and selective perception: An investigation of managers' belief structures and information processing. *Academy of Management Journal, 31*(4), 873–896. doi:10.2307/256343

Watson, E. E. & Schneider, H. (1999). Using ERP systems in education. *Communications of the AIS, 1*(2), Article No.3.

Yen Teoh, S., & Pan, S. (2008). Understanding the influences of social integration in enterprise systems use. *Journal of Enterprise Information Management, 21*(5), 493–511. doi:10.1108/17410390810904256

Yigitbasioglu, O. (2015). The role of institutional pressures and top management support in the intention to adopt cloud computing solutions. *Journal of Enterprise Information Management, 28*(4), 579–594. doi:10.1108/JEIM-09-2014-0087

Yin, R. (2003). *Case study research design and methods* (3rd ed.). Thousand Oaks, CA: Sage.

Zhu, Y., Li, Y., Wang, W., & Chen, J. (2010). What leads to post-implementation success of ERP? An empirical study of the Chinese retail industry. *International Journal of Information Management, 30*(3), 265–276. doi:10.1016/j.ijinfomgt.2009.09.007

KEY TERMS AND DEFINITIONS

Enterprise Information Systems (EIS): Any kind of information system that improves the functions of enterprise business processes by integration.

Top Management (TM): The person(s) with overall responsibility for the organization.

Top Management Characteristic (TMC): Leadership or technical attributes related to the success of EIS implementations.

Chapter 2
Managing IT to Innovate and Strategize in the Higher Education Sector:
Role of Enterprise Systems

Prithvi Jyoti Bhattacharya
Higher Colleges of Technology, UAE

ABSTRACT

The current dynamic business landscape has compelled all organizations worldwide to innovate and strategize continually to remain competitive and achieve an edge in the market. Information technology has often made lofty promises to assist such endeavors. This chapter explores the management of large, integrated, and packaged software suites collectively called enterprise systems (ES) to enable innovation and strategic decision making in organizations. In particular, this chapter explains how a globally renowned and highly ranked university manages its enterprise-systems-based IT platform to offer new services, develop new processes, and make astute strategic decisions – all in an attempt to retain and improve its position in the global higher education market.

INTRODUCTION

In recent times, Information Technology (IT) has been reported to play a strategic role in organizations, and a variant of IT called 'Enterprise Systems' are increasingly becoming the favored IT platform around the world. Enterprise Systems (ES) are large-scale, packaged, application systems that can be used to streamline and

DOI: 10.4018/978-1-5225-5279-6.ch002

integrate business processes, and enhance information levels within the organization as well as with its business partners. ES has become a generic term that includes a number of systems like Enterprise Resource Planning (ERP), Customer Relationship Management (CRM) and other similar systems. These systems were originally adopted to improve operational efficiency. However, all organizations that have adopted Enterprise Systems hope to get more out of these systems than just improving operational efficiency; they want to be able to strategize and innovate, and thus compete better in the market. This case study, using a highly ranked and world-renowned University based in Australia, attempts to explore how Enterprise Systems can enable organizations to innovate and strategize.

Background

UniCo (real name not disclosed for confidentiality reasons) is a public university located in Australia. It was founded in the 19th century and is one of the oldest Universities in Australia. It is a member of Australia's "Group of Eight" universities and the Universitas 21. The University time and again ranks among the top universities in Australia and the world. UniCo has one of the highest financial endowments of any university in Australia. UniCo has consistently ranked first or second on the key national research indicators set up by the federal government to allocate public funds for research and training infrastructure in the nation. The university has close to 36,000 students, and more than 7,300 staff members, both academic and professional. Figure 1 shows a partial view of the high level organization chart of UniCo.

UniCo was originally established to offer degrees to advantaged students at a standard that would be at par with that of Oxford. In subsequent times, the University took up research, public service, and cultural initiatives in line with its spirit as a public institution. Today, UniCo retains its public spirit, but is now a massive complex organization facing the insecurity of being between the extremes of regulated responsibility towards the public and market-driven private income. It aims to strike a balance between a traditional vision of teaching and research with the more recent objectives to meet professional, economic, and community demands.

The University should focus its development in a limited number of strategic priority areas in which it has outstanding and demonstrated strength and capacity and critical mass, while providing broad opportunities in other areas which do not warrant the same priority in core resource allocation – a broad base with a sharp edge. (Academic Staff Member, Faculty of Medicine, Dentistry and Health Sciences)

Figure 1. Organization structure (partial)
(Source: UniCo Website)

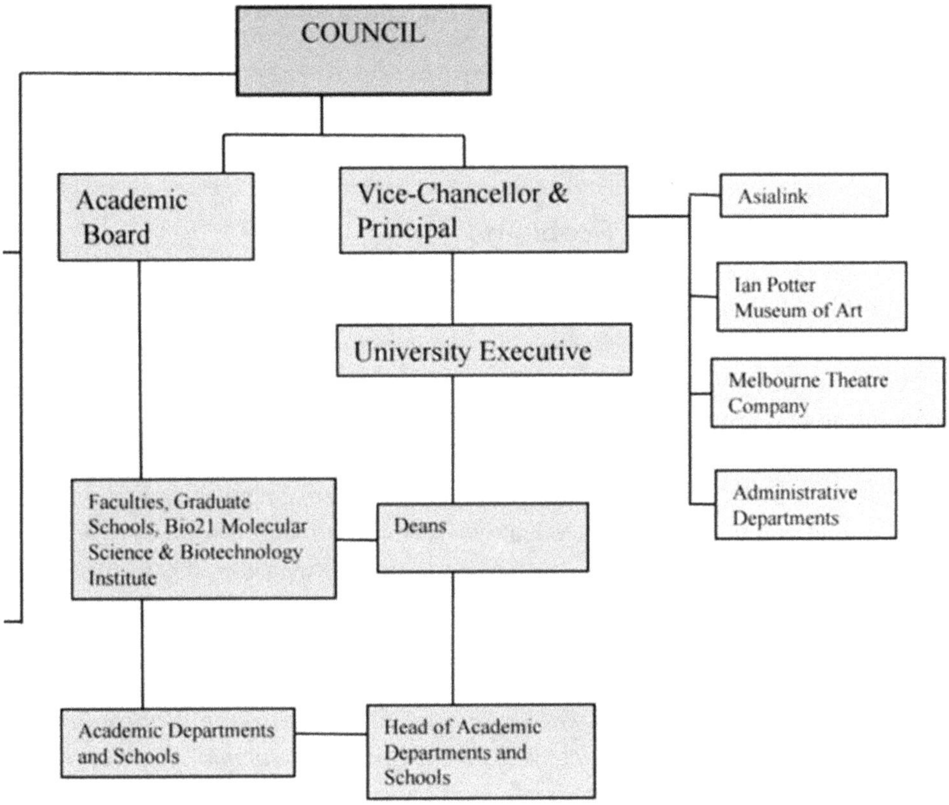

A US-style BA-BSc that enabled students to develop more background knowledge and discover their interests might be wise. (Academic Staff Member, Department of History and Philosophy of Science/School of Population Health)

To realize the vision to be 'one of the finest in the world', UniCo devised the 'Triple Helix', with three key priorities for the institution – research, teaching and knowledge transfer. As the name suggests, the triple helix is made up of three strands intertwined with each other. The first strand, Research, involves the systematic generation of new knowledge, birth, and growth of new ideas and experimentation with new techniques. The second strand, Learning and Teaching, involves a body of ideas, is informed by available research, and implants inquiry as way of thinking. The final strand, knowledge transfer, involves many dimensions of interaction between academia and society at large.

The approach embraces the notion of a triple helix: a public-spirited institution, defined by tightly-bound strands of research, teaching and knowledge transfer, offering its students, staff and community one of the finest universities in the world. (Vice Chancellor)

MAIN CHALLENGES FACED BY THE ORGANIZATION

Issues, Controversies, Problems

Like most institutions of this sector, UniCo was challenged by the current trends in the higher education sector globally – increased pressure to publish more research, an enormous rise in enrolments of local and international students, diminishing financial support from the government, and increasing dependence on fee-paying students to finance core activities. To achieve the mission of being 'one of the finest' universities globally, UniCo needed an internal discipline of regular evaluation of its activities and outcomes. It also needed the flexibility to shift resources to address indicators of academic performance, and fine-tuning the measures that faculties and departments use to make strategic decisions.

When it comes to making decisions about, communicating or facilitating our research, we do not operate in a mode that enables a hypertext organisation. Data about our research is compartmentalised across different University communities trapped in its use for specific purposes. (Academic Staff Member, University Systems Project)

Our system of promotion does not place enough emphasis on contributions to knowledge transfer. (Academic Staff Member, Faculty of Engineering)

To address these challenges, UniCo looked at the IT infrastructure to assess how well it supports or enables the strategy of the institution. The University operated using a plethora of IT systems namely Merlin, GeneSys and MUFAS. Merlin was the student administration system, *MUFAS* was an in-house built, standalone finance system, and Genesys was a standalone HR System which also handled the payroll. The University IT environment was devolved and fragmented, and this resulted in unproductive delivery of information services. So a decision was made to fundamentally change this business model to a more strategic and unified one. Figure 2 below shows this intended change.

The University commenced a review of its IT Infrastructure to assess its fitness to support the 'Agenda' of 1996. This review produced a report that identified issues

Figure 2. The IT Architecture at UniCo
(Source: UniCo Website)

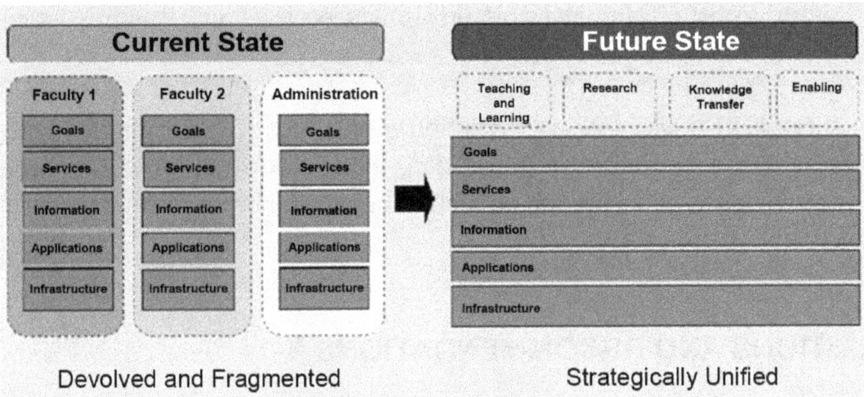

with the existing IT infrastructure. In accordance with the strategy of UniCo, the review report identified that a new IT platform was required to support the following:

- Automation of workflow for different business processes, from leave approvals to payments for procurements
- Putting new, better, re-engineered business processes in place
- Extension of business processes to include external stakeholders like affiliates and suppliers
- Implementation of accrual –based accounting instead of cash-based accounting
- Enhancement of budgeting through techniques like Activity Based Costing
- Conducting business to business transactions electronically (B2B E-Commerce), especially online procurement
- Provision of Self-service facilities for students and staff members
- Management of career and employee relations
- Increasing the visibility of cross disciplinary research
- Attracting world class academics and researchers
- Linking of academia to industry based on demand and supply of knowledge and skills
- Assessing the health of a faculty, department or affiliate to make decisions about them
- Identifying, assessing and exploiting different sources of funding
- Better decision making for career management of academics based on performance in terms of publications, attracting grants etc.

- Improving the quality of grants applications, thus enhancing the competitive position of UniCo in the Higher Education market
- Ability to better plan and cost investment projects and ventures based on accurate information of past performances

In short, UniCo was seeking a solution to a number of issues in the areas of operational business processes. In addition, it was also seeking ways to identify its own strengths as an institution, as well as opportunities in the higher education market that it can potentially serve.

SOLUTIONS AND RECOMMENDATIONS

As an attempt to address the above mentioned challenges and requirements, the approach that UniCo chose was to adopt a specialized kind of an IT system: An Enterprise System. The benefits of this approach to address the challenges facing the organizations has been widely found in academic as well as industry literature. Enterprise Systems promise to enable organizations with both operational efficiency as well as other more 'strategic' benefits to equip them to compete better in the market. Such 'strategic' benefits range from enabling product innovation, integrating mergers and acquisitions, implementing new and unique processes and enabling astute strategic decision making (Shang & Seddon 2002; Davenport 2004; Weill & Ross 2009; Staehr et al. 2012; Mostaghel et al. 2015). Several large organizations around the world like ExxonMobil, Microsoft, Canadian National Railways, Colgate-Palmolive, Kraft Foods, BHP Billiton and many others have taken the approach of adopting Enterprise Systems to address the strategic challenges. Many higher education institutions have also followed the same path (Motiwala et al. 2012; Ruivo et al. 2014). A recent study provided insight on our limited understanding of how digital and ES platforms attain innovation, through a study involving 189 organizations (Sedera et al 2016). Recent studies support that the use of enterprise systems (ES) enhances buyer-supplier integration (BSI). More specifically, ES enhances BSI indirectly, mediated by inventory management capabilities (IMC). This shows that Enterprise System not only enable integration within the adopting organization, but also beyond it (Lorenzo et al. 2017).

UniCo invested in an Enterprise System from the vendor, Oracle. This investment was driven by more than achieving operational efficiency; the organization aimed to introduce new services and new ways of doing things to support its strategy of becoming the top research intensive university in Australia and among the best in the world. UniCo had the Oracle E-Business suite of Enterprise Systems licensed to address the above-mentioned objectives. The decision was made to adopt a new, packaged Enterprise System to manage the Finance, Human Resources, and Research requirements of the University. The chosen ERP software package was Oracle E-Business Suite Release 11. This system was internally termed as 'THEMIS'. The project started in 2001 and it took UniCo twelve months to decide the setting up of the system. This period involved activities like vendor experiences, demonstrations, and gathering business requirements.

We decided to go forward and have integrated administrative systems; THEMIS has Oracle Finance, Oracle HR, and a component called Oracle Contracts that we have customized to build our assets reporting module with all the finances in it. I was a part of the implementation team, which started in 2001. (Manager, Financial Systems and Research Reporting)

The Implementation Project for the Enterprise System

Once decided that UniCo will implement 'THEMIS', the Enterprise System from Oracle, UniCo set up a dedicated project for this purpose with an estimated timeframe of 3 years and a budget of approximately 1 million Australian dollars for the implementation. UniCo established a project team of 100+ members that consisted of a wide range of people with different but complementary skills:

- Subject Matter Experts (SMEs) from the different functional areas of business namely Human Resources, Finance and Research Administration of UniCo
- Technical staff members like User Interface designers, programmers, testers, database administrators and IT infrastructure staff from UniCo.
- External consultants from a consulting firm with specialist knowledge in Oracle ERP systems

Given the project team consisted of members from UniCo as well as external consultants, a clear statement of work was created that clearly defined the scope of work to be done and deliverables to be produced from the two parties, UniCo and the external consulting firm. All team members reported to a project manager who was assigned on a full time basis for this project.

The project team was divided into three streams, one for each functional areas of business namely Human Resources, Finance and Research Administration. Each of the streams had members from both UniCo as well as the external consulting firm. Each stream had a 'stream leader', who was a member of UniCo, and all the project steam members reported to him. Each stream leader reported to the project manager.

In addition, UniCo set up a steering committee led by the Chief Information Officer and consisting of senior managers representing the different functional areas involved in the implementation, namely Human Resources, Finance and Research Administration. The objective was to make the implementation project a 'business' project, not just an 'IT' project. A conscious decision was made to involve the representatives of the business functional areas from the very beginning.

A traditional waterfall model was applied whereby analysis was followed by blueprinting and design, which was followed by configuration of the Oracle E Business Suite as well as customization through additional coding. This was followed by extensive testing and, finally, release management activities for rolling out the system across UniCo. The roll out of the system was done in phases built around the three core modules implemented: Human Resources, Finance and Research Administration. Prior to the Enterprise System implementation, UniCo had three different systems for each of the aforementioned three functional areas. Therefore, the roll out strategy used was a phased cutover, whereby each module was implemented and the corresponding legacy system was turned off, one by one.

The Finance module was the first module implemented and it went live in 2003, but UniCo was still running the current in-house built system for HR and payroll then. The HR component of the Oracle E-Business suite came on board in 2004, and then Research component the year after in 2005.

The following sections explain the details of how the Enterprise Systems was used to address the challenges faced by the organization.

Integrating Data and Systems With the Enterprise System

UniCo used the Oracle E-Business Suite to integrate its Finance, HR, and Research systems- three isolated, standalone systems. There were also a lot of conversations around the integration between Finance and Human Resources (HR); Finance went in production first. Finance uses the concept of 'job category', for a position to do the financial delegations. But, HR uses the concept of 'job positions' for other purposes. So there were a lot of discussions around which concept would be used for the delegations and what effect that would have on the system.

The THEMIS System allows UniCo to consolidate the different divisions, different departments, so that it can have a look at either a consolidated view, or a more detailed view. The THEMIS System will basically provide the financial forecasts for expenditures and revenue; anything associated with finance gets in there.

UniCo has looked into having an integrated approach for all its web-based content too. UniCo has about 400,000 separate web pages with all kinds of information – it was a big endeavour to try to put everything into one place. UniCo implemented what it calls a "Management Information Library" or MIL. The MIL, as the name suggests, is a library in essence. It is a consolidated or a central place to put all kinds of documents - interactive and static. Interactive documents are documents that get the latest updated data, while static documents are things like government reports, static University reports, the University Budget Philosophy, the University Handbook and so on.

Now we have an end-to-end integrated system, a single system. So it reduced redundancy of data collection and improved the accuracy of data. (Development Manager, IT Services)

It comes down to why you use a single ERP system versus the best isolated systems. If you were using three or four different systems, then you'd have to manually use some sort of a mechanism to consolidate it and then look at it, whereas this system allows doing that in one go. Obviously, different fields, or different accounts will generate a stream of revenue, but it's all in the one system. (Manager, Finance & Planning)

Basically, the THEMIS System allows us to consolidate the different divisions, different departments, different areas to input, consolidate, and then we can have a look at a consolidated view, or we can just dive into the more detailed view, as well. (Manager, Finance & Planning)

Furthermore, UniCo has a number of subsidiary organizations like the University Press, Business School, NOSSAL Global Institute, and so on. These are known as its affiliates. When it does its consolidated financial statements, all of these affiliations are brought to account; if they form an affiliation with the university, they form a part of its consolidated financial statements. Some of them are called 'semi-autonomous bodies', and they have their own boards. So they have their own profit and loss statements that they have to provide independently. But they still are a part of the University.

It depends on the agreement that the university has with those affiliates. So some it will, some it won't. (Manager, Human Resources Systems)

I would say that the way that the ERP works is flexible enough to allow for the different types of companies that we deal with because we have many different types - including trusts and the University Press. They are completely separate organizations. Some have their own ABN, some trade under our ABN. (Deputy Manager, Accounting Operations)

UniCo has used its Enterprise Systems to integrate with other organizations using the 'Accounting string' or the 'Chart of Accounts' used by the Enterprise System.

This is a series of segments (numbers) to define an account. For example, the first segment is the company number; this is 01 for the University, and a range of other numbers depending on what the affiliate is. The second segment is the budget unit; like 600 for the faculty of science. Then there is a segment called section that relates to the budget unit; that is the child unit of the budget unit, like the department of Chemistry. The budget unit can have several sections in it.

There are further plans of integrating other systems with THEMIS. For example, the recruitment process involves identifying a vacancy, defining job descriptions, advertising a position, collating responses, short listing candidates, arranging interviews, recording feedback, evaluating candidates, selecting the candidates and making offers. All this is currently done outside of THEMIS, in a system called Page Up. Page Up in essence, is the system for people who have applied for a position at UniCo, but have not yet been hired. It is only when the candidate is offered and he/she accepts the position that he/she is recorded in THEMIS as an employee and thereafter anything to do with this person is through THEMIS. The plan is to replace PageUp and use THEMIS from the early stages of hiring employees.

We are looking at integrating the recruitment system with THEMIS. At the moment there's quite a bit of duplication because a lot of the data that they are keyed in Page Up, they are re-entered into THEMIS. (Development Manager, IT Services)

Optimizing Processes With the Enterprise System

The 'Procure to Pay' process in the Accounts Payable area was streamlined using the Finance module of the Enterprise System. The predecessor of the THEMIS Finance module, called MUFAS, did not provide automation of the processes. For example, when a supplier's invoice came into the University, sometimes it would get lost or someone might store it somewhere and one could never find the original.

It [MUFAS] used to be a very manual sort of system. There wasn't a lot of online approvals and a lot of the processes was pretty much offline, as well as paper-based. We have definitely improved beyond the old system. (Development Manager, IT Services)

UniCo also optimized the way corporate credit card transactions were processed and accounted for. This was done using the iCMS component of the ERP system. This component optimized the process by making it a simple, two-step process. The first step is to be electronically processed by the employee (or a delegate) incurring the expenditure, and the second step is to be processed by the shared service centre for charging the appropriate budget unit that should bear the expenditure. This process was "miles ahead" of the previous manual process that was in place.

Business processes have also been optimized in the area of Human Resources. An integral part of the ERP system is the HR system, including the payroll system. Payroll is a process that happens fortnightly for all staff. There are two categories of staff at UniCo – a) casual staff and b) continuing or fixed term staff. THEMIS is used to pay members belonging to both of these categories. A key contribution that Enterprise Systems made within HR was the enablement of the concept of 'Self Service'. Prior to the system, the lifecycle of an employee, from hiring to termination, was managed outside the system, with only the key events being registered in the systems. With THEMIS in place, lifecycle events like applying for leave, performance appraisals, promotions, undertaking training courses were fully automated by giving control of these events to the employee and his/her supervisor(s).

It [THEMIS] is used for performance appraisal. There are a range of life cycle events, from once somebody starts at the university to things like taking leave, being promoted, going on professional development or training - THEMIS is used for all of those things. (Human Resources Manager, Law School)

Two additional areas where the business processes were streamlined is petty cash flow and student loans. This was one area which used to be manual, run using spreadsheets; it wasn't integrated with the Accounting System.

We didn't have a very good view of it [petty cash]. So that's one thing that we have done now, make the information and process online so that gave you real time information for control. (Development Manager, IT Services)

The process of registering for loan is still paper based but after that it gets into the system so we can send out reports of installments or schedules of payments for students. We do follow-ups like reminding if they fall behind and manage that whole side. And in terms of turnaround, to get a loan out to them, the system provides sufficient information like guarantor information quite readily. The whole process improved. (Development Manager, IT Services)

Another key business process in the area of HR, was optimized using the new Enterprise System. The major optimization here was the fact that UniCo could run a pay without closing down the system. In the previous payroll system, Genesis, UniCo had to shut down the system from 2:00 to 5:00 pm while the pay was run.

Yet another area where business processes have been streamlined is in the area of research administration. Prior to THEMIS, UniCo used to have manual forms where one could enter anything; validating data was very time consuming and, sometimes, not possible at all. With the research component of THEMIS in place, there are a lot more consistent lists of values that users must choose from. So, one can't really get it wrong easily. In addition, there is an online validation process.

There's still a way to go with trying to ensure the integrity of the data that's entered initially but certainly we're not experiencing the same kinds of problems that we did prior to this system. With submitting a contract or a grant, at the end it validates what you've just entered. So if you've forgotten to enter something that's mandatory, it will tell you and you can't proceed prior to that. (Research Systems Business Analyst, IT Services)

Also, in a lot of ways UniCo purchased an Oracle product, which in itself has been reasonably heavily customized to its needs. Many of such issues are intended to be addressed by adopting the next release (Release 12) of the Oracle E-Business Suite which is coming up. One of the key things is looking at whether or not the next implementation is going to be as customized. However, some degree of flexibility in process design is to be maintained, given the diverse nature of the different parts of the organization.

We do a very different type of business here in Architecture, Building and Planning faculty, than they do in Science [faculty] and than that they do in Education [faculty]. Medicine faculty is predominantly research driven. We're predominantly teaching driven; so different load of teaching and different profile. The systems and processes of the university have to have the flexibility to adjust to those differences. (Business Services Manager, Faculty of Architecture Building and Planning)

'Informating' With the Enterprise System

UniCo uses Oracle's Enterprise Technology, in conjunction with Microsoft's easy-to-use interface to build analytics capability to enhance its informating platform. UniCo uses the Oracle Enterprise System as its backbone for capturing and disseminating information. However, it also uses an Oracle Data Warehouse that is fed data from the E-Business Suite. At the very top it uses Microsoft's tools to analyze and present the information.

The Enterprise System enables reporting; there are a number of ways UniCo can generate reports. One is through an Oracle-based 'Discoverer' application; developers can develop reports and then release them into various THEMIS 'responsibilities', or roles with different access levels, that can be used by authorized staff members. The other option of creating reports is to create in THEMIS what is called a 'concurrent program'; it is essentially a script that one can run that produces reports.

THEMIS has been seen to have increased the information levels across the organization, especially compared to the systems that were being used prior to its adoption.

Certainly the capacity of the system to pull out information is better than the old system, Genesis. Finance is definitely stronger on the data you can pull out, and HR is weaker. The strength of reporting in HR is nowhere near as well developed as finance and that's probably a conscious decision on the part of the university. (Business Services Manager, Faculty of Architecture Building and Planning)

However, the platform will be only effective if the necessary data is entered accurately into the system. For example, for researchers it is important that they enter and store their qualifications, what publications that they have had to date, what projects they are involved in, which department the staff member or the researchers belong to, and the like. All of that information is stored in THEMIS.

For example, with things like your qualifications, some of your background information, cultural background, that's really up to the discretion of the individual. So the facility is there, the capability is there; it's really up to the people to enter that. So from that perspective it is real time as soon as you enter it; it's in the system so everybody can see it. (Development Manager, IT Services)

The university is certainly trending towards developments to manage high level information, in the Management Information System where the senior members of the university see the university's future. This includes the development of a Management Information Library (MIL) which is built on the Oracle platform itself. The MIL provides a high level, consolidated, central place to put all kinds of documents - interactive and static.

Addressing the Challenges Using the Enterprise System

It has been found in the literature that Enterprise Systems can enable a number of benefits for adopting organizations. Many large organizations have adopted Enterprise systems and benefitted significantly from the operational efficiency they provide (Sumner 1999; Shang and Seddon 2002; Spathis and Constantinides 2003; Davenport 2004; Motiwalla et al 2012). In addition, it has been found that such systems can enable Innovation through new products and processes, as suggested by Rajagopal (2002), Bradford (2003), Swanson and Pang (2005) and King and Burgess (2006) and Srivardhana (2007). It has also been reported that such systems facilitate Strategic decision-making, as suggested by Bligh and Turk (2004), Spathis and Constantinides (2004), Rom and Rohde (2006), Goodhue et. al (2002) and Rigby and Ledingham (2004).

Automation and More Efficient Business Processes

UniCo enabled by its Enterprise systems platform was able to run its operations a lot more efficiently by automating most of its processes. These benefits included usual benefits from Enterprise Systems like reduction in errors in data entry, faster transaction processing times, fewer employees. UniCo was able to run their operations more efficiently after their adoption of the system. These benefits, collectively called 'operational efficiency', have been achieved by a vast majority of adopters of Enterprise Systems.

Innovative Services: Attracting Global Talent and Implementing Self-Service Model

The Enterprise System enabled UniCo to undertake innovation by delivering new products/ services and process. UniCo launched a new product called 'Find an Expert' for prospective researchers, industry partners and community in general. It is an interactive online tool that allows people to search for researchers with a given topic of research or area of interest. This product is driven by the information about all researchers, their research interests, research projects undertaken, relevant publications and subjects taught stored centrally in the THEMIS system. The concept of this product thus originated from the fact that all this useful information is stored in the Enterprise Systems in an integrated manner and can be extracted to provide this useful service.

The ES enabled all of the research-related information to be put together into a web-based format. So anyone can look for an 'Expert' on a topic or field of his or her interests by searching the system online. The web-based application was built on top of the ES to actually deliver the functionality or the service to the public and within the university and research communities.

This functionality is extended to the affiliates of the University too. For example, in Mental Health and Research and other similar affiliate organizations, the research output of affiliates is actually input into the system in a similar way to what other research staff data is put in. So, this 'Find An Expert' service allows those subsidiaries, and affiliate type organizations to access information easily.

UniCo, enabled by the Enterprise Systems platform, was able to undertake process innovation. Such process innovation was based on two key elements enabled by the ES – a) Accrual Accounting and b) Electronic Authorizations.

With THEMIS in place, UniCo introduced a different method of accounting to what it was using prior to implementing the system. In other words, they shifted from cash-based accounting to accrual-based accounting. This was a major process change and impacted the entire financials of UniCo. The THEMIS system had built-in support for accrual based accounting and UniCo could innovate by simply implementing the finance module of the system. So it can be said that the Enterprise System helped UniCo innovate the foundation for all finance related business processes.

The THEMIS system has certainly facilitated the opportunity to use accrual accounting. For example, when UniCo does Purchase Orders for purchasing assets or for large purchases, the system captures details of the planned commitment of funds at the Purchase Order stage rather than at the Accounts Payable Invoice stage.

Another key contribution of the THEMIS system was electronic authorization. The Oracle E-Business Suite had an underpinning workflow component for approvals and authorizations. This meant that UniCo could introduce electronic authorization. This helped UniCo innovate the foundation for many business processes. An example is the introduction of electronic workflows for approvals for invoices for procuring goods and services.

One of the key new processes enabled by the ES is 'I-Procurement' - procuring goods and services from trusted vendors online. This process is a part of Accounts Payable wherein a purchase requisition is raised for goods. Then a vendor is selected from a list of pre-selected vendors in the market and a Purchase Order is placed. Once goods and invoice are received, goods are checked and the invoice is forwarded for payment. The Accounts team usually makes the payment by the due date. There are several approvals required in this process. Prior to THEMIS, this process was completely manual and involved paper-based forms and human intervention at every step of the process. THEMIS introduced 'I-Procurement' to redesign the 'Procure to Pay' process. The entire process was to be executed using workflows using a standard browser, similar to purchasing books online. This involved connecting to the systems of external parties (like vendors), generating all required documents like Purchase Orders, and requesting/granting approvals at different stages of the process. 'I-Procurement' was essentially a new process spanning across and beyond the boundaries of UniCo to fulfill its procurement needs in a completely automated manner. This process was implemented entirely using their THEMIS system.

A second new process that the Enterprise System enabled was 'Self Service', in a number of areas. These processes were predominantly paper-based and not automated prior to the adoption of Oracle E-Business suite. This new 'Self Service' processes were implemented in a number of functional areas, ranging from Human Resources (HR) to Research Administration.

Certainly prior to the THEMIS implementation we did rely on a greater number of forms. With THEMIS, we were then able to move a number of transactions to a self-service model. As a staff member if you want to apply for some restricted number of things, you would go into THEMIS and do it via self-service. Then that request would travel to your supervisor and you'd be in the system and be approved. (Human Resources Manager, Law School)

When the HR module of THEMIS came on board, self-service elements were introduced for maintaining personal and professional details, leave applications, and others. Employees could now retrieve useful documents like Pay slips and 'Pay As You Go' Tax Summaries at the end of financial year using THEMIS.

THEMIS also enabled Performance Development Review (PDF) to be done online, as a self-service. Time sheeting of casual staff members through this self-service model is a new process enabled by THEMIS. The university has a very significant casual staff workforce, and the ability of self-entry and approval by an authorized officer rather than reliance on administrative staff at the departmental level has reduced the workload. Prior to THEMIS, it was all processed by administrative staff. Most departments would actually have an off-system timesheet that staff would fill out, it would then go to the administrative person who would process the hours into the system, and then it would get approved. But the payroll module of the THEMIS system has been automated for the casual staff members whereby they submit their own timesheets for approval to their supervisors, who then approve it, and forward to Central HR services, who then finally process the payment - all using the Enterprise System. In addition, the system implemented other HR processes like performance appraisal, taking leave, being promoted, undergoing trainings and so on – THEMIS is used to process all of this. The system also does have other advantages in automatic calculation of overtime which help UniCo manage compliance and maintain good relations with staff.

Electronic authorizations were extended to research administration too. UniCo developed a self-service method to administering research so that there is not much paperwork going from departments and faculties to central administrative divisions. A key area of self-service in research is Ethics Applications. Research Students, who need to deal with humans or animals for their research, need to apply for Ethics clearance before they commence their study. 'Self-Service' was implemented for this process by enabling the students to apply for ethics clearance online and submit for approval to different people up a hierarchy. The application would move seamlessly through the approval hierarchy and a notification will be sent to the student once it is approved. This is completely a new way of doing things by putting the research student in charge of the application- all one using the THEMIS system.

Before THEMIS, pieces of paper had to be signed off, physically, by heads of departments and deans. And then it kind of, by the time it actually got the research office for example, there's a lot of time passed. We tried to minimize this by introducing this online system because it could all be then done electronically. So workflow and streamlining the processes and making them more efficient and having them online I would say are the big objectives there. (Research Systems Business Analyst, IT Services)

So Enterprise Systems enabled innovation at UniCo through the following new products/ services and processes:

- **'Find An Expert':** A new service for identifying researchers.
- **'I-Procurement':** A new process for procuring goods and services.
- **'Self-Service':** A new service for employees and researchers.

Strategic Decision Making for Long Term Growth

The Enterprise System enabled UniCo to make strategic decisions. A part of the Enterprise Systems platform is the MIS – Management Information System. This is a data warehouse that feeds all the data from THEMIS. The system can collate data from research, finance and HR and do management -style strategic reports out of that.

The data in the data warehouse pulls the data from different sources and stores them in an aggregated manner. These are rolled on as 'Information Cubes' in the system. There are cubes for staff, students, courses, research components etc. These cubes are analyzed to make decisions on these entities. For example, which courses are profitable, which staff member draws maximum grants and so on.

UniCo used Enterprise Systems enabled platform to make strategic decisions on the financial front. It has allowed the organization to have figures allowing the decision makers to make financial decisions. UniCo has built a customized project module that it uses for budgeting using the Oracle toolset, used for business decision-making purposes. The reports provided from the Finance module would tell the executive of the university what areas are costing them the most. For example, about four years ago, a review was done of the Arts faculty, and the number of the subjects that were run in the Arts faculty was reduced, because some subjects had extremely low enrolment numbers. This decision was enabled by the data and analysis using the Enterprise System.

The financial reporting at UniCo is basically faculty-based, and there is a series of reports set up for the faculty to be able to run their own reports. All of those reports are set up in Discoverer; a part of the Oracle Enterprise platform. The majority of the reports are standard reports that can be run for particular projects as well as for a faculty.

THEMIS enables financial forecasting – a key component of strategic decision making. Using this, UniCo does a budget to actual variance analysis on a monthly basis for most faculties. It also has quarterly forecast budgets; it can change their budget based on forecast. Essentially, the budget gets uploaded into the THEMIS System, and then the actual financials go through the system. The system allows the finance managers to look directly and quickly at the forecasting data and compare it to the actual. In terms of making strategic decisions, if one was to look at the business direction, the forecast information is readily available. It is that availability of information that allows UniCo to make strategic decisions. With isolated systems, the availability of data becomes difficult and very manual. If the forecast is going well, then one has to make decisions about where to spend the money. If, for example, one is not getting the revenue that one wants, but the costs are still the same, a simple decision would be where to save costs, or how to increase revenue? With Enterprise Systems in place, in the long-term, one could have a look at the trends and analyze them in terms of how the finances are doing.

Another key area where UniCo makes strategic decisions is around workforce management or staffing. The university is essentially a service industry and has a very high salary component to total expenditure. So staffing decisions, particularly long-term ones, are very expensive decisions. These include building a faculty profile, building across multiple academic disciplines, and also ensuring the proper profile of seniority and even age of staff members, A key element here is ensuring that there are enough staff members at lower levels to certainly provide career growth into more experienced staff.

The system has enabled information about staff across a number of parameters. These range from absenteeism reports to their performance review reports. Performance Review is a process which has to happen at least annually for all staff within certain categories. The deans or heads of departments need each of these reports, once a year. The system also has information about research publications by academic staff members- a key indicator of performance of academic staff.

The Oracle-based THEMIS suite is quite good from a management perspective with how it provides for management reporting which we certainly didn't have with our former system. Certainly the new system has a lot stronger management reporting; the capacity to draw information out of a broad range of areas (Business Services Manager, Faculty of Architecture Building and Planning)

Another area where strategic decisions need to be made is research. UniCo has benefitted from using the Enterprise Systems in making research-related decisions. The system shows the funding outputs of each faculty and each department. The Research module of THEMIS has every researcher's output in terms of income from grants, and what they have published. It's absolutely essential for every academic staff member to put in their publications, their funding and grants, and contract. This information is aggregated to work out what the university's key strengths are, and also, at going ahead to look at potential funding.

The University has spent a lot of time strengthening in the last couple of years, significant amount of effort building a coherent set, a strategic base at the top end of our programs providing academic leadership, providing administrative leadership, ensuring we have the strength in our academic programs are very important to us. And I guess looking at our staffing profile from our systems enables us to make those decisions of where we need to invest in staff for the future. (Human Resources Manager, Law School)

Again, the ES also enables UniCo to make decisions about its offerings to the market- courses of study that are offered to students. This involves understanding what courses are popular, and how much it costs to put a course together. Part of what it is doing at the moment is popularity polls using the Enterprise Systems. This is a market research where people say, "What under-graduate degree?" This helps in developing UniCo's 'New Generation' degrees. A costing is done by the faculties to say "if you put a degree together, this is how much you have to charge for it".

Using Enterprise Systems, UniCo is able to do trend analysis on particular courses and subjects. To answer a question like 'how much revenue does a course or a subject generate?' the Enterprise System would use its integrated Finance and HR modules traverse the data back and forth to generate the required reports.

The finance systems have been used to determine how much it actually costs to run those subjects, because if it's costing the university a lot of money, they have to take a step back and say what benefit to the community is it providing at such cost to the university? Business decisions have to be made about whether removing those subjects would impair the education in that area going forward. (Manager, Financial Systems and Research Reporting)

In addition, international students being a major source of revenue for UniCo, it uses Enterprise Systems to make decisions on the regions/countries in the world that UniCo attracts students from. This helps in marketing activities for drawing more international students and thus additional revenue. So Enterprise Systems enabled strategic decision making in UniCo through the following:

- Profitability of faculties / departments
- Key research strengths
- Most valuable academic staff members
- Most popular courses/subjects
- Prominent countries/regions where international students come from

So we've got a [data warehouse] cube on staff, we've got one on students and we've got a couple that's in the Research space, like publications. So based on that, we can do analysis such as student load per staff member, which courses are profitable or popular based on student numbers and enrolment and the like. (Development Manager, IT Services)

We have one of the most advanced research systems in the world. We've had groups go on trips around the world to find even Harvard and places like that to find out what they're doing, have come back saying that we are miles ahead. We are certainly well advanced of any of the universities in Australia. (Manager, Financial Systems and Research Reporting)

FUTURE DIRECTIONS

The journey of UniCo with Enterprise Systems is not finished yet; challenges still remain that are yet to be addressed. The recruitment process is still not integrated with the Human Resource processes of the organization. Several faculties still continue to use their 'legacy' system to appoint part-time or casual employees. Researchers are reluctant to update their publications and grants information on time, as they see little personal benefits in these tasks. There are cultural and political barriers on taking strategic actions based on performance of certain courses or programs of study. However, it may be necessary to have a broader look, beyond the Enterprise System, to address these issues.

As further research, the intention is to follow up this case study by revisiting the case after a few years to explore any further use of the Enterprise Systems to address other or subsequent challenges faced by the organization, or take advantage of any emerging opportunities. Another interesting aspect for future research could be to explore any challenges and issues raised by the use of the Enterprise System itself in adopting organizations, and approaches to address them.

CONCLUSION

This chapter presents a case study on a globally renowned public University (referred to as UniCo) based in Australia and its journey of using its IT systems to achieve its operational and strategic objectives. The study raises a few key issues that one should focus the attention to in the analysis of the case; these issues can be broadly classified into two categories or sets. On the one hand, rising costs of operations and diminishing financial support from the government created a set of operational challenges. On the other hand, increased pressure to publish more research and establishing and maintaining a competitive edge to attract more competent local and international students and researchers created another set of strategic challenges for UniCo.

The case analyses these challenges the organization was facing and how Enterprise Systems was adopted to address those challenges. In particular, the case illustrates how Enterprise Systems assisted in not only improving operational efficiency through integrated systems and optimized processes, but also helped the organization in establishing and maintaining a competitive position in the higher education market through two other means namely, (a) innovating and (b) strategizing. The case thus illustratively explains how Enterprise Systems, through integrated data and systems, optimized processes and real-time information and its analysis, can help organizations achieve operational efficiency, ability to undertake product/service and process innovation, and make better strategic decisions. However, challenges still remain to be addressed, with further use of IT, and even beyond it. These include issues such as (a) dis-integration of the recruitment process with the Human Resource processes (b) several faculties and departments still continuing to use their 'legacy' system to appoint part-time or casual employees (c) reluctance of researchers to update their publications and grants information on time and (d) cultural and political barriers on taking performance-based strategic actions on certain courses or programs of study.

ACKNOWLEDGMENT

I am thankful to my colleagues and guides Prof Peter Seddon and Prof Rens Scheepers for their continuous help and support.

REFERENCES

Amid, A. (2014). Organizational Levels Model for Measuring the Effectiveness of Enterprise Resource Planning System (Case Study TUGA Company, Iran). *Universal Journal of Industrial and Business Management*, *2*, 25–30.

Bligh, P., & Turk, D. (2004). *CRM unplugged: releasing CRM's strategic value*. John Wiley & Sons.

Bradford, M., & Florin, J. (2003). Examining the role of innovation diffusion factors on the implementation success of enterprise resource planning systems. *International Journal of Accounting Information Systems*, *4*(3), 205–225. doi:10.1016/S1467-0895(03)00026-5

Davenport, T. H., Harris, J. G., & Cantrell, S. (2004). Enterprise systems and ongoing process change. *Business Process Management Journal*, *10*(1), 16–26. doi:10.1108/14637150410518301

Goodhue, D. L., Wixom, B. H., & Watson, H. J. (2002). Realizing business benefits through CRM: Hitting the right target in the right way. *MIS Quarterly Executive*, *1*(2), 79–94.

King, S. F., & Burgess, T. F. (2006). Beyond critical success factors: A dynamic model of enterprise system innovation. *International Journal of Information Management*, *26*(1), 59–69. doi:10.1016/j.ijinfomgt.2005.10.005

Lorenzo, O., Bjorn, C., Oksana, K., & Angel, D. (2017). Integration through orchestration: the interplay between enterprise systems & inventory management capabilities. *Journal of Enterprise Information Management*.

Mostaghel, R., Oghazi, P., Beheshti, H. M., & Hultman, M. (2015). Strategic use of enterprise systems among service firms: Antecedents and consequences. *Journal of Business Research*, *68*(7), 1544–1549. doi:10.1016/j.jbusres.2015.01.049

Motiwalla, L. F., & Thompson, J. (2012). *Enterprise systems for management*. Upper Saddle River, NJ: Pearson Education.

Rajagopal, P. (2002). An innovation—diffusion view of implementation of enterprise resource planning (ERP) systems and development of a research model. *Information & Management*, *40*(2), 87–114. doi:10.1016/S0378-7206(01)00135-5

Rigby, D. K., & Ledingham, D. (2004). CRM done right. *Harvard Business Review*, *82*(11), 118–130. PMID:15559450

Rom, A., & Rohde, C. (2006). Enterprise resource planning systems, strategic enterprise management systems and management accounting: A Danish study. *Journal of Enterprise Information Management, 19*(1), 50–66. doi:10.1108/17410390610636878

Ruivo, P., Oliveira, T., & Neto, M. (2014). Examine ERP post-implementation stages of use and value: Empirical evidence from Portuguese SMEs. *International Journal of Accounting Information Systems, 15*(2), 166–184. doi:10.1016/j.accinf.2014.01.002

Sedera, D., Lokuge, S., Grover, V., Sarker, S., & Sarker, S. (2016). Innovating with enterprise systems and digital platforms: A contingent resource-based theory view. *Information & Management, 53*(3), 366–379. doi:10.1016/j.im.2016.01.001

Shang, S., & Seddon, P. B. (2002). Assessing and managing the benefits of enterprise systems: The business manager's perspective. *Information Systems Journal, 12*(4), 271–299. doi:10.1046/j.1365-2575.2002.00132.x

Spathis, C., & Constantinides, S. (2003). The usefulness of ERP systems for effective management. *Industrial Management & Data Systems, 103*(9), 677–685. doi:10.1108/02635570310506098

Srivardhana, T., & Pawlowski, S. D. (2007). ERP systems as an enabler of sustained business process innovation: A knowledge-based view. *The Journal of Strategic Information Systems, 16*(1), 51–69. doi:10.1016/j.jsis.2007.01.003

Staehr, L., Shanks, G., & Seddon, P. B. (2012). An explanatory framework for achieving business benefits from ERP systems. *Journal of the Association for Information Systems, 13*(6), 424.

Sumner, M. (1999, April). Critical success factors in enterprise wide information management systems projects. In *Proceedings of the 1999 ACM SIGCPR conference on Computer personnel research* (pp. 297-303). ACM. doi:10.1145/299513.299722

Swanson, E. B., & Wang, P. (2005). Knowing why and how to innovate with packaged business software. *Journal of Information Technology, 20*(1), 20–31. doi:10.1057/palgrave.jit.2000033

Weill, P., & Ross, J. W. (2009). *IT savvy: what top executives must know to go from pain to gain*. Harvard Business Press.

KEY TERMS AND DEFINITIONS

Enterprise Systems (ES): A large-scale, packaged, application software systems that can be used to streamline and integrate the business processes of an organization and considerably improve information and knowledge levels within the organization as well as with its customers and suppliers.

Informate: To provide access to real time data and the ability to analyze the organizational performance from the inside as well as the behaviour of external stakeholders like customers and suppliers.

Innovation: An activity that adds unique value for the customers, enables competitive advantage, and generates value for shareholders. The dimensions along which an organization can undertake innovation are product and process.

Integrate: To harmonize the data, processes, and systems within the organization as well as beyond the boundaries, extending to customers, suppliers, business partners, and other stakeholders.

Optimize: To standardize the processes (both front-end/customer facing and back end/administrative) by aligning them with so-called "industry best practices" or enhance processes to meet the strategic needs of the organizations that yield competitive advantage.

Process Innovation: The *critical* analysis and *radical* redesign of work flows and business processes to achieve *dramatic* improvements in important measures of performance.

Product Innovation: The development of new products, changes in design of established products, over time.

Strategic Decision Making: Involves making intentional choices or responses about concerns that materially affect the survival prospects, well-being, and nature of the organization.

Strategy: A systematic approach to place the organization in its environment to ensure its continued growth and make it immune to surprises.

Chapter 3
Synergy for Sustainability in the Upcoming Telecommunications Revolution:
The Case of a Developing Economy

Abdul Rafay
University of Management and Technology (UMT), Pakistan

Arsala Khan
University of Management and Technology (UMT), Pakistan

ABSTRACT

This chapter examines the merger of two cellular companies (MOBILINK and WARID) for sustainability in the telecommunication sector of Pakistan. During 2011-2013, WARID faced the news of the possible sales of the company due to falling revenues, constant poor network quality, and lower network coverage in rural areas. In 2014, all telecom players participated in the auction for 3G/4G licenses, but WARID did not participate due to its technology neutral license (TNL). Important decisions were taken in 2014 like launching of 4G/LTE services in major cities, US$500 million investment, increase of tower sites, opening of new regional sales offices. These decisions along with presence of TNL made WARID an attractive target for merger with MOBILINK. In 2016, the formal merger was finalized for benefits like synergies in CAPEX/OPEX, fastest 4G network, network reach to rural areas, roll-out of new services like the internet of things (IoT) and mobile banking. The merger proved successful. During the first quarter of 2017, the company generated PKR 38.7 billion in revenues, up from the same period a year before.

DOI: 10.4018/978-1-5225-5279-6.ch003

PROLOGUE

In July 2016, the Amsterdam-based telecom company VimpleCom Ltd (VCL), Global Telecom Holding (GTH), together with WARID Telecom Pakistan and Abu Dhabi Group of UAE announced to merge MOBILINK and WARID, creating Pakistan's next generation digital telecommunications provider. Earlier in November 2015, these groups agreed to combine their telecommunication businesses. Prior to this, MOBILINK and WARID had received approvals from Competition Commission of Pakistan (Appendix – I), Pakistan Telecommunication Authority (Appendix – II), Securities and Exchange Commission Pakistan, local and international creditors and State Bank of Pakistan (Technologytimes.pk, 2016; Attaa, 2016). Finally, the merger process of MOBILINK and WARID achieved a crucial milestone after the Islamabad High Court approved the amalgamation of the two companies in December 2016 (Thenews.com.pk, 2016).

On the occasion of this announcement, Jean-Yves Charlier, Chief Executive Officer of VimpelCom, said:

VimpelCom continues to make significant progress against its transformation program. This market consolidation represents a further step in our strategy to strengthen

our position in the markets in which we operate. As the VimpelCom Group focuses on an ambitious digital transformation, we are investing in revolutionizing our technology core and product portfolio. We look forward to bringing innovative new products and services to our customers in Pakistan.

Chairman of Abu Dhabi Group, His Highness Sheikh Nahayan Mabarak Al Nahayan, said:

I am very happy that we have arrived at this milestone today which marks a unique moment for the Pakistan telecom sector. It is very positive news for Pakistan's economy. The Dhabi Group remains committed to Pakistan and its economic progress and I would like to thank all the relevant authorities for their support in closing this transaction. I am confident that together, the combined businesses will continue to deliver excellence to their customers while contributing positively to the business environment in Pakistan.

Adeel Bajwa, Chairman Board Executive Committee of WARID Telecom, expressed his gratitude:

This is a milestone in the Telecommunication history of Pakistan. Such steps will go a long way towards enhancing investor confidence in the country and the progress of Cellular Industry. We are very excited about the opportunities this affords us for better serving our customers.

THE BACKGROUND

Telecommunication Sector of Pakistan

The telecommunication sector is one of the best developed technology sector and fastest growing segments of Pakistan's economy. Due to high growth rate, this industry has become the back-bone of Pakistan's economy. It is providing many employment and investment opportunities due to high revenues generated via huge number of users of telecom services. Along with opportunities, this industry also provides vast facilities to communicate and connect globally in better, reliable and modern manner. Pakistan is still an unsaturated market and with the falling cost of handsets there are plenty of new subscribers to go for. But eventually as in saturated markets, if mobile operators want to avoid simply competing on price, they will have to compete on superior services, innovative features and ease of use. Due to current developments of market convergence among Internet, cable and telecom industry there is a need of regulatory restructuring for creating a platform for timely relocation towards next-generation-networks (NGNs) like 4G and beyond (Xia, 2011). For a country like Pakistan, it is necessary to analyze its market regulations and policies and other technological advancements for sustainable development. In developing countries like Pakistan, market analysis yields many factors creating an impact on 3G[1]/4G[2] networks including the institutional aspects as well as indirect aspects that have impact on regulatory policies (Xia, 2012).

In 2017, Pakistan won the GSMA[3] award and ranked third in the list of mobile telecom operator groups. Currently the government is working on a project to provide at least one mobile tower per 100 users in every village of Pakistan by 2030. The government has also started preparations for the introduction of 5G[4] by 2021. Pakistan would be the first country in South Asia to test 5th generation cellular services for faster connectivity. 5G in mobile broadband technology will be launched with an effort from the whole technology industry with firms such as Qualcomm, ZTE, Intel and many others have decided their immense contribution (Yasin, 2017).

The SIM[5] verification drive by PTA[6] during 2014-2016 caused a significant drop in cellular connections in the Telecom Industry of Pakistan. The campaign came in the wake of growing terrorist activities, street crimes and money extortion as mafias used illegally acquired SIMs. To get rid of them, the government and the

PTA undertook a biometric[7] verification campaign and PTA cancelled millions of suspected fake and illegal cellular connections after their users failed to prove their identity via biometric verification. But in 2017, the cellular industry recovered on the back of a fast growing digital economy and popularity of social networking. As on 30th June 2017, the headcount of mobile phone users swelled to all time high 139.75 million in Pakistan (Siddiqui, 2017). Subscription of new mobile connections, however, slowed down with a growth of just 5%. Slowdown was attributed to the compulsory biometric verification, restriction on SIM connections to a maximum of five for a subscriber compared to unlimited connections in prior years, near-saturation in major cities and slow pace of penetration by cellular firms in semi-urban and rural areas. In the year 2017, 12.55 million new 3G/4G connections were acquired. The growth in mobile phone internet connections has also slowed down, but the number of broadband connections touched 42.08 million. During 2017, over 30% of the cellular phone used internet on the phone as well. Use of cellular internet has gone up with a significant rise in the use of apps like Skype and WhatsApp and increase in business activities like e-commerce, e-banking, mobile phone-led branchless banking and dispatch of money via mobile services.

There were 5 cellular companies working in Pakistan; MOBILINK, WARID, TELENOR, ZONG and UFONE. MOBILINK leads the chart with 39 million customers and WARID has a total of 11 million customers. After the merger of MOBILINK and WARID, MOBILINK is able to widen the lead with more than 50 million post-merger customers. At the end of June 2017, TELENOR was closing in the gap with 38 million subscribers; ZONG has 25 million customers while UFONE stood with 19 million customers. Mobile teledensity[8] reached 72.52% (Jabri, 2017).

How WARID Uplifted Its Profile to Become Attractive for Merger With MOBILINK

WARID launched in Pakistan in May 2005. WARID was backed by the Abu Dhabi Group, one of the largest groups in Middle East and the single largest foreign investor group in Pakistan. The company was established as a joint venture between the Abu Dhabi Group and Singtel, a Singaporean telecommunications company. Singtel invested USD 758 million (30% of equity) in WARID, which was later purchased by the Abu Dhabi Group. At that time, many viewed the purchase of remaining stake as part of the Abu Dhabi Group strategy to support WARID's continued growth and to enhance its market position.

WARID implemented a new modern corporate identity as a result of the dynamic changes taking place in the telecom industry in Pakistan. With a reflection of a new strategy, WARID aimed to perceive not only as a telecommunication operator of voice services, but also as a universal provider of comprehensive communications services to both residential and business customers. The company also aimed to reflect the changes in telecom industry of Pakistan by taking its customers on board to the latest technological advancements in the communication sector. WARID developed and executed affinity partnership agreements with reputed organizations. This unique approach enabled WARID to effectively drive sales, improve retention and enhance brand awareness. Till 2011-12, WARID was offering state-of the art telecommunication services at 7000 destinations in Pakistan.

The Problems

WARID's share in the telecom sector had also fallen from a massive 18 million users in 2009 to 12.5 million users in 2013 owing mainly to lower network coverage in the rural areas of Pakistan. As a result WARID had been facing declined revenues. 2012-2014 had been a roller coaster ride, starting from the rumors that the company was up for sale which affected the company's business and led not only to the loss of potential customers but also the existing customers. The financial analysts and specialists were confirming the company's windup keeping in view its performance. The reasons behind the company's possible fall were that it had strong footing in the urban areas but didn't tap the large consumer base in the rural areas of Pakistan.

Financial Overview

Profitability analysis[9] of WARID exhibited a net loss for the years 2010-2012. The accumulated loss was reduced by 13.5% in 2011 but again mounted by 37.3% in 2012. Regarding ROA[10] the company was following a negative trend of -8%, -7.3% and -10% for years 2010, 2011 and 2012 respectively. Loss per share was following the same trend. Return on equity (ROE) was also deteriorating. Revenues of the company showed growth of 5.88% for 2010-2011 and 9.1% for 2011-2012. If revenue was increasing and profitability was declining, the reasons must be higher costs. Gross profit margins of the company depicted an increasing trend whereas net profit margins were not only negative but stagnant during the period. The problem lied with operating expenses of the company that left the company with net accumulated losses. A 17% increase in financial charges[11] during 2011-2012 was also a hampering element. In the nutshell, the situation was quite alarming and the company was required to make major changes in its capital structure.

The liquidity position[12] of WARID continued to suffer since 2010. This trend was persistent although current assets[13] were increasing. Current assets and current liabilities[14] were following a fluctuating trend e.g. in 2011 decrease in current assets was 7.55% but in 2012 it was improved by 130%. Current liabilities grew by 153% in 2011 and by 11.27% in 2012. Current ratio[15] in 2010 was approximately 0.28

which implied that there were PKR 0.28 to meet currently maturing obligations of PKR 1. In 2011 and 2012 it was 0.1 and 0.21 respectively which clearly exhibited that WARID's liquidity position was a cause of concern. One reason for this was short term borrowings along with associated obligations that had been perpetually increasing. Another stringent test of short term liquidity, Acid test ratio[16], also depicted about similar results i.e. 0.15, 0.075 and 0.18 for years 2010-12 respectively. Due to continuous deteriorating liquidity position, it was presumed that it might cause trouble for the company.

The company's financial leverage[17] is measured by debt equity ratio[18]. In the case of WARID, total debt has increased by 9% from 2011 to 2012 while for 2010-11 it was only 1%. Difference between debt to equity ratio of 2010 and 2011was not that substantial but there was a clear growth in this ratio for year 2011-12. A debt ratio of 7.58 for 2012 means that for PKR 1 of equity financing, PKR 7.58 were provided by creditors. Long term debt to equity ratio was 1.88, 0.6 and 1.17 for years 2010, 2011 and 2012 respectively. Depending upon the debt ratio, it raised concerns about the company's debt servicing capacity[19] in the long run. It is important to note that WARID was operating with high leveraged (too much debt verses equity) capital structure with the trend in ratios brought about largely due to changes in current liabilities of the company over the year.

This decline in revenues along with a few administrative issues caused Singtel having 30% share in WARID to sell off its share for as low as USD 150 million and exit the scene. During 2011-12 WARID faced the news of the possible sale of the company (Brohi, 2013). The consequences of the news stirred both internal (employees) and external (suppliers and customers) stakeholders. Uncertainty reached at its peak during the next couple of months with constant media hype that WARID is looking for a prospective buyer (Ahmed, 2013). People were shifting to other networks given the constant poor network quality of the company especially in the major cities. Due to the selling news the management was in no position to revise or issue any new tariff plans.

The company was still in a state of confusion when in July 2013 PTA issued an SOP[20] to all telecom service providers notifying them that from July 31, 2013 all unverified SIMs would be blocked. Starting from August 1, 2013 SIMs would only be issued and activated through biometric verification which meant that SIMs that were not sold from authorized business centers or franchises would not be activated. The SOP further notified that telecom service providers should instruct their customers to use SIMs issued against their own CNICs[21] and biometrics; all other SIMs should either be transferred or verified to their own CNIC or would be blocked. This was another big task that management had to face along with the proposed sale of the company. The situation demanded immediate attention so that the company might not lose its customers base.

Due to these factors coupled with the critical financial position of the company telecom service providers backed off from the company buyout. Pakistani Telcos[22] were of the view to bid for WARID but once the picture became clear, they backed off.

The Crucial Meeting

Multiple meetings were being organized by management to decide the future course of action regarding the spectrum license[23] auction. All were aware of the fact that in midst of this situation the company needed a strategy to come back and reclaim its position but the biggest question was how? Most of managers were agreed with the proposed action plan to directly switch to LTE[24] network from GSM[25] network. The managers tried to convince the Board of Directors (BOD) of the possible strength and existing resources of WARID and that this would actually help the company to retrace its position in the market and claim profits. In this regard, a high level meeting was held in 2014 to discuss in detail various core and non-core issues like choice of 3G or 4G/LTE, budgetary issues, supply chain establishment, inbound/outbound logistics, operational strategy, PTA SIM verification policy, high employee turnover and finally decision making with mutual consent. Extracts from minutes regarding relevant discussions and decisions are summarized below (Rafay & Khan, 2016):

3G or 4G/LTE?

The issue regarding whether to launch 3G or 4G LTE was a critical one. The decision made at this very moment could have long term effects on the company's revenue and most importantly its profitability in the future. All the big players in the market had all auctioned for 3G/4G licenses in which WARID did not participate. It could directly jump to LTE service because it had a Technology Neutral License (TNL)[26]. The launch of 3G service by these players was widely appreciated by the customers and many had started using them. WARID was under pressure regarding which technology to opt for as its customers had started to switch to other networks that had launched 3G. Deciding to launch 3G meant that the company would have to go with a head on war with the players in the market which were bigger and had a huge customer base. This would mean that WARID would still miss out on opportunity to attract more customers and profit for them.

The increased activity on social networking sites and the constant need of users to stay online has led Telcos to have remarkable bandwidth[27]. The need for providing high speed, improved coverage, mobility and soaring data rate required a shift to 4G network for WARID. The shift in 4G LTE networks provide telecom users with increased data rate (up to hundreds of megabits per second) and support for the latest technologies like video conferencing, voice chats (Duan, Huang, & Warland,

2015) etc. The 4G upgrades are not an easy transition even with the provision of required spectrum and other exclusive rights. Any telecom operator considering switching to 4G network has to follow IMT- Advanced standards[28]. 4G networks are supported on IP[29] packet switching[30] in place of circuit-switching technology[31] and employ MCM[32] like OFDM[33] method.

The cost of a 4G upgrade was usually very high but with the passage of time it decreased as the process and service matured. Mainly, two key factors are significant for a mobile operator while making the upgrade decision: 4G upgrade cost and user switching cost. In competitive markets like Pakistan with multiple operators, an early upgrade would mean larger market share and improvement in quality of service would reimburse the high upgrade cost (Duan, Huang, & Warland, 2015). Sometimes in competitive situations like these, other operators delay the upgrade to avoid huge upgrade costs and severe competition pressures. In view of foregoing, shift to 4G means handling much technical issues that need to be properly addressed before the transition. In order to adopt 4G/LTE network, WARID would be required to have greater network and subscriber intelligence that would result in quality of service and network proficiency (Kamrujjaman, & Alam, 2014).

The idea to launch LTE services had more problems attached to it. The minimum amount of spectrum required for LTE was 2x10 MHz[34], while WARID had just 2x8.8 MHz in 1800 MHz band, which meant that it could offer LTE services only to a limited number of subscribers. To be able to provide full speed LTE, WARID would have to buy two additional spectrums. Previously, it only had the ability to provide LTE at the 1800 MHz band, it now would have to purchase the 2100 MHz band and 2400 MHz band because 1800 MHz band is only supported by limited phone sets. Moreover, WARID would have to free-up its 1800 MHz band for data services which is currently being used for voice services. It meant that company would have to shift 900 MHz band with additional investment. Not to forget, WARID has 2x5 MHz spectrum from 900 MHz band which might get congested if entire voice traffic is shifted to it.

After hours of debate among board members, management decided with mutual consent to launch 4G/LTE services in major cities of Pakistan instead of 3G. This decision was mainly to the fact that 4G/LTE was the latest technology and that WARID wanted to gain competitive advantage over its rivals who were all providing 3G services. The new strategy focusing on latest technological advancements was a decision to bring the company out of influx of past years issues. The management at WARID believed that by launching this latest technology it was investing in the future since almost all the latest phones have LTE support in them. It would also attract the niche market of high end rich consumers who were interested in high speed data technology and yet at the same time were willing to pay for it. However, to be able to provide 3G alongside with 4G/LTE, WARID would then have to purchase three additional bands that would have added to the project cost without any significant return.

Budget Analysis and Allocation for 4G/LTE

Despite the edge that WARID has over its competitors for having a TNL which saved money by not participating in the 3G/4G bidding held by the PTA, WARID had required a lot of cash to carry out the proposed upgrade. WARID was enjoying the industry's largest post-paid user base and the highest ARPU of around PKR[35]260, well over the industry average of PKR199 as on June 2014. According to official resources, this was due to its loyal corporate clientele that alone generated an ARPU of PKR1400. After carrying out a consumer survey the management realized that customers were very satisfied with their service and wanted WARID to upgrade its network to the latest 3G/4G technology. Given this scenario, WARID had maintained its 12.5 million customer base which meant that it could win back its customers through coverage of rural areas as well.

After consensus, the management decided to develop a budget for the expansion and up-gradation of its existing network. The company planned to invest USD500 million which would be largest investment in LTE services by any telecom company. This budget was established for the next five years. The Chairman approved the budget, out of which USD100 million were allocated to upgrade the network and procurement of LTE equipment (Baloch, 2015). The budget was divided in a series of sections that included a section for the upgrade to 4G/LTE and also to increase its tower sites by 2000 to enhance the total coverage area. The estimated cost of setting up one tower was around USD 150000-200000 which meant that a significant part of the budget had to be allocated for the network expansion.

Supply Chain Establishment

The amplification effects of the supply chain management majorly arose from the manufacturing industry. However, in a service supply chain, the amplification process yields that the workload is a more suitable measure for amplification effects rather than the inventory levels. Furthermore, the most effective countermeasure for amplification is to improve the quality within all the service chain (Akkermans & Vos, 2003). Quality within supply chains can be enhanced through improved capacity management and inventory management which also help in achieving both the cost cutting as well as service related objectives of the company (Jammernegg & Reiner, 2007). When the service provider is able to comprehend the needs of its consumer with the appropriate offerings it yields to enhanced joint demand chain that results in greater demand chain efficiency and consumer satisfaction (Heikkilä, 2002). One of the critical issues faced by WARID was to manage the supply chain. WARID should be able to offer to wide range of customers and contexts.

After consensus, the BOD decided to launch the 4G/LTE service at once which meant that all cities had to have the necessary technical and customer support services. This was a big issue given the fact that WARID was operating on a tight budget and did not have enough resources to cater to all the cities at once. In order to launch the 4G/LTE service in Pakistan with proper coverage and effect, WARID decided to divide the entire project into multiple phases. This meant that R&D department at WARID had to come into play once again. It was decided to gather data that included demographics and other information such as socio-economic class of the people living in the major cities of Pakistan. A detailed analysis of the statistics helped the management to decide that Phase 1 of the launch should only include the metropolitan cities of Pakistan including Karachi, Lahore and Islamabad. After successful implementation the company shall work on the Phase 2 of the project which included a total of 10 neighboring cities around the metropolitan.

Inbound and Outbound Logistics

The inbound logistics for WARID included SIM cards, designing packages, printing brochures or leaflets for packages, roaming and network signals. As suggested by managers, WARID decided to manufacture its SIMs in a local software house. Previously SIMs used to be manufactured in China, Malaysia, Korea and Singapore and then later on imported to Pakistan. The outbound logistics included calls, Short Messaging Services (SMS), video chats and video conferencing. The distribution network of WARID comprised of local distributers that deliver SIMs to customer care centers and franchises. The warehouses were situated at the zonal offices of WARID. It was also decided to open around 19 regional sales offices throughout the Pakistan that shall operate under their respective zonal office.

Operational Strategy

The operational strategy at WARID was always to provide quality services at competitive prices for improved customer experience. In this regard, WARID decided to invest in 4G/LTE networks. After reasonable arguments, WARID management decided to work in liaison with some strategic partners thus, putting emphasis on its core competencies through the influence of the competitive strength of the strategic partner. In view of this, WARID decided to enter into an agreement with Ericsson[36] for networking services and requisite equipments for up-gradation.

PTA SIM Verification

The up-gradation and expansionary phase in WARID had a big anchor attached to it which would not let the process go on so smoothly. The government of Pakistan wanted the Telcos to biometrically verify all SIMs sold by them or they would get blocked.

Following options were proposed to retain the customer loyalty.

- Advertise heavily and try to educate people to get their SIMs verified
- Invest in a procedure that involved getting the biometrics from the customer at his/her residence.
- Establish verification facilities at franchises, business centers and convenience stores.
- Offer free incentives (airtime, SMS and mobile data) after verification

WARID, being as a responsible telecommunication organization, decided to implement Biometric Verification System (BVS) at different outlets all over the country. It was decided to install over 10,000 devices at costs USD300-400 per device. These devices shall be installed across Pakistan in compliance with PTA's directives pertaining to mandatory installation of BVS. Doing so shall ensure that no one could be able to purchase new SIM using fake CNIC. During its 1st phase of development, WARID decided to introduce the Billing Validation Services in Karachi, the largest city of Pakistan and then expanded its implementation to other places across other cities covering 12 business centers and 358 franchises. In its 2nd development phase, WARID shall install over 2000 devices in same cities. In its 3rd phase WARID shall install over 10,000 devices across Pakistan.

High Employee Turnover Rate

WARID has been experiencing a very high employee turnover rate from the past year. This problem was associated with a number of reasons. The news of WARID being put up for sale has also added to this problem. Employees were leaving WARID because everyone was of the view that WARID would be sold and that they would lose their jobs. Moreover, the recent up-gradation of the cellular industry to 3G has created many new job openings for employees in the industry. A lot of employees were leaving WARID because they were getting offers from other operator companies. Other employees who did not leave demanded an increase in their remuneration to match the offers that they were receiving from other companies. It was suggested to introduce new pay scales with better incentives for the employees to keep them motivated. He was of the view that their services shall be required to handle tasks related to the new launch of 4G/LTE. Moreover, best people for the job are those who are already in the same industry. Management agreed with the suggestions and asked the HR personnel to make clear to the employees that the decision to sale WARID has been overturned by the BOD and the pay packages shall be revised within few months.

The decisions taken in the above said meeting played a key role in revival of WARID. At one time the Abu Dhabi Group was thinking to sell WARID and exit Pakistan, however the top management decided the other way around and ready to take on the challenge and get back in the race. WARID got a new direction to provide best cellular experience to its current and prospective customers through the efforts of the management and the decisions made by them to tackle the problems the company had been facing.

THE MERGER

Initially no one really cared if the merger would become a reality, until December 2016 when the customers heard the news of formal merger. MOBILINK completed the acquisition of 100 percent of WARID's shares, while Abu Dhabi Group's shareholders acquired 15 percent of the shares of MOBILINK. Then these shares were decomposed, where VimpelCom retained 85 percent shares of the merged company while Dhabi Group got 15 percent of the merged company. The merged company benefited from VimpelCom's one billion dollars investment in infrastructure building.

Brief Profiles of Counterparties

VimpleCom

VimpelCom is an international communications and technology company, headquartered in Amsterdam, and driven by a vision to unlock new opportunities for customers as they navigate the digital world. Present in some of the world's most dynamic markets, VimpelCom provides more than 200 million customers with voice, fixed broadband, data and digital services. VimpelCom's heritage as a pioneer in technology is the driving force behind a major transformation focused on bringing the digital world to each and every customer. VimpelCom offers services to customers in 14 markets including Russia, Italy, Algeria, Pakistan, Uzbekistan, Kazakhstan, Ukraine, Bangladesh, Kyrgyzstan, Tajikistan, Armenia, Georgia, Laos, and Zimbabwe. VimpelCom operates under the "Beeline", "WIND", "Djezzy", "MOBILINK", "Kyivstar", "banglalink" and "Telecel" (Waridtel.com, 2016).

Global Telecom Holding (GTH)

GTH, which is 51.9 percent owned by VimpelCom Limited, is a leading international telecommunications company operating mobile networks in high growth markets in Africa and Asia, having a total population under license of approximately 401 million as of 31 March 2016. GTH operates mobile networks in Algeria ("OTA"), Pakistan ("MOBILINK"), and Bangladesh ("Banglalink"). GTH reached more than 86 million customers as of 31 March 2016. GTH is traded on the Egyptian Exchange under the symbol (GTHE.EY), and on the London Stock Exchange, its GDRs[37] are traded under the symbol (GLTD:LI) (Waridtel.com, 2016).

Abu Dhabi Group

Dhabi Group / Dhabi Holdings manages the investments of His Highness Sheikh Nahyan Mubarak Al Nahyan and his affiliates, which span various sectors and geographies, particularly: Telecommunication (WARID & Wateen, Pakistan), Banking & Financial Services (Bank Alfalah, Pakistan), Real Estate, Energy (Oil & Gas / Geothermal), Healthcare and Construction in South Asia (Pakistan), Eastern Africa (Uganda), GCC (mainly UAE) and the Caucasus/Black-Sea Region (Georgia, Turkey) (Waridtel.com, 2016).

MOBILINK

MOBILINK is operated by Amsterdam-based Vimplecom, the international communications and technology company. MOBILINK's turnaround over the past two years (2016-2107) has shown significant improvements in service revenue growth (from -5% to +12% year on year), ARPU (from -11% to +15% year on year) and increased EBITDA[38] margins (from -3 p.p.[39] to +4 p.p year on year).

Prospective Benefits

Following were the expectations of both companies at the time of merger:

- Financial Benefits
  - MOBILINK and WARID will benefit from VimpelCom's USD one billion investment into building one of the largest and most ambitious IT infrastructures in the industry.
  - The transaction is expected to create CAPEX[40] and OPEX[41] synergies of around USD500 million with an annual run rate of more than USD100 million of Free Cash Flow (FCF)[42] improvements.

  - The combined revenue of both companies for the 12 months to March 2016 amounted to USD1.4 billion.
- Technological Benefits
  - Pakistan's digital development will be elevated to the next level, further reducing the digital divide.
  - Over 50 million customers in Pakistan will benefit from high-speed mobile telecommunications and a best-in-class digital mobile network.
  - MOBILINK, a provider of the largest 2G and 3G networks in the country, along with WARID's 4G/LTE service, will become Pakistan's fastest 4G network provider. With MOBILINK and WARID customers both on board, the two organizations will strive to minimize discrepancies in their cellular service (Shairani, 2016).
  - The combined MOBILINK and WARID entity will be the leading telecommunications provider of 2G, 3G and Long Term Evolution (LTE) services in Pakistan, providing higher quality national voice and data coverage, faster downloads, and a wider portfolio of products and services. MOBILINK only has a 2G and 3G coverage, and WARID only has 4G LTE. Both MOBILINK and WARID customers can benefit from an integrated 4G service operating under the new entity. In areas of absence of either network, a 3G or 4G service will be made available to customers.
  - MOBILINK serves a wider coverage reach while WARID's signal coverage is smaller in comparison. Different network capabilities of both companies would help each other regarding better signal coverage.
  - The joint operator shall plan to overcome little gaps in communication in Pakistan's rural factions. In order to resolve limited signal coverage in outskirts and rural areas, there would be a plan to revolutionize communication and form a digitalized generation. The merger shall provide customers with sound network coverage reach from urban to sub-urban and rural areas.
  - The Group's new digital systems will enable faster roll-out of new local products and services, particularly in the areas of communications, the Internet of Things (IoT)[43], and mobile financial services.
  - Customers often complain that Pakistan has poor online banking facilities, and the combined entity is paying attention. It plans on offering easy-to-operate and user-friendly online banking services on mobile. This will be implemented through a mobile banking and mobile financial services feature.
  - Mobile entertainment shall be a primary focus. The merged company shall put focus to introduce high definition audio and video sharing on

mobile by enhancing the existing connection. In this regard, it is decided to sign a contract of one billion dollars with Ericsson to recondition hardware and technology.

- The focus of the merger is also to come up with a high definition video and audio calling that needs a few major customer service upgrades too.
- Both companies will revamp their packages in combination to produce executive packages for customers. The new entity will claim to offer premium quality service.
- Merger would also lead to a unified SIM recharge system, MOBILINK customers will also be able to recharge their SIMS using a WARID card, and vice versa (Khan, 2016).
- The combined entity shall upgrade SMS delivery by making it quicker and faster; a popular facility amongst mobile phone users.

POST MERGER SCENARIO

MOBILINK officially confirmed in February 2017, that it has given up its brand name and will now be called as "JAZZ" after its merger with WARID.

Financial Results for FY 2016 (Q4) of the Merged Entity

For the year 2016, merged company generated a total of PKR 38.7 billion in revenues during Q4 2016, up from PKR 26.8 billion during same period last year. For the entire fiscal year 2016, the company posted revenue of PKR 135.6 billion, up from PKR 104.2 billion during previous year. Reported service revenue increased by 43% during the quarter, while organically, excluding WARID, it increased by 16%. Mobile Financial Services revenue continued to show robust growth at 34%. Data revenue, excluding WARID, organically increased by 62%, mainly due to successful data monetization initiatives, including attractive bundle offers and the unification of the tariff portfolio, together with continued 3G network expansion. WARID's data was consolidated starting July 01, 2016.

Underlying EBITDA margin, excluding both restructuring costs related to the performance transformation programme and integration costs related to the WARID transaction, was 40.3% in Q4 2016, slightly lower year-on-year due to the consolidation of WARID during 2016. CAPEX increased to PKR 10.1 billion in Q4 2016 while the full year CAPEX to revenue ratio decreased to 16.6% and full year operating cash flow margin was 26%. In November 2016, the company declared, for the first time after 11 years, a gross dividend of PKR 5 billion (approximately USD 50 million) to its shareholders.

The company offered 3G to WARID customers in 30 cities while 4G was offered to MOBILINK customers in 17 cities. Interconnect, site sharing activities and marketing synergies reached PKR 8.2 billion of run-rate of synergies in Q4 2016. MOBILINK's customer base increased 14% YoY in Q4 2016 (Attaa, 2017a).

Financial Results for FY 2017 (Q1) of the Merged Entity

During first quarter of 2017, the company generated PKR 38.7 billion in revenues, up from PKR 36.8 billion — for both WARID and MOBILINK — during the same period a year ago. The revenue growth was supported by all revenue streams; in particular, data revenue grew by 29% YoY due to growth in data customers. Total data revenues reached PKR 5.2 billion during the quarter.

Underlying EBITDA margin, excluding PKR 0.6 billion of restructuring costs related to both performance transformation and the WARID integration, was 43.4% in Q1 2017, improving by almost 4 percentage points year on year. CAPEX increased to PKR 3.6 billion in Q1 2017 while the LTM[44] CAPEX to revenue ratio decreased to 17.7% in Q1 2017 and the operating cash flow margin was 34%.

The company closed the quarter with a whooping 52.5 million customers, up from 48.3 million a year ago, showing 8.7% percent growth. At the end of the first quarter, 3G was offered in more than 350 cities while 4G/LTE was offered in over 50 cities. The company's ARPU stood at Rs. 231.1 per user per month while voice usage in minutes per user per month were 609. The average data usage per user per month remained at 465MBs (Attaa, 2017b).

The initial financial results and increasing customer base show clearly the success of merger.

DISCLAIMER

The published chapter shall be meant only for academic discussion and shall not be intended to serve as endorsement, source of primary/secondary data or illustrations of effective and ineffective management.

ACKNOWLEDGMENT

The authors are thankful to all published sources and management of relevant companies for kind support and permission to use company information for the chapter.

REFERENCES

Ahmed, A. (2013, June 26). The Abu Dhabi Group plans to sell its 100% stake in WARID Telecom Private Limited (WTPL) – AHL Research. *Investor Guide 360.* Retrieved from http://investorguide360.com/latest-economic-news/abu-dhabi-group-plans-to-sell-its-100-stake-in-WARID-telecom-private-limited-wtpl-ahl-research/

Akkermans, H., & Vos, B. (2003). Amplification in service supply chains: An exploratory case study from the telecom industry. *Production and Operations Management*, *12*(2), 204–223. doi:10.1111/j.1937-5956.2003.tb00501.x

Attaa, A. (2016). *PTA okays Mobilink-Warid Merger with Conditions.* Retrieved from https://propakistani.pk/2016/05/23/pta-okays-mobilink-warid-merger-with-conditions/

Attaa, A. (2017a). *Post Merger: Jazz Posts 44% Growth in Revenues During Q4 2016.* Retrieved from https://propakistani.pk/2017/02/27/post-merger-jazz-posts-44-growth-revenues-q4-2016/

Attaa, A. (2017b). *Post Merger: Jazz Posts 44% Growth in Revenues During Q4 2016.* Retrieved from https://propakistani.pk/2017/05/11/jazz-posts-5-4-revenue-growth-q1-2017/

Baloch, F. (2015, January 23). One year later: Warid Telecom comes back with a bang. *The Express Tribune.* Retrieved from http://tribune.com.pk/story/826035

Brohi, B. (2013, June 27). WARID Telecom up for sale. *Dawn News.* Retrieved from http://www.dawn.com/news/1021149

Duan, L., Huang, J., & Walrand, J. (2015). Economic Analysis of 4G Upgrade Timing. *Mobile Computing. IEEE Transactions on*, *14*(5), 975–989.

Heikkilä, J. (2002). From supply to demand chain management: Efficiency and customer satisfaction. *Journal of Operations Management*, *20*(6), 747–767. doi:10.1016/S0272-6963(02)00038-4

Jabri, P. (2017). *Mobile phone users cross 140.5mn with 72.52pc tele-density.* Retrieved from http://www.brecorder.com/2017/07/31/362290/mobile-phone-users-cross-140-5mn-with-72-52pc-tele-density/

Jammernegg, W., & Reiner, G. (2007). Performance improvement of supply chain processes by coordinated inventory and capacity management. *International Journal of Production Economics*, *108*(1), 183–190. doi:10.1016/j.ijpe.2006.12.047

Kamrujjaman, Md., & Alam, Md. N. (2014). *Feasibility of Migration from 3G to 4G including Comparative Study*. Northern University Bangladesh. Retrieved from https://www.researchgate.net/publication/271647287

Khan, W. (2016). *Mobilink-Warid Merger: Mobilink to Warid calls will now count as on-net*. Retrieved from http://techrave.pk/2016/10/mobilink-warid-merger-mobilink-warid-calls-will-now-count-net/

Rafay, A., & Khan, A. (2016). Shift towards Next Generation Networks (NGNs) for sustainability: Evidence from an emerging Economy. *Journal of Cases on Information Technology*, *18*(3), 1–12. doi:10.4018/JCIT.2016070101

Shairani, K. T. (2016). Mobilink and Warid are now one happy family — but what's in it for customers? *Dawn News*. Retrieved from https://www.dawn.com/news/1272471

Siddiqui, S. (2017). *Mobile phone subscribers reach close to 140m*. Retrieved from https://tribune.com.pk/story/1471997/mobile-phone-subscribers-reach-close-140m/

Technologytimes.pk. (2016). *CCP approves Mobilink merger with Warid*. Retrieved from http://technologytimes.pk/post.php?title=CCP+approves+Mobilink+merger+with+Warid

Thenews.com.pk. (2016). *Islamabad High Court allows merger of Mobilink, Warid*. Retrieved from https://www.thenews.com.pk/print/173740-Islamabad-High-Court-allows-merger-of-Mobilink-Warid

Waridtel.com. (2016). *VimpelCom and Dhabi Group announce completion of Mobilink and Warid transaction*. Retrieved from http://www.waridtel.com/media-center/press-release/archive/381

Xia, J. (2011). The third-generation-mobile (3G) policy and deployment in China: Current status, challenges, and prospects. *Telecommunications Policy*, *35*(1), 51–63. doi:10.1016/j.telpol.2010.12.008

Xia, J. (2012). Reprint of: Competition and regulation in China's 3G/4G mobile communications industry—Institutions, governance, and telecom SOEs. *Telecommunications Policy*, *36*(10), 798–816. doi:10.1016/j.telpol.2012.10.002

Yasin, A. (2017). Pakistan to be the first country in South Asia to test 5G services. *Dawn News*. Retrieved from https://www.dawn.com/news/1326401

ENDNOTES

1 Third Generation (3G) is a mobile communications standard that allows mobile phones, computers, and other portable electronic devices to access the internet wirelessly.

2 Fourth Generation (4G) is the broadband wireless mobile communications that will supersede the third generation (3G).

3 GSM Association (GSMA) is a trade body that represents the interest of mobile operators worldwide.

4 Fifth Generation (5G) aims at higher capacity than current 4G, allowing a higher density of mobile broadband users, and supporting device-to-device, ultra reliable, and massive machine communications.

5 Subscriber Identity Module (SIM) is an integrated circuit chip that store the International Mobile Subscriber Identity (IMSI) number and its related key, which are used to identify and authenticate subscribers on mobile telephony devices such as mobile phones and computers.

6 Pakistan Telecommunication Authority (PTA) is state regulator of Pakistan responsible for the establishment, operation and maintenance of telecommunications in Pakistan.

7 Biometric verification is any mean by which a person can be uniquely identified by evaluating one or more distinguishing biological traits including fingerprints, hand geometry, earlobe geometry, retina and iris patterns, voice waves, DNA etc.

8 Teledensity is the number of telephone connections for every hundred individuals living within an area.

9 Profitability analysis is the evaluation of a company's return on investment and involves identification as well as measurement of profitability sources.

10 Return on Assets (ROA) shows the percentage of how profitable a company's assets are in generating revenue.

11 Financial charges include interest charges on borrowings, commitment fees and other charges paid by the borrower for availing the loan facility.

12 Liquidity position of a company refers to the ability of meeting its short term financial obligations.

13 Current Assets are all assets that can reasonably expect to be converted into cash within one year. Generally these include cash and cash equivalents, accounts receivable, inventory, marketable securities etc.

14 Current liabilities are a company's debts or obligations that are due within one year and include short term debt, accounts payable, accrued liabilities and other debts.

15 Measuring current ratio of a company leads to the ability of that company to satisfy its current liabilities with the available current assets.

16 Acid-test or quick ratio measures the ability of a company to use its cash to retire its current liabilities immediately.

17 Financial leverage refers to the use of debt to acquire additional assets.

18 The debt-equity ratio implies the amount of Debt a company is using to finance its assets. It helps to assess a company's long term financing structure and credit risk.

19 Debt servicing capacity is the ability to make payments of the principal and interest on outstanding loans.

20 Standard Operating Procedure (SOP) is a set of step-by-step instructions created to carry out routine operations and their purpose is to achieve efficiency, quality output and uniformity of performance, while reducing miscommunication and failure to comply industry regulations.

21 Computerized National Identity Card (CNIC) is issued by National Database and Registration Authority (NADRA) of Pakistan and serves as an identification document to authenticate an individual's identity as the citizen of Pakistan. This is machine-readable and carry facial and fingerprint information.

22 Telco refers to Telecommunication Company.

23 Spectrum license for mobile telecoms define the frequency band, geographic reach and duration of usage rights, fees and charges, network coverage, roll-out requirements, and technical or service parameters.

24 Long Term Evolution (LTE) is a technical process for high-speed data for phones and other mobile devices.

25 Global System of Mobile Communication (GSM) is a set of protocols for second generation (2G) cellular network.

26 Technology Neutral License (TNL) allows operators free to evolve their services as the technology advances without the use of any non-interfering technology in the specified band.

27 In telecom bandwidth is the frequency range between lowest and highest attainable frequency usually measured in hertz. The hertz (symbol Hz) is the unit of frequency and is defined as one cycle per second.

28 International Mobile Telecommunications-Advanced (IMT-Advanced) standards are issued by the International Telecommunication Union (ITU) for 4G mobile phone and Internet access service.

29 The Internet Protocol (IP) is the method by which data is sent from one computer to another on the Internet. Each computer (known as a host) on the Internet has at least one IP address that uniquely identifies it from all other computers on the Internet.

30 Packet switching is a digital networking communications method that groups all transmitted data into suitably sized blocks, called packets. Packet switching increases network efficiency, robustness and enables technological convergence of many applications operating on the same network.

31 Circuit switching is a methodology of implementing a telecommunications network in which two network nodes establish a dedicated communications channel (circuit) through the network before the nodes may communicate.

32 Multi-carrier modulation (MCM) is a method of transmitting data by splitting it into several components, and sending each of these components over separate carrier signals.

33 Orthogonal Frequency Division Multiplexing (OFDM) is a technique for transmitting large amounts of digital data by splitting the radio signal into multiple smaller sub-signals that are then transmitted simultaneously at different frequencies to the receiver. OFDM reduces the amount of crosstalk in signal transmissions.

34 The megahertz (MHz) is equal to one million hertz (1,000,000 Hz) and is commonly used to express microprocessor clock speed. The clock speed typically refers to the frequency at which a processor is running.

35 PKR is the abbreviation of Rupees; official currency of Pakistan. In August 2017, USD1 = PKR105 (Approx)

36 Ericsson is a Swedish multinational corporation that provides communication technology and services.

37 Global Depository receipts (GDR) is a certificate issued by a depository bank, which purchases shares of foreign companies and deposits it on the account.

38 Earnings before Interest, Tax, Depreciation and Amortization.

39 Percentage points (p.p)

40 CAPEX stands for Capital Expenditures

41 OPEX stands for Operational Expenditures

42 Free Cash Flow (FCF) represents the cash that a company is able to generate after spending the money required to maintain or expand its asset base. It is a measure of a company's financial performance, calculated as operating cash flow minus capital expenditures.

43 The Internet of Things (IoT) is the inter-networking of physical devices, vehicles, buildings, and other items embedded with electronics, software, sensors, actuators, and network connectivity which enable these objects to collect and exchange data.

44 Last Twelve Months (LTM).

APPENDIX 1

CCP Conditions for Approval of Merger

In January 2016, Competition Commission of Pakistan (CCP) approved the merger of WARID Telecom and MOBILINK subject to the following conditions:

Spectrum Concentration

At present, the total spectrum allocated to five mobile operators in the country is 108 MHz, out of which the merged company will have a combined spectrum of 37.2 MHz (combined spectrum of MOBILINK and WARID) after the merger, which would be 34.44% of total spectrum that's allocated to telcos.

Apart from the overall size of the spectrum allocated to an operator, each frequency band carries unique properties in terms of its spectral efficiency and propagation properties on wireless mediums, making certain frequencies more attractive than others.

CCP said that merged company will hold the largest portion of the 900 MHz frequency band i.e. 12.4 MHz, which can be used for both GSM (2G/3G) and LTE networks. Not to mention, no further allocation of spectrum is possible in 900MHz band. Merged company will also have 14.8MHz in 1800MHz band — that also allows for LTE deployments — that accounts for 21.49% of total spectrum allocated in this band. Then in 2100MHz band, the merged company will have 10MHz of spectrum that's usable for 3G.

With respect to spectrum shares of merged company, the Commission said that it is necessary to remedy the disproportionate allocation.

In this regards, PTA is asked to develop a spectrum sharing framework, after which in a case when new operator — that wants to operate in the country — comes in and there's no additional spectrum available then merged company will be allowed to negotiate the terms with new entrant to share the spectrum with it.

PTA is asked to finalize the spectrum sharing framework with-in six months.

Merged company, however, is restricted from entering into any spectrum sharing or trading arrangements with Telenor Pakistan due to structural link between Telenor ASA and VimpelCom unless approved by the PTA and the Competition Commission of Pakistan.

Wholesale Access to Mobile Virtual Network Operators (MVNOs)

Competition Commission of Pakistan recommended PTA to carry out a reassessment of its current MVNO framework in consultation with all stakeholders including mobile operators and to take reasonable steps to encourage and facilitate MVNO entry in Pakistan. Commission recommended PTA that it may consider imposing a wholesale access obligation on all mobile operators currently operating in Pakistan on a non-discriminatory basis.

Once PTA is able to develop this new framework on MVNOs, merged company is binded to provide wholesale access rights to MVNOs, including but not limited to network access, call origination, call termination, international roaming for voice/ SMS and data services, and access to portability database.

Such access to wholesale call origination shall be granted on fair and reasonable commercial terms and conditions.

Interconnection

Competition Commission of Pakistan, in its decision, binded merged company to continue providing interconnection on the same terms to all other mobile network operators in accordance with the Reference Interconnection Offer (RIO) approved by PTA.

Competition Commission offered all mobile operators to approach the commission in case of any anti-competitive behavior.

Infrastructure Sharing

Ongoing Infrastructure Sharing Agreements between WARID/MOBILINK and other mobile companies in cases where merged company will plan to discontinue its sites should be re-negotiated.

Not to mention, almost 4,000 sites are supposed to be decommissions when MOBILINK and WARID will be merged into one company.

In such renegotiation, merged company will do one of following:

- If there's only one tenant on a shared-site, merged company will allow such guest operator to buy all fixtures, tower or any other equipment on site on fair-value and reasonable commercial terms and conditions with-in a period of 30 days from initial notice
- If there is more than one tenant on a Site, then merged company will auction all fixtures, tower or any other equipment on such Sites subject to a fair value

reserve price within the period of thirty (30) days from initial notice. Upon successful sale, the acquiring mobile company will continue to host the other guests on the same terms and conditions.

- If parties cannot agree on the fair value then price would be determined by an independent valuer mutually agreed between the parties who shall carry out the valuation within fifteen (15) days from reference.
- If the sale is not completed within the thirty (30) day period of reference then guest operator(s) would vacate the Site
- However, for a period of five months from the date of Order, merged company will not request an mobile operator that is a guest on its Sites to leave the site without granting one of the two options noted

Joint Control / Cross Holding

Just because VimpelCom is partly owned by Telenor ASA, and that Telenor Pakistan is fully owned by Telenor ASA, while MOBILINK is fully owned arm of VimpelCom, CCP said that merger must not raise any cross-holding or joint control issues in local market.

Commission said that merging parties will appoint an independent third-party-reviewer (TPR) to ensure compliance on current merging companies' cross-holding with Telenor Pakistan and to ensure the compliance of CCP's earlier decisions at the time when VimpelCom had acquired Orascom.

Commission, in its decision, said that this term will cease in case Telenor ASA sells its shares in VimpelCom, which it has intended lately.

APPENDIX 2

PTA Conditions for Approval of Merger

In May 2016, PTA approved the merger of WARID Telecom and MOBILINK subject to the following conditions:

- MOBILINK will accept and own all liabilities of WARID Telecom with respect to all the licenses that WARID was issued by PTA.
- MOBILINK and WARID will not change names and brand names without 30 days' prior written notice to PTA.
- The Parties shall not reduce total number of interconnection circuits (E1s) allocated by them to other operators including LL, LDI, cellular licensees,

without prior approval of PTA. Interconnect capacities or their pricing shall not be used in a manner to impede other operators access to the customer of the Merged Entity.

- The determination is subject to clearance of all outstanding dues by the Parties.
- PTA would be free to take any legal action if information provided by MOBILINK and WARID to regulator for the approval application is fabricated, incomplete or false.
- The Parties shall continue to operate as separate legal entities till the amalgamation order.
- If the Parties intend to do on-net calls amongst each other before the amalgamation order, they shall submit proposal for PTA's approval.
- The Parties shall not merge their networks and shall not decommission any network element. BTSs may be decommissioned with 30 days prior notice to PTA. PTA may restrict decommissioning of any BTS, if so considered by PTA in alignment with condition 201 and 202 of the CCPs decision on merger.
- MOBILINK and WARID will continue to submit their annual audited financial statements as separate entities till the amalgamation order.
- The agreements between the Parties and franchises/authorized retailers / sellers shall remain unchanged for a term of 6 months from the date of issuance of the NOC.
- The Parties will inform all the consumers and public at large through advertisements in at least two leading daily Newspapers, i.e., one Urdu and one English, of major circulation for three consecutive days regarding the merger within seven working days of the date of amalgamation order.
- Rationalization / re-arrangement/ redeployment of Base Transceiver Station (BTS) sites may occur as a result of merger. Merged Entity to ensure efficient deployment of microwave backhaul frequency links after the merger and return redundant and unused microwave spectrum scarce resource to Frequency Allocation Board (FAB) as coordinated and determined by FAB.
- All BTS sites of MOBILINK and WARID shall continue to remain operational and will not be dismantled for a minimum of 4 months' period from the date of NOC.
- The Merged Entity shall give first right to guest user (currently sharing BTS site with MOBILINK / WARID) to acquire BTS sites rendered surplus. In case, the guest user refuses to acquire surplus BTS sites, the Merged Entity shall apply a judicial and transparent process / method to transfer ownership of BTS sites to other telecom licensees including tower / infrastructure licensees in order to avoid possible dismantling of BTS sites.

- Merged Entity shall ensure that record of all de-commissioned BTS sites and backhaul links is updated with FAB and PTA.
- Merged Entity shall ensure that Quality of Service (QoS) being experienced by MOBILINK customers for 3G does not deteriorate with the joining of WARID's customers as a result of merger. Similarly, the QoS being experienced by WARID's LTE customers should not deteriorate.
- The Merged Entity's pricing/ tariff offers shall not be in any way anticompetitive, unfair or burdensome.

MOBILINK and WARID filed unconditional acceptance of the above said terms and conditions of PTA.

Section 2

Innovation in Information Technology

In this section of *Technology Management in Organizational and Societal Contexts*, our focus turns to how organizations can foster innovation through information technology. In particular Petra Nylund and co-authors' story of software developer PlantFactory addresses two important topics: home bias and the dynamics of competition.

First, the PlanetFactory story introduces the topic of "home bias," the propensity of innovators to carryout activates in geographic proximity to a firm's location. A double edge sword, home bias can help innovators reduce cost and gain a deep knowledge of customers. However, home bias can keep innovators away from the lead users and other partners they need to work with in the early stages of product development. The PlanetFatory setting of Cabrianes, Spain is certainly far afield from Silicon Valley and other Internet startups.

Second, the authors' analysis speaks to the dynamics of competition. After analyzing PlanetFactory's position using Porter's classic five forces approach and the static concept of "competitive advantage", they introduce D'Aveni's concept of the "Age of Temporary Advantage." Indeed, with the speed of competition in a digital age, competitive advantage can only be held for a limited time.

Chapter 4
Home Bias in Innovation Ecosystems:
Relying on Local Users for Knowledge

Petra A. Nylund
University of Vic - Central University of Catalonia, Spain

Núria Arimany-Serrat
University of Vic - Central University of Catalonia, Spain

Xavier Ferràs-Hernández
University of Vic - Central University of Catalonia, Spain

José Antonio Corral-Marfil
University of Vic - Central University of Catalonia, Spain

ABSTRACT

Innovation ecosystems enable small- and medium-sized enterprises to compete with innovations that include the knowledge of others. Integrating external knowledge in innovation is however a complex process, and there are many pitfalls to consider. This chapter analyses a small company that leverages its innovation ecosystem to compete in a dynamic global industry characterized by rapid technological change. The company faces the challenge of creating and commercializing a breakthrough innovation for digital education, with educational content for teachers and easy-to-use authoring tools for modifying this content. As the company attempts to benefit from lead-user experiences, home bias towards knowledge from local users appears as an unexpected hurdle. The analysis of the case encompasses the firm, the industry, and the ecosystem levels, and indicates key concepts for innovation in ecosystem settings.

DOI: 10.4018/978-1-5225-5279-6.ch004

INTRODUCTION

Innovation ecosystems enable small and medium-sized enterprises to compete with innovations that include the knowledge of others. Integrating external knowledge in innovation is however a complex process, and there are many pitfalls to consider. This case study analyses a small company which leverages its innovation ecosystem to compete in a dynamic global industry characterized by rapid technological change. The company faces the challenge of creating and commercializing a breakthrough innovation for digital education, with educational content for teachers and easy-to-use authoring tools for modifying this content. As the company attempts to benefit from lead-user experiences, home bias towards knowledge from local users appears as an unexpected hurdle. The analysis of the case encompasses the firm, the industry, and the ecosystem levels, and indicates key concepts for innovation in ecosystem settings.

ORGANIZATION BACKGROUND

PlanetFactory was established on January 1, 2000, in the garage of Bernat Casanovas in Cabrianes. The company specialized in creating educational applications and experienced very rapid initial growth. In 2007, the global financial crisis brought with it cost cuts in Spanish public administration. These cuts were especially hard on PlanetFactory, which relied on the public sector for 90 percent of its turnover. In an extremely critical moment, Bernat and his team decided to transform the company to minimize its exposure to fluctuations in the demand of intermediaries that they had suffered as a service company. They wanted to be able to reach the end user directly, and therefore planned to create a competitive proprietary product. Hence, the first objective they set was to develop an innovative product based on internal R&D. Once they had developed the product, the second objective would be to internationalize the company. The financial situation of the company is summarized in Tables 1 and 2.

The PlanetFactory team had also learned the business of customized services should not depend so much on a single industry, i.e. that of public administration (Corral-Marfil et al., 2015). They made a great effort to find customers in the publishing industry. As they worked with publishers, they detected a market opportunity. The team observed a need to create content in the educational business. They also found that those who developed content at the publishers lacked technical skills. Therefore, they decided to create a product that was extremely easy to use.

The team sensed that tablets would replace PCs completely, which would make all content based on the Flash program obsolete. In 2010, PlanetFactory decided to

Table 1. Balance sheet for PlanetFactory 2007-2013

Assets	2013	2012	2011	2010	2009	2008	2007
A) Assets (non-current)	654,296	603.656	447.723	460.878	475.121	484.799	56.344
B) Assets (current)	421,498	322.366	317.160	279.912	372.671	364.091	473.875
Total assets	1,075,794	926.022	764.883	740.790	847.792	848.890	530.219
Liabilities and Owners' Equity	2013	2012	2011	2010	2009	2008	2007
A) Owner's Equity	668,021	562.261	456.920	439.094	400.394	322.897	172.741
B) Liabilities (non-current)	464,009	291.246	196.697	207.433	221.443	232.460	11.463
C) Liabilities (current)	47,163	72.514	111.266	94.264	225.956	293.533	346.015
Total Liabilities and Owners' Equity	1,179,193	926.022	764.883	740.790	847.792	848.890	530.219

Table 2. Income statement for PlanetFactory 2007-2013

	2013	2012	2011	2010	2009	2008	2007
A) Operating profit	-434	67.535	27.342	55.269	107.790	212.460	153.568
B) Financial expenses	10,029	-5.966	-4.296	-3.669	-12.408	-6.538	-497
C) Profit before tax (A + B)	9,595	61.569	23.046	51.600	95.381	205.922	153.071
D) Income tax expenses	-4,970	43.772	-5.220	-12.900	-17.884	-55.766	-39.911
E) Profit for the year (C + D)	4,625	105.341	17.826	38.700	77.497	150.155	113.160

produce a multi-platform authoring tool to create content for both tablets and PCs. In 2013, they begun sales of MakerPF.

SETTING THE STAGE

Early in the morning of January 8th, 2017, Bernat Casanovas, CEO and founder of PlanetFactory, was pacing his office, ignoring the view of the hills around the small village of Cabrianes, Catalonia. His employees would soon be back after the Christmas holiday and would want to know where he was taking the company in the near future. After years of working on the new product MakerTeacher for the creation and commercialization of a digital school environment, it was make-or-break time. Bernat had to decide whether to continue investing in MakerTeacher based on the last test results.

The current range of products and services of the company was divided into three areas: authoring tools, educational applications, and services.

MakerPF

MakerPF was an intuitive authoring tool that allowed publishers to quickly and easily create, develop and publish interactive educational content without programming or technical support. With a user-friendly software, the same people who had conceptualized an idea could create content for multiple platforms by the click of a button. Most of the active MakerPF licenses had been sold to Spanish customers and nearly all were in the possession of publishing houses such as Oxford University Press, McGraw Hill Interamericana, Grupo Planeta, and Anaya. The sales of MakerPF were made by sales representatives who visited the publishers in person. The company was working on solving some customer service issues such as the time difference with the U.S., language differences with Brazil, and cultural differences concerning Nigeria.

MakerPF had two direct competitors: Kitaboo, a tool by Hurix Systems, and Raptivity, made by the Harbinger Group. The two competing enterprises were from India and considerably larger than PlanetFactory. According to Bernat Casanovas, the competitive advantage of PlanetFactory consisted in providing a more user-friendly tool.

Make It

Make It was a simplified version of MakerPF. Make It was an authoring tool for tablets, as opposed to MakerPF which was used on PCs. Make It was aimed at teachers and students and combined multiple dynamic screens, an integrated teaching method, and professional presentation techniques in one extremely user-friendly application. Make It was sold directly to consumers and PlanetFactory had found the commercialization of this product much more cumbersome than that of MakerPF.

Educational Apps

The company offered 20 educational apps for children. Most were sold through the App Store, four through Google Play, and two exclusively on Samsung devices. In 2017, the counter on the web site of PlanetFactory showed over 7,000,000 app downloads mainly from the U.S., the UK, and a number of Asian countries.

CASE DESCRIPTION

The company had the ability to create a comprehensive solution for a digital education model, including both content and tools for creating and modifying contents in a learning environment. The new solution was based on the current successful authoring tools of the company which targeted publishers and school children. The part of the project that worried Bernat Casanovas was the commercialization. On one hand, he knew that the company had to look for new products and new markets to stay on the path of rapid growth, and a product aimed at teachers would be perfect to complement the current offering. In addition, Bernat considered it a huge potential market. Moreover, teachers were a complex market of which PlanetFactory did not have much knowledge. The company was still small and had never had a powerful sales department. It had been very tough to get this far, and Bernat was not sure whether he wanted to risk all he had achieved.

The Market Situation

The interest of some governments in digitizing schools was becoming increasingly apparent. Governments around the world were developing a digital agenda in the area of education. Many countries, e.g. Brazil, Colombia, Holland, Scotland, and Indonesia, were implementing the widespread use of tablets in schools. The case of South Korea was emblematic: The country had invested $2.4 billion in a program to replace textbooks with tablets in all schools by 2016 (Stallard & Cocker, 2014).

In most countries, both students and new teachers beginning their careers were digital natives. There was a growing interest among teachers and students in digital and collaborative training models, and PlanetFactory had received questions and requests from teachers regarding their authoring tool. New providers such as Kahn Academy were providing free Internet instruction which was being used within and outside traditional school environments (Thompson, 2011). In some cases, technology was incorporated into courses through computers, portable media devices with Web 2.0 technology, micro-blogging, and social media networks (Schnackenberg, 2014). Other courses incorporated gamification and visual technologies with the goal of increasing student motivation and engagement (Villagrasa et al., 2014).

Emerging countries opted for digital distance learning for demographic and logistical reasons. For instance, Malik (2012) analyses the initiative to adopt e-learning at an institute of higher learning in India, and Zhang (2007) reports the case of a distance course for students located in more than 60 cities around China.

In other countries, cultural movements and educational trends supported the direct involvement of parents in children's learning, learning from home (e.g. with the Montessori methodology) or learning at home (i.e. home schooling). Digital innovation was also beginning to be viewed as a core component of the education curriculum (Fichman, Dos Santos, & Zheng,, 2014). Digital learning and authoring tools were therefore gaining significant relevance. End users of e-learning were very interested in authoring tools, mainly for mobility and parameterization possibilities. Educators needed to create their own content and customize content created by others. For example, González et al. (2005) describe how mobile technologies were used to teach physics, by means of developing mobile applications and the use of smart phones. There was a definite increase in demand for customized learning.

Tablets were replacing personal computers worldwide. The technological development was tightly aligned with the educational progress. Large companies such as Apple, Samsung, and Microsoft were developing programs targeting schools. With the great potential of digital learning, these companies perceived a promising business opportunity in the educational market. Moreover, due to the gradual standardization of mobile devices, manufacturers began to differentiate from competition by incorporating quality content in their devices, including apps, programs, and games.

In the education industry there were many teachers interested in authoring tools. End users of e-learning are highly interested in authoring tools, mainly because of its mobility and parameter setting options (Weiss, 2011). Both teachers and students need to create their own content or modify existing content. The shift from students as consumers to students as creators is a key trend in education: "There is a growing trend on university campuses in which students are doing more content creation and design, across the spectrum of disciplines." (Johnson et al. 2014, p. 14).

A New Product Idea

MakerPF had excess functionality compared to their needs, and therefore too high a price for the teacher market. PlanetFactory was sure that the tablet, not the PC, would be the tool of choice for digital education. For this reason, PlanetFactory wanted to make it easy to transfer content created on a tablet or PC to tablets. Transferring content from PCs to tablets was technically complex and required a developer's license. The publishing houses had trained technicians, but teachers had neither the training nor the tools to produce and publish content. Bernat Casanovas saw an enormous market in the creation of content by teachers. He wanted to create a new authoring tool adapted to the needs of teachers and a kiosk that would organize, distribute, and sell digital content on different devices and platforms.

The new product had the working title MakerTeacher and would consist of a new authoring tool and a kiosk. MakerTeacher would be positioned half way between the Make It and the MakerPF. Similarly to Make It, it would be very easy to use and would incorporate templates, but like MakerPF it would enable content creation with computers for increased comfort in adapting teaching materials. The created content could be visualized on tablets, smart phones or computers.

Content created with the new authoring tool, would be transferred to the kiosk, in order to make them available to students world-wide. The kiosk would allow for the organization, distribution, and commercialization of digital content in diverse devices and iOS, Android, and web based platforms. It would consist of a multiplatform cloud repository of contents.

The new product would be aimed at teachers of primary schools around the world. This was a vast market compared to the current market of the company that consisted of the publishers who bought MakerPF. To try to access a market as large and dispersed, PlanetFactory was considering targeting groups of schools and educational communities instead of providing the product school by school, or country by country. First, several religious congregations with strong presence in Latin America, appeared as ideal candidates: the Jesuits, Marist Brothers, Opus Dei, and others. Second, another point of entry to groups of schools that PlanetFactory considered were national and regional governments. The company had already been in contact with the governments of Catalonia, Colombia and Brazil. Third, other possible partners for commercializing the new product included Learning Management Systems (LMS) companies such as Moodle, Blackboard, or Schoology. The LMS providers had access to the global market and direct access to schools. The only issue with this route was that there were many different LMS and PlanetFactory had to ensure compatibility with each.

There was yet another factor that had to be taken into account: The education industry was undergoing a stage of revolutionary change. The change arose from globalization, with participatory teaching methods that highlighted entrepreneurial skills spreading from leading countries such as Finland. The lecture was a dying practice. The change was also related to new needs associated with the rapid pace of knowledge and the generation of new skills. Finally, the immense possibilities offered to teachers by digital education contributed to the new era of education. In such a dynamic moment, it was impossible to predict the paradigm that would characterize the future of the industry.

CURRENT CHALLENGES FACING THE ORGANIZATION

Bernat Casanovas contemplated what his team had expressed in the last meeting. His business partner, Gemma Malet, wanted to create the new product MakerTeacher in cooperation with existing customers:

The best would be to reach agreements with major publishing houses who agree that our authoring tool could be a way of commercializing their content.

Bernat had disagreed with Gemma:

If we are to join forces with a publisher and their content, we would need to find one that's very open to change, and most aren't. Also, the schools and teachers interested in digital learning don't look favorably on publishers. They tend to generalize and see publishers as a reminiscence of the past. So, if we would go with a publisher, besides getting the other publishers against us, we could lose the trust of schools and teachers.

The CTO, Jordi Puigdellívol, worried about the cost of 273,000 euros and the size of the project.

I think its suicide. We are too small to enter a market of this size. The project is too complex for our company. We don't have enough resources to manage it.

Bernat was not sure if entering the teacher market would bring PlanetFactory closer to a successful future or if it would be the end of the company he had created. To move the innovation-process forward, the three agreed to finalize a prototype, and test in on a group of lead users. Based on the result of the test, they would revise their strategy and decide whether to move on and invest further in the launch of the product.

Home Bias Toward Local Users

Firms increasingly need to collaborate to achieve competitive advantage (Dyer & Singh, 1998). Therefore, firms increasingly innovate together with other organizations e.g. corporations, small and medium enterprises, entrepreneurs, venture capitalists, universities, and other institutions (Chesbrough, 2003). Such loose networks of organizations and individuals that interact and coevolve to create and capture value from innovation are denominated innovation ecosystems (Moore, 1993, Iansiti & Levien, 2004, Adner & Kapoor, 2010).

PlanetFactory was acutely aware of the need to constantly input external knowledge in the innovation process for MakerTeacher. Bernat Casanovas was regularly reading international news about technological advancements, and saw San Francisco as the epicenter of the innovation ecosystem of PlanetFactory. As a prototype for MakerTeacher was ready, Bernat sought user input to validate the design. In 2016, the team decided to test the prototype by introducing it for free in several schools in the nearby town of Manresa. The teachers were enthusiastic about the product and the team prepared to make minor adjustments before launching the product to the rest of Catalonia, followed by a gradual international expansion. However, as user statistics from the Manresa teachers became available, the team was disappointed to learn that the teachers who had shown such great interest in MakerTeacher were not actively using it at all.

Testing a product with lead users is a common methodology for assessing viability. However, it appears the Manresa teachers were not actually lead users. Lead users have strong current needs for the product in question (Von Hippel, 1986). The systematic identification of lead users allows the company to select users that are ahead of the market trend and who can retrieve high expected benefits from using the product (Bilgram et al., 2008). Manresa is not a city with exceptionally high use of digital contents in educational settings, and teachers are therefore not likely to develop needs ahead of the market. Potential lead users of MakerTeacher would probably already use much digital content and would harbor an urgent need for customizing existing content and for creating content of their own.

The ideal characteristics of lead users were not prioritized when choosing a test population for MakerTeacher. Instead, user selection was based on proximity to the city of Manresa. Geographic proximity is known to reduce coordination costs for interactive learning and innovation (Boschma, 2005). PlanetFactory could access the local education system through personal contacts and their reputation as a local business helped them gain trust with teachers and administrators. The proximity of entities within the ecosystem is a salient predictor of knowledge output (Whittington et al., 2009), and working with local users would help PlanetFactory interact more with users to gain a deeper understanding user needs. The high costs for identifying, evaluating and dealing with foreign partners often make companies work with local entities (Rangan, 2000). It is thus fully understandable that PlanetFactory was subject to home bias selecting the users for testing MakerTeacher.

Home bias is the propensity to carry out activities near the firm's main location (French & Poterba, 1991). Companies do tend to rely excessively on knowledge from nearby locations in their innovation processes (Jaffe and Trajtenberg, 1999; Bacchiocchi and Montobbio, 2010). Besides reducing costs for gaining the knowledge, relying on local knowledge helps companies integrate that knowledge as it has been generated in the ecosystem in which they are embedded (Frost, 2001). Through relying on users from nearby Manresa, the MakerTeacher team could gain a deep understanding of user needs as they share a belief system with the users, and do not have to cross cultural barriers. Home bias thus allows the partners to interact in a setting where they are comfortable with the context and the cultural codes. They then avoid the psychological challenges entailed in cross-cultural interaction (Molinsky, 2007). The drawback in the case of PlanetFactory was that although they involved users in the innovation process, these users where not lead users and their use did thus not show the full potential of the product. In spite of involving the innovation ecosystem to develop MakerTeacher, PlanetFactory was not able to benefit from external knowledge due to home bias. SMEs generally experience more difficulties in benefiting from open innovation than do larger firms (Brem et al. 2017). To progress beyond the recent test results, the company would have to see past this home bias.

TEACHING NOTE: SOLUTIONS AND RECOMMENDATIONS

The case considers the situation of a small company without the support of strong local ecosystem partners, which is competing in a dynamic global industry characterized by rapid technological change. The case is apt for courses on technology management, innovation, competitive strategy, or marketing management.

Our initial analysis concerns how to approach the dilemma of the CEO Bernat Casanovas: Does PlanetFactory have the required capabilities for entering the teaching market, trying to sell applications and content to schools and teachers? Through which channels can they enter the market? We begin by reviewing the commercial strength of PlanetFactory, which is deficient in the words of the CEO, and to discuss possible strategies for commercialization. We discuss whether PlanetFactory could enter the market alone, selling directly to schools, or if it needs another strategy. The discussion then turns to financial issues. What are the financial requirements to address such a market? Can PlanetFactory afford the level of risk associated with the investment?

The case yields three key questions that form the core structure of our analysis:

- Does PlanetFactory have the capacity required for entering the teaching market?
- What is the expansion strategy of PlanetFactory?
- Should PlanetFactory enter the teaching market? How?

Each of the key questions is analyzed and discussed in the text below.

Expansion Strategy

We study which is the expansion strategy of PlanetFactory, and whether it should be focused on sales, finance, or bigger strategic questions. The essence of strategy can be synthesized in three points (Rumelt, 2011):

- An objective analysis of reality.
- A proposal for differential action. The strategy is essentially the way PlanetFactory differs from its competitors.
- A consistent and coherent action plan.

To settle on a strategy for expansion, we must thus consider the extent to which PlanetFactory has made an objective analysis of reality that allows the company to develop a proposal for differential action, and whether it can deploy a consistent and coherent action plan to profit from this difference. The core of the matter is hence what the differential competences of PlanetFactory are and why the company should be able to succeed in the context of digital education. In short, we need to analyze how PlanetFactory can create, develop and perfect competitive advantage in this new market. Considering these components of an expansion strategy, we can conclude that the company has little of an elaborated strategy. The case presents a dilemma the answer to which can only be found through a strategic analysis and, if appropriate, through the formulation and execution of the selected strategy. To develop a financial analysis and strategy formulation we need to carry out an internal analysis of the company and an external analysis of the industry, which we do in the following sections.

Internal Analysis

As a first approach to the strategic dilemma and to understanding of the target industry, we carry out an internal analysis based on the business model canvas (Osterwalder and Pigneur, 2010) for the proposed product. The nine building blocks of the business model canvas are key partners, key activities, key resources, value proposition, customer relationships, channels, customer segments, cost structure, and revenue streams.

- **Key Partners:** Lead users emerge as a key partner group (Von Hippel, 1986). Lead users give feedback and collaborate during the introductory stages of a product motivated by the influence they can exert on product development (Pollock & Hyysalo, 2014). One possibility of real lead users for MakerTeacher is home schooling communities. Another possibility is to associate the lead user to an educational methodology such as Montessori. PlanetFactory could reach exclusive agreements with certain methodological segments of the education industry, e.g. regarding IT support for the implementation of teaching methods. Certain geographic markets may more digitalized and innovative schools, and hence have a higher concentration of lead users.

- **Key Activities:** The key activities are the development of the solution and the introduction of this breakthrough innovation on the market. The first activity falls well within the experience of PlanetFactory and is not expected to cause any problems, whereas the market introduction requires breaking new ground and creating new capabilities.
- **Key Resources:** The key resources of the model include creativity, technology, experience, talent, and flexibility. From the point of view of the resource-based theory, a core competence should be unique, rare, and inimitable. The core competences are exclusive attributes that enable the firm to generate real competitive advantage (Collis & Montgomery, 2008). PlanetFactory must thus ensure that the identified key resources are really core competences, and not commodities available to several competitors.
- **Value Proposition:** The value proposition is centered on the ease of use of the solution and on the contribution to the improvement and customization of content, leading to better teaching materials and shorter preparation times.
- **Customer Relationships:** The customer relationships take a complex structure including buyers, prescriptors, and users. In the digital education market, the buyer may not coincide with the prescriptor of the product or with the user. The buyer may be a school or a governmental entity, prescribed by teachers or parents' associations, with the teacher, and even the students, as users.
- **Channels**: The main channels for market entry include direct sales, sales via intermediaries e.g. publishing houses, electronic sales, etc. Sales via publishing houses appear convenient since nearly all of the current MakerPF customers were publishing houses. However, to reach the mass market, PlanetFactory needs to grab the electronic sales channel.
- **Customer Segments**: The aim of the new business is to capture the under-served segment between publishing houses and students: The teachers. PlanetFactory currently has a complex product directed to the publishing-house segment as well as a simple product directed to the student segment. The teacher segment requires more authoring functionality than the student segment, but also a simpler product than that directed towards the publishing houses.
- **Cost Structure**: Table 3 outlines the financial analysis. With a recent debt increase of 147.000, the debt-to-equity ratio is being stretched. Further investments should be accompanied by a continued increase in sales and profits. The radical increase of the net result in 2012 improves the economic situation and also puts the company in a better position for embarking on a project of the budgeted proportions.

- **Revenue Streams:** Pricing will be completely different if the company manages to obtain revenue streams in the form of bulk licenses to governments and associations, compared to individual teacher purchases. The focus must therefore be entirely on revenue streams from large groups.

External Analysis: The Digital Education Industry

The business model canvas serves as a first analysis but does not provide an answer to the dilemma of PlanetFactory. The tool has helped us visualize the business model, but does not identify how PlanetFactory can generate and protect competitive advantages. We therefore turn to the external environment and make an analysis of the industrial environment. Is digital education an attractive industry? Does it provide us with a business opportunity? We use Michael Porter's five competitive forces model (Porter, 2008) to analyze the digital education industry.

- **Threat of New Entrants:** Is it easy for newcomers to enter the digital education industry? What are the requirements for entry? Financial capacity? Proprietary technology? Creativity? Small entrants can successfully enter markets with large incumbents when they target very specific niches (Markman & Waldron, 2014). It appears that if PlanetFactory can enter this market, any competitor with the same technological and creative capabilities can enter. What happens if bigger competitors decide to enter using greater capacity for prescription, financing, R&D, or marketing? What would happen if the Learning Management System (LMS) providers such as Moodle, Blackboard, or Schoology decide to enter the market? The threat of new entrants does make the industry less attractive, and PlanetFactory would need to partner with other potential entrants to ensure a sustainable position as competition intensifies.

Table 3. Financial analysis of PlanetFactory 2008-2013 ('000 euros)

	2008	2009	2010	2011	2012	2013
Net equity	323	400	439	457	562	668
Debt	526	447	302	308	364	511
Assets	849	848	741	765	926	1179
Sales	600	531	264	280	428	513
Net result	150	78	39	18	105	106

- **Threat of Substitutes**: If PlanetFactory develops this new product, can it be easily replaced by other similar products? In the excitingly dynamic sector of mobile applications, although it may be difficult to conceptualize, new paradigms are continuously emerging and replacing existing technology. In the changing digital environment, there is a clear threat of substitute products. The competitive advantage of one product can rapidly be surpassed by an emerging substitute (D'Aveni, Dagnino, & Smith, 2010).
- **Bargaining Power of Buyers**: Who makes the purchasing decision and under what criteria? In the analysis of the business model canvas, we discussed that the purchasing process relies on multiple agents and it can therefore be expected to be complex and even geographically fragmented. The key buying criteria may be contingent on the characteristics of different countries or regions. However, it is possible that the basic criterion for purchase is the price / quality ratio, and the reputation of the product with teachers, the trust generated by the product, and its usability and compatibility with other tools.
- **Bargaining Power of Suppliers:** To what degree is PlanetFactory dependent on its suppliers? We consider that they are technology suppliers who offer their software as a complement to hardware sales, which generates another key question: which will be the future technological standard? Apple's closed iOS platform or Google's open Android platform? The battle for platform dominance is still raging in this industry (Eisenmann, Parker, & Van Alstyne, 2011), and PlanetFactory's preference for iOS can be expected to limit future choices.
- **Industry Rivalry**: Although the market is still green, with low barriers to entry and high threat of substitutes, industry rivalry can be expected to become important. However, given the possibility of institutional buyers, it is also possible that the market will eventually become an oligopoly of a few brands that will impose a standard accepted by the educational community. The market is currently void, and it may contain few players. In this situation, PlanetFactory would have a business opportunity.

After deepening the understanding of the industry, we can conclude that indeed there is a window of opportunity in a vast market that is not yet directly accessed by authoring tools. However, PlanetFactory has no clear elements of differentiation and is therefore not strong in this industry. How can PlanetFactory generate a competitive advantage? The opportunity lies in being the first-mover and get entrenched in the budding market through building the appropriate relationships. Our challenge now moves from creating a competitive strategy to focusing on the strategy for introducing a breakthrough innovation.

Towards Solving the Dilemma: How to Introduce a Breakthrough Innovation

What are the requirements for introducing a breakthrough innovation in a market that does not explicitly ask for the innovation, but in which there is a latent opportunity? According to literature:

- Competition in dynamic environments with rapid, technological change creates fierce battles between industry giants and small start-ups (i.e. David vs Goliath competition). On the Internet, being smaller can be a competitive advantage for companies that are able to move and scale quickly. Technology foments the reconfiguration of markets, and small start-ups have more strategic flexibility to capture the opportunities of technological change and new market segments (Yoffie & Cusumano, 1999). Firms must understand that their core competences are, how to build new dynamic capabilities, and how to exploit them in high-velocity markets (Wirtz et al., 2007; Helm & Gritsch, 2014). A salient dynamic capability in these markets is that of creating radical product innovations (Slater, Mohr, & Sengupta, 2014).
- Once the innovation has been introduced, PlanetFactory's focus should shift towards creating barriers to entry for competitors. For this purpose, there are numerous strategies associated with first-movers: investment in R&D to maintain technological superiority of second investment in brand gained notoriety, generating greater understanding and experience of the market, or creating an standard that generates lock-in effects for customers who learn to use a certain authoring tool and are than locked in by this skill (Suarez & Lanzolla, 2005)
- In a rapidly changing industry, competitive advantages can be transient (Gunther McGrath, 2013), and PlanetFactory needs to create a process for renewing the competitive advantage based on continuous innovation to adapt to the frequent and rapid changes in the market. Participants may perceive the importance of not endangering the ease-of-use of the product when renewing its functionality. User-support services instead reinforce usability as a competitive advantage and also provide useful feedback for the continuous innovation of the product.

Once the innovation has been introduced to the market, the speed of growth is key to maintaining a competitive position and gaining economies of scale for financing continued development of the product.

The Transformation of the Industry

To conclude the analysis of the case, we make reflections regarding the future of the industry. Porter's five competitive forces is a somewhat static method that was initially conceived in a less dynamic environment. The education industry is however undergoing a major transformation driven by the advent of digital technology and the globalization of markets:

- Methodologically, the lecture format is getting outdated. In particular, Socratic teaching is made obsolete by the expansion of Anglo-Saxon methodologies, oriented towards critical thinking, generating collective knowledge in the classroom, and working in teams and projects. Innovative educational methods are gaining attention e.g. flipped classroom, games, and creative classroom (Peres & Mesquita, 2016; Sankey & Hunt, 2014).
- Conceptually, knowledge has a tendency to become obsolete quicker as we generate new knowledge to replace it. Skills and capabilities last longer. Education is thus increasingly designed to consolidate skills such as learning, knowledge creating, and entrepreneurial skills. New tools will be needed to foment this new direction of education.
- Technologically, today's students, especially at higher levels, have real-time connection with sources of information and audiovisual material, sometimes of higher quality than that offered by the teacher. This access is dramatically reconfiguring and emphasizing the role of teacher.

Digital technology is beginning to radically transform the education industry in a way similar to the situations of industries such as music, movies, publishing, photography, etc. Our analysis shows that PlanetFactory can develop the required capabilities for taking advantage of this transformation and launching a breakthrough innovation. The team would however have to overcome the existing home bias towards knowledge from local sources, and test the prototype on a population of real lead users.

REFERENCES

Adner, R., & Kapoor, R. (2010). Value creation in innovation ecosystems: How the structure of technological interdependence affects firm performance in new technology generations. *Strategic Management Journal*, *31*(3), 306–333. doi:10.1002/smj.821

Bacchiocchi, E., & Montobbio, F. (2010). International Knowledge Diffusion and Home-bias Effect: Do USPTO and EPO Patent Citations Tell the Same Story? *The Scandinavian Journal of Economics*, *112*, 441–470.

Bilgram, V., Brem, A., & Voigt, K. I. (2008). User-centric innovations in new product development—Systematic identification of lead users harnessing interactive and collaborative online-tools. *International Journal of Innovation Management*, *12*(03), 419–458. doi:10.1142/S1363919608002096

Boschma, R. (2005). Proximity and innovation: A critical assessment. *Regional Studies*, *39*(1), 61–74. doi:10.1080/0034340052000320887

Brem, A., Nylund, P. A., & Hitchen, E. (2017). Open Innovation and Intellectual Property Rights: How do SMEs benefit from patents, industrial designs, trademarks and copyrights? *Management Decision*, *55*(6), 1285–1306. doi:10.1108/MD-04-2016-0223

Chesbrough, H. W. (2003). *Open Innovation: The New Imperative for Creating and Profiting from Technology*. Cambridge, MA: Harvard Business School Publishing.

Collis, D. J., & Montgomery, C. A. (2008). Competing on resources. *Harvard Business Review*, (June-August): 2008.

Corral-Marfil, J. A., Arimany-Serrat, N., Ferràs, X., & Nylund, P. A. (2015). PlanetFactory: Competing under rapid technological change. *Journal of Cases on Information Technology*, *17*(2), 1–14. doi:10.4018/JCIT.2015040101

D'Aveni, R. A., Dagnino, G. B., & Smith, K. G. (2010). The age of temporary advantage. *Strategic Management Journal*, *31*(13), 1371–1385. doi:10.1002/smj.897

Dyer, J. H., & Singh, H. (1998). The relational view: Cooperative strategy and sources of interorganizational competitive advantage. *Academy of Management Review*, *23*(4), 660–679.

Eisenmann, T., Parker, G., & Van Alstyne, M. (2011). Platform envelopment. *Strategic Management Journal*, *32*(12), 1270–1285. doi:10.1002/smj.935

Fichman, R. G., Dos Santos, B. L., & Zheng, Z. (2014). Digital innovation as a fundamental and powerful concept in the information Systems curriculum. *Management Information Systems Quarterly*, *38*(2), 329–353. doi:10.25300/MISQ/2014/38.2.01

French, K., & Poterba, J. (1991). Investor diversification and international equity markets. *The American Economic Review*, *81*(2), 222–226.

Frost, T. (2001). The geographic sources of foreign subsidiaries' innovation. *Strategic Management Journal, 22*(2), 101–124. doi:10.1002/1097-0266(200101)22:2<101::AID-SMJ155>3.0.CO;2-G

González, M. Á., Martín, M. E., Llamas, C., Martínez, Ó., Vegas, J., Herguedas, M., & Hernández, C. (2015). Teaching and learning physics with smartphones. *Journal of Cases on Information Technology, 17*(1), 31–50. doi:10.4018/JCIT.2015010103

Gunther McGrath, R. (2013, June). Transient advantage. *Harvard Business Review.*

Helm, R., & Gritsch, S. (2014). Examining the influence of uncertainty on marketing-mix strategy elements in emerging business to business-export markets. *International Business Review, 23*(2), 418–428. doi:10.1016/j.ibusrev.2013.06.007

Iansiti, M., & Levien, R. (2004). Strategy as Ecology. *Harvard Business Review, 82*(3), 68–78. PMID:15029791

Jaffe, A. B., & Trajtenberg, M. (1999). International Knowledge Flows: Evidence from Patent Citations. *Economics of Innovation and New Technology, 8*(1-2), 105–136. doi:10.1080/10438599900000006

Johnson, L., Adams Becker, S., Estrada, V., & Freeman, A. (2014). *NMC Horizon Report: 2014 Higher Education Edition.* Austin, TX: The New Media Consortium.

Malik, K. (2012). Inception to dissolution: E-learning initiative at an Indian institute. *Journal of Cases on Information Technology, 14*(3), 35–56. doi:10.4018/jcit.2012070103

Markman, G. D., & Waldron, T. L. (2014). Small entrants and large incumbents: A framework of micro entry. *The Academy of Management Perspectives, 28*(2), 179–197. doi:10.5465/amp.2011.0112

Molinsky, A. (2007). Cross-cultural code-switching: The psychological challenges of adapting behavior in foreign cultural interactions. *Academy of Management Review, 32*(2), 622–640. doi:10.5465/AMR.2007.24351878

Moore, J. (1993). Predators and prey: A new ecology of competition. *Harvard Business Review, 71*(3), 75–86. PMID:10126156

Osterwalder, A., & Pigneur, Y. (2010). *Business model generation: a handbook for visionaries, game changers, and challengers.* Hoboken, NJ: John Wiley & Sons.

Peres, P., & Mesquita, A. (2016). Flipping the classroom to gain time: A pedagogical innovative model. *Journal of Cases on Information Technology, 18*(4), 36–52. doi:10.4018/JCIT.2016100103

Pollock, N., & Hyysalo, S. (2014). The business of being a user: The role of the reference actor in shaping packaged enterprise system acquisition and development. *Management Information Systems Quarterly*, *38*(2), 473–496. doi:10.25300/MISQ/2014/38.2.07

Porter, M. E. (2008, January). The five competitive forces that shape strategy. *Harvard Business Review*. PMID:18271320

Rangan, S. (2000). Search and deliberation in international exchange: Microfoundations to some macro patterns. *Journal of International Business Studies*, *31*(2), 205–222. doi:10.1057/palgrave.jibs.8490902

Rumelt, R. (2011). *Good strategy, bad strategy, the difference and why it matters*. Deckle Edge.

Sankey, M., & Hunt, L. (2014). Flipped university classrooms: Using technology to enable sound pedagogy. *Journal of Cases on Information Technology*, *16*(2), 26–38. doi:10.4018/jcit.2014040103

Schnackenberg, H. L., Vega, E. S., & Simard, D. A. (2014). Paradigm shift: Introduction of a social media network and web 2.0 technology into a college classroom environment. *Journal of Cases on Information Technology*, *16*(2), 1–12. doi:10.4018/jcit.2014040101

Slater, S. F., Mohr, J. J., & Sengupta, S. (2014). Radical product innovation capability: Literature review, synthesis, and illustrative research propositions. *Journal of Product Innovation Management*, *31*(3), 552–566. doi:10.1111/jpim.12113

Stallard, C. K., & Cocker, J. (2014). *Education technology and the failure of American schools*. Lanham, MD: Rowman & Littlefield.

Suarez, F., & Lanzolla, G. (2005, April). The half-truth of first-mover advantage. *Harvard Business Review*. PMID:15807045

Thompson, C. (2011, July 15). How Khan Academy is changing the rules of education. *Wired Digital*.

Villagrasa, S., Fonseca, D., Redondo, E., & Duran, J. (2014). Teaching case of gamification and visual technologies for education. *Journal of Cases on Information Technology*, *16*(4), 38–57. doi:10.4018/jcit.2014100104

Von Hippel, E. (1986). Lead users: A source of novel product concepts. *Management Science*, *32*(7), 791–805. doi:10.1287/mnsc.32.7.791

Weiss, C. (2011). *E-Learning End Users Have Spoken – time for the industry to listen*. Retrieved June 12, 2015, from http://elearninfo247.com

Whittington, K., Owen-Smith, J., & Powell, W. (2009). Networks, propinquity and innovation in knowledge-intensive industries. *Administrative Science Quarterly*, *54*(1), 90–122. doi:10.2189/asqu.2009.54.1.90

Wirtz, B. W., Mathieu, A., & Shilke, O. (2007). Strategy in high-velocity environments. *Long Range Planning*, *40*(3), 295–313. doi:10.1016/j.lrp.2007.06.002

Yoffie, D. B., & Cusumano, M. A. (1999). The competitive dynamics of Internet time. *Harvard Business Review*, *77*(1), 70–80. PMID:10345393

Zhang, Y. (2007). A teaching case for a distance learning course: Teaching digital image processing. *Journal of Cases on Information Technology*, *9*(4), 30–39. doi:10.4018/jcit.2007100103

KEY TERMS AND DEFINITIONS

Breakthrough Innovation: A technology that creates a new behavior or a new industry.

Digital Education: Education using digital materials.

First-Mover Advantage: A competitive advantage achieved from being first to enter a market.

Home Bias: A propensity towards activity near the home location of the firm.

Innovation Ecosystems: Loose networks of organizations oriented towards innovation.

Lead-Users: Users with strong, early needs for a product or service.

Technology Management: The organization of technology development and use.

Transient Advantage: A short-lived competitive advantage.

Section 3
Security

This section of *Technology Management in Organizational and Societal Contexts* has a single chapter devoted to information security. The number of chapters, however, does not represent the importance of the topic. Indeed, the promise of the digital revolution currently stands in peril due to security challenges by hackers and rogue governments. For the Internet to be a highway of information and payments, security challenges have to be overcome. In this chapter, Daya Sagar Guptaand and G. P. Biswas present a technical presentation on securing cloud storage using an algorithm with a homomorphic frame.

Chapter 5
On Securing Cloud Storage Using a Homomorphic Framework

Daya Sagar Gupta
Indian Institute of Technology Dhanbad, India

G. P. Biswas
Indian Institute of Technology Dhanbad, India

ABSTRACT

In this chapter, a cloud security mechanism is described in which the computation (addition) of messages securely stored on the cloud is possible. Any user encrypts the secret message using the receiver's public key and stores it. Later on, whenever the stored message is required by an authentic user, he retrieves the encrypted message and decrypts it by using his secret key. However, he can also request the cloud for an addition of encrypted messages. The cloud system only computes the requested addition and sends it to the authentic user; it cannot decrypt the stored encrypted messages on its own. This addition of encrypted messages should be the same as the encryption of the addition of original messages. In this chapter, the authors propose a homomorphic encryption technique in which the above-discussed scenario is possible. The cloud securely computes the addition of the encrypted messages which is ultimately the encryption of the addition of the original messages. The security of the proposed encryption technique depends on the hardness of elliptic curve hard problems.

DOI: 10.4018/978-1-5225-5279-6.ch005

INTRODUCTION

To secure network communication, there exists two methods of cryptographic systems: symmetric key encryption and asymmetric key encryption. Till 1976, only symmetric-key cryptographic systems were known, in which both sender and receiver of an authentic message use the same key for encryption and decryption respectively. Some structures, like the one-time pad, attained information theoretic security, which indicates that the structure is protected even against adversaries with unlimited computing power. However, information theoretic security includes a high economy in terms of the key size and need randomness. Additionally, symmetric-key cryptography requires substantial key-management.

Diffie & Hellman (1976) proposed an asymmetric key cryptosystem (also known as public-key encryption) known as the Diffie-Hellman (DH) key exchange protocol which could be used to securely exchange a symmetric key between two users. They also presented the conception of a trapdoor one-way function. The property of a one-way function f is that it is easy to calculate $f(x)$ from x, but hard to calculate x from $f(x)$. These trapdoor functions become the backbone of asymmetric key cryptography. The idea of Public-key encryption is to generate a couple of keys that are mathematically associated with each other, comprising of a private key SK and a public key PK. The sender of a message uses the public key PK of the receiver to encrypt his message and the receiver then uses his private key SK to decrypt the incoming message. The public key is known to everyone and the corresponding private key is kept secret. The security of the corresponding private key depends on the hardness of one-way functions because finding the private key from the public key information is as hard as reversing a one-way function. Public key encryption (PKE) provides a key advantage over symmetric cryptography.

PKE plays an important role in the field of information security. This paper uses a PKE technique to ensure the security of the designed protocol. The proposed protocol is based on the homomorphic encryption. The idea of homomorphic encryption was first proposed by Rivest, Adleman & Dertouzous (1978) as a notion of *privacy homomorphism.* A public key encryption technique which includes the homomorphic property: $E(m_1 \, o_M \, m_2) = E(m_1) \, o_C \, E(m_2)$ is termed as homomorphic encryption. In general, a PKE has three algorithms: *keyGen* which generates a pair of key (public key and private key), *encrypt* which encrypts the message using the public key and *decrypt* which decrypts the message using the private key. Homomorphic encryption also includes these three conventional algorithms with the inclusion of an efficient algorithm *evaluate* which takes cipher texts $c_1, c_2 \ldots\ldots c_n$ and public key as inputs and produces a valid encryption of some function f on messages $m_1, m_2 \ldots\ldots m_n$ i.e.

$$E(f_M(m_1, m_2 \ldots\ldots m_n)) \leftarrow f_C(c_1, c_2 \ldots\ldots c_n)$$

The proposed scheme is based on the elliptic curve cryptography (ECC). The elliptic curves play a very important role in the field of cryptography. The security of elliptic curve cryptography is much better than that of the RSA cryptosystem. Their scheme deals with the properties of an elliptic curve. The proposed protocol is based on the bilinear property designed for elliptic curves. ECC depends on the difficulty provided by elliptic curve operations like addition operation of the points on elliptic curves. Elliptic curve cryptography is nothing but a kind of PKE with a pair of keys i.e. secret and public keys. On the bilinear map, the Computing Diffie-Hellman Problem (CDHP) is difficult, but the Decision Diffie-Hellman Problem (DDHP) is easy. Miller (1985) and Kblitz (1987) independently proposed the security of elliptic curve cryptosystem algorithm which depends on the discrete logarithm problem of elliptic curves. A number of ECC-based cryptosystems like Nobelis et al. (2012), Gupta & Biswas (2017a), Gupta & Biswas (2017b), Gupta & Biswas (2017c), Munir & Mohammed (2012) etc. are proposed in the literature.

In this paper, the authors are going to present a cryptographic technique which is based on the difficulty of elliptic curve Diffie-Hellman problem and Bilinear Diffie-Hellman problem as discussed in Boneh D & Franklin (2001), Gupta & Biswas (2015a), Gupta & Biswas (2015c) and others. The main work is done to secure the cloud storage. To do so, the authors use the homomorphic encryption technique in their proposed work. To implement the proposed protocol, the authors use four algorithms*: keyGen, encrypt, decrypt* and *evaluate*. The *keyGen* is used to generate the public parameters and a secret key. The public parameters are publicly known to others, whereas the secret key is secret to the authenticated user. *Encrypt* algorithm encrypts the secret message which is then stored on a remote server called the cloud. This encrypted message is not understandable to the remote server, *i.e.* the message is securely stored in the cloud so that the original message is not visible to the remote server. The *decrypt* algorithm decrypts the encrypted message to get the original message. This *decrypt* algorithm is only used by the authenticated user so that only the true user can get the original message. The *evaluate* algorithm is used for a special purpose. This algorithm is used to show the homomorphic property of the proposed protocol. In this paper, user requests to the remote server for the addition of stored messages stored on cloud storage. The cloud system calculates the addition of messages stored and fulfills the request of the client to return the addition calculated. After getting the calculated addition of message from cloud system, the user decrypts the message and finds the addition of the original messages.

Related Works

Benaloh (1987) proposed a technique to verify Secret-Ballot election. The Benaloh' scheme uses the homomorphic property in which the result of an election is verified by each user. Naccache & Stern (1998) proposed a PKE technique which describes a cryptosystem which is based on the hardness of computing higher residues modulo a composite integer. The scheme is described by two versions: the first scheme is *deterministic* which is practically oriented and the second scheme is *probabilistic* which uses the homomorphic property to encrypt the message. Okamoto & Uchiyama (1998) proposed a novel and probably secure PKE technique with some interesting point. The technique is a probabilistic encryption scheme which is based on the homomorphic property. Pascal Paillier (1999) proposed a public-key cryptosystem Based on Composite Degree Residuosity Classes. This paper describes a secure computational problem which is known as the Composite Residuosity Class Problem, and using this hard problem, Paillier proposed three public-key encryption schemes in which two schemes are homomorphic probabilistic encryption schemes computationally comparable to RSA. Sander, Young & Yung (1999) proposed a circuit-private additively homomorphic scheme. The above referred all contribution includes a homomorphic property based on the addition wherever Elgamal (1984) proposed a technique which is based on the multiplicative homomorphism. This contribution was based on the difficulty of computing discrete logarithms. Boneh, Goh & Nissim (2005) presented a cryptosystem which allows both addition and multiplication homomorphic property. In addition, they included the computation of the 2-DNF formula in their contributed paper. Cramer et al. (2001) proposed a scheme for multiparty computation (MPC). Their MPC is based on the homomorphism. They also showed that their MPC protocol with n players is secure against the active adversary and has the better complexity. Their proposal uses the threshold cryptosystem. This circuit can handle an arbitrary circuit. Damgard & Jurik (2003) proposed a PKE scheme which is derived from the Paillier cryptosystem. This scheme copies the attractive properties of homomorphism from Paillier encryption. They also found two new interesting properties other than Paillier properties: the first property describes that each and every participating user can deal with the same modulus which is used to generate a public key as well as a private key. In the second property, the construction of a threshold decryption scheme is efficient. The messages of arbitrary length can handle efficiently by this scheme, even though the public key and the secret key are of fixed size.

Melchor et al. (2008) proposed a homomorphic encryption scheme based on the lattice. In their scheme, the encryption of the addition of two message is equal to the addition of the encryption of same messages. Armknecht & Sadeghi (2008) showed that the algebraic homomorphic encryption (AHE) scheme which supports both addition and multiplication is still missing. The previous AHEs presented are insecure, inefficient and only supports addition or multiplication. Thus, they proposed an efficient homomorphic encryption which supports both addition and multiplication and exhibits the probable security. Their AHE is based on the problem of the coding theorem. Gentry (2009) proposed a fully homomorphic encryption (FHE) to solve the problem of AHE. In his scheme, FHE means that any arbitrary function may be computed on encrypted file without having the decryption key. The arbitrary function *f* must be an efficiently computable function. His scheme is applicable in many real-life scenarios such as search engine queries, searching in the encrypted data stored on a cloud storage. He started his work with designing somewhat homomorphic encryption and completed with the FHE. Van Dijk et al. (2010) used modular arithmetic to propose an FHE scheme. Their scheme is worked over the integers. They used Gentry (2009)'s somewhat "bootstrappable" homomorphic encryption scheme to construct an FHE. Their scheme works on the addition and multiplication over the integers instead of ideal lattices. They also claim that their scheme is simple and efficient. Coron et al. (2011) showed that the scheme presented by Van Dijk et al. (2010) has a large public key size, thus, it comes with a higher cost. Coron et al. (2011) reduced the size of the public key and proposed an FHE scheme. Their scheme also works on the addition and multiplication over integers and costs low in comparison with the public key size of Van Dijk et al. (2010). They also proved that their scheme is as secure as Van Dijk et al. (2010). They also claim that their scheme exhibits simplicity and is implemented using simple arithmetic operations. Erkin et al. (2012) proposed a homomorphic encryption scheme to generate the private recommendations for the online services. In their proposal, they claim to protect the private information from the service provider. They proposed the encryption of private data in such a manner that the computation on encrypted data is equal to the encryption of same computation on the original data. Lopez-Alt et al. (2012) proposed an *on-the-fly multiparty computation* (MPC) assisted by a cloud which has a highly computational capability but is untrusted. They designed their MPC protocol using the homomorphic encryption. The remote cloud may do any arbitrary dynamically chosen computation on the stored encrypted data but unable to do the decryption on that data. To do so, they proposed a *multi-key* FHE based on the NTRU cryptosystem.

Aguilar-Melchor et al. (2013) presented a survey on recent advances in homomorphic encryption. Their survey was focused on the cryptographic constructions as well as software engineering tools related to the FHE. They also provide some practical implementations on that. Gentry et al. (2013) proposed a learning with error (LWE) based FHE and claimed that their scheme is comparatively simple. They used a new method for their proposal and called this method as *approximate eigenvector.* Their scheme mostly use the operations of matrix multiplication and addition, thus it is easy to understand and simple to implement. Brakerski & Vaikuntanathan (2014) presented an efficient FHE based on the (standard) LWE. They showed that there may be an LWE-based "somewhat homomorphic" encryption which uses a *re-linearization* technique. They also proposed a *dimension-modulus reduction* technique by which the size of cipher text is reduced and the complexity of decryption algorithm is shortened. Ames et al. (2015) developed a scheme for a cloud based health monitoring system based on homomorphic encryption. Their proposal is based on the Branching Programs. They also proved that their scheme is simulated in real ECG data. Gupta & Biswas (2015b) proposed an ECC-based homomorphic encryption. In their scheme, the computation on the cloud storage is secure. The cloud may compute the addition of the stored files where the computed addition is same as the addition of original encryption files. The cloud stores and computes the files without knowing the secret key. They also claim the security of their protocol using the properties of an elliptic curve. Kocabas & Soyata (2015) proposed a cloud computing technique for medical issues and also removes the privacy concern about the service provider. They used the FHE to secure the health monitoring data stored on the cloud and ease the computation on that stored health related data without knowing the actual data. Wei (2016) proposed a homomorphic encryption without bilinear pairing. Their scheme uses proxy re-encryption and used for the multi-party computation. However, their scheme is a "somewhat homomorphic" encryption as it only computes infinite addition but only one multiplication. They also provided a security model for their proposal based on the subgroup decision problem and did a formal security analysis. Gupta & Biswas (2017a) proposed an ECC-based homomorphic encryption for the cloud storage security. Their proposal computes both addition and multiplication operations for homomorphism. They claim their scheme could be implemented using the properties of an elliptic curve and bilinear mapping. They also provided a security analysis for their scheme based on the hard assumption of ECDLP and BDHP. Ibtihal & Hassan (2017) proposed a homomorphic encryption scheme for a mobile cloud computing (MCC). They provided security for outsourcing images.

Road Map of the Paper

Next, the Background section defines some preliminaries definitions about the elliptic curve, bilinear pairing, identity based encryption (IBE), and some related computational problems. Next, the proposed scheme is laid out. The security analysis of the proposed scheme is described in the following section. Finally, the authors conclude the paper.

BACKGROUNDS

This section gives a brief idea about the properties and definitions that have been used in the proposed protocol. The authors start this section with the definition and property of elliptic curves and show their functionalities. Next, they put some background on bilinear pairing and define the homomorphic encryption. Finally, some assumptions based on the elliptic curves and pairing is given.

Elliptic Curve

The points of elliptic curve include an Abelian group with addition as a binary operation. An elliptic curve over real number is defined by the following algebraic equation:

$$y^2 = x^3 + ax + b$$

where x, y, a and b are real numbers. A special type of elliptic curve is used in cryptography to secure the network communication for which $4a^3+27b^2\neq0$. The operation on elliptic curve points is defined as follows:

Let X and Y be two points on an elliptic curve, the following different methods are applied to add these two points according to their positions on the elliptic curve:

1. If Points X and Y both have different x and y coordinates, i.e. $X = \left(x_1, y_1\right)$ and $Y = \left(x_2, y_2\right)$, the straight line connecting X and Y intercept the elliptic curve on a point which is $-Z$. The reflection of this point about x-axis will be addition of points X and Y *i.e.* $Z=X+Y$
2. If point X overlaps i.e. we have to find $Z=X + X$, in this case, a tangent to point X is drawn. This tangent intercepts the curve on point $-Z$. Then taking reflection of this point about x-axis, we get $Z=2X$
3. If we are adding two points X and $-X$ which is inverse to each other, the intercepting point of these two points, when connected as a straight line is at Infinity. This point is called a point of Infinity O which is the additive identity of the group, i.e. $X + (-X) = O$.

A special type of operation in the elliptic curve is scalar multiplication, in which a point of the elliptic curve is multiplied by a scalar number. Suppose X is a point on an elliptic curve and k is a scalar then kX is calculated as add up the given point X, k time.

Bilinear Pairing

The bilinear mapping is used as a function to map an element of a group to an element of the different group. Let two groups G_1 and G_2 of order q for some large prime q. The group G_1 is an additive group, whereas G_2 is a multiplicative group. A bilinear map $\hat{e}: G_1 \times G_1 \rightarrow G_2$ between these two groups G_1 and G_2 must satisfy these following properties:

- **Bilinearity:** $\hat{e}(aX, bY) = \hat{e}(X, Y)^{ab}$ for all $X, Y \in G_1$ and all $a, b \in Z_q$
- **Non-Degeneracy:** All pairs *in* $G_1 \times G_1$ does not map to the identity element in G_2.
- **Computability:** $\hat{e}(X, Y)$ must be easy to compute and efficient for any $X, Y \in G_1$.

The map satisfying all above-described properties is said to be an admissible bilinear map. In their case, the authors use the bilinear mapping that maps two points of an elliptic curve to an element in the multiplicative group, i.e. G_1 is an elliptic curve group whereas G_2 is a multiplicative group.

Homomorphic Encryption

The homomorphic encryption is used to encrypt a message so that the authors can use an operation on the encrypted message to get encryption of the same operation on the original message. Homomorphic encryption is defined by following four algorithms:

- **Keygen (t):** It takes a security parameter t, and outputs a pair of keys, i.e. public key (PK) and private key (PR)
- **Encrypt (PK, m):** It takes public key and message as input and outputs an encrypted message c
- **Decrypt (PR, c):** It takes cipher text and private key as input and returns plain text as output
- **Evaluate (PK, c_i, f):** This algorithm takes the public key, a set of cipher text and a function to be computed on cipher texts as input and outputs a cipher text which is the encryption of the same function calculated on the corresponding original plain texts

Hard Assumptions

Some of the hard problems which are difficult to solve are given below:

- **Elliptic Curve Discrete Logarithm Problem (ECDLP):** Given X, Y $\in G_1$, it is hard to find k $\in Z_q^*$ such that Y = kP.

- **Bilinear Diffie-Hellman Problem (BDHP):** Given X, aX, bX, cX for any a, b, c $\in Z_q^*$, it is hard to compute ê (X, X) ^{abc}.
- **Computational Diffie-Hellman Problem (CDHP):** Given (X, aX, bX) $\in G_1$ for any a, b $\in Z^*_p$ computation of abX is hard to the group G_1.

PROPOSED PROTOCOL

In this section, the authors are going to propose their homomorphic encryption scheme. As mentioned earlier, the technique uses the properties of an elliptic curve to secure the communication. Their technique includes four algorithms as mentioned before. The description of these algorithms is given here:

keyGen(t)

As the name suggested, the algorithm is used to generate the key pair. The algorithm is laid here:

1. This random algorithm takes a security parameter $t \in Z^+$ and generates the system parameters as follows:
   a. **q_1:** A t-bit random prime
   b. **q_2**: A t-bit random prime
   c. **n:** q_1q_2
   d. **G_1:** Additive group of finite points on an elliptic curve of a finite order n
   e. **G_2:** Multiplicative group of a finite order n
   f. **ê:** $G_1 \times G_1 \rightarrow G_2$: An admissible bilinear map that takes two points of G_1 as input and produces an element of G_2
2. Let P be a base point in group G_1 and g be a generator in group G_1. Calculate $Q = q_2P$.
3. The public key PK is $(n, G_1, G_2, \hat{e}, g, P, Q)$ and the private key is q_1.

Encrypt (PK, M)

This algorithm is used to encrypt the secret message which is used to generate a cipher text of original plain text. The generated cipher texts are then stored on the cloud system for later retrieval. The algorithm steps are laid here:

1. The message space M must have the integers $M = \{0, 1, 2, 3. . . N\}$ where N is always less than q_2 i.e. $N < q_2$.

2. 2. Select a random integer $x = \{0, 1, 2, 3... (n-1)\}$ and encrypt a message m to get a cipher text as follows:

$$C = (mP + xQ) \in G_1$$

C is a cipher text of message m.

Decryption

The algorithm decrypts the cipher text to obtain original message. This algorithm is run by the authentic user because he has the private key *PR*. The steps involved in this algorithm are given below:

1. 1. To decrypt the cipher text C, user calculates,

$$D = \hat{e}(C,P)$$

$$= \hat{e}\ ((mP + xQ),P)$$

$$= \hat{e}\ ((mP + xq_2P),P)$$

$$= \hat{e}(P,P)^{m+xq_2}\ (m + xq_2 < n)$$

$$= g^{m+xq_2}$$

(Since $\hat{e}(P,P) = g$ as degeneracy)

$$= g^m g^{xq_2}$$

2. 2. Now, user calculates,

$$D^{q_1} = (g^m g^{xq_2})^{q_1}$$

$$= g^{mq_1} \text{ (Since } q_1 q_2 = n \text{ and } g^n = 1\text{)}$$

$$= (g^{q_1})^m$$

3. Since m is small, decryption can be done by pre-computation table of powers of g^{q_1}.

Evaluate (PK, C_i, +)

In this algorithm, an addition operation is done on a set of cipher texts C_i *(i= 1, 2 …)*. The addition operation is distributed on modulo n. Let C_1 and C_2 be two cipher texts of two plain texts m_1 and m_2. The *evaluate* algorithm returns with modulo addition of these two cipher texts *i.e.* $C_e = C_1 + C_2$. The computed C_e must be the encryption of $(m_1 + m_2)$ mod n due to the homomorphic property.

Homomorphic Property

In this section, the authors illustrate the homomorphic property of their proposed protocol. Let an authenticated user request for the addition of cipher texts C_1 and C_2 to the remote server. Cloud system, then calculates the addition of stored encryptions *i.e.* $C_e = C_1 + C_2$ using evaluate algorithm and returns C_e to the user. The user then decrypts C_e using decrypt algorithm to get $(m_1 + m_2)$ mod n. According to the encryption algorithm $C_1 = m_1 P + x_1 Q$ and $C_2 = m_2 P + x_2 Q$. Since C_1, C_2 ϵ G_1, the addition is

$$C_e = C_1 + C_2$$

$$= (m_1 P + x_1 Q) + (m_2 P + x_2 Q)$$

$$= (m_1 + m_2) P + (x_1 + x_2) Q \epsilon\, G_1$$

This addition C_e is done by cloud system and sent to the authenticated user. The user decrypts this additive cipher text to get addition of original message *i.e.* $(m_1 + m_2)$.

Implementation of Proposed Scheme

The proposed protocol uses the operations on points of a standard elliptic curve described in the above sections. These operations can be easily computed using Koblitz (1987) and Miller (1985). Now, in their scheme, decryption algorithm is based on the bilinear map discussed in the above sections. This map is firstly introduced by Boneh and Franklin (2001). The computation of bilinear map is done using Weil and Tate pairing.

SECURITY ANALYSES

This section provides the security analyses of the schemes proposed in section 3. Their proposed schemes are based on the elliptic curve discrete logarithm problem (ECDLP) and Bilinear Diffie-Hellman problem (BDHP) which are hard to solve. In presented scheme, the private key q_1 is only known to the authenticated user so that the encrypted message is only decrypted by him. The cloud system only can see the encrypted message. The server is unable to decrypt the original message since it does not have the private key to decrypt the original message.

CONCLUSION

In this paper, the authors have presented a scheme which provides a better way of calculating on encrypted data. The proposed scheme described a quite simple way to secure the messages which are to be stored on a remote server. The encrypted message is stored on a remote server in such a manner that server is not able to decrypt the original message. The server can only do computation on encrypted files. The homomorphic property of their encryption scheme computes the addition of encrypted files so that an authenticated user can easily decrypt this addition to get the addition of original files. The authors also showed that their proposed scheme is secured under possible attacks. The security is concerned with the hard problem on an elliptic curve.

REFERENCES

Aguilar-Melchor, C., Fau, S., Fontaine, C., Gogniat, G., & Sirdey, R. (2013). Recent advances in homomorphic encryption: A possible future for signal processing in the encrypted domain. *IEEE Signal Processing Magazine*, *30*(2), 108–117. doi:10.1109/MSP.2012.2230219

Ajtai, M. (1996). Generating hard instances of lattice problems. In *Proceedings of the twenty-eighth annual ACM symposium on Theory of computing* (pp. 99-108). ACM. doi:10.1145/237814.237838

Ames, S., Venkitasubramaniam, M., Page, A., Kocabas, O., & Soyata, T. (2015). Secure health monitoring in the cloud using homomorphic encryption: A branching-program formulation. In *Enabling Real-Time Mobile Cloud Computing through Emerging Technologies* (pp. 116–152). IGI Global. doi:10.4018/978-1-4666-8662-5.ch004

Armknecht, F., & Sadeghi, A. R. (2008). A New Approach for Algebraically Homomorphic Encryption. *IACR Cryptology ePrint Archive, 2008*, 422.

Benaloh, J. D. C. (1987). *Verifiable secret-ballot elections*. Yale University.

Boneh, D., & Franklin, M. (2001). Identity-based encryption from the Weil pairing. In Advances in Cryptology—CRYPTO 2001 (pp. 213-229). Springer Berlin Heidelberg. doi:10.1007/3-540-44647-8_13

Boneh, D., Goh, E. J., & Nissim, K. (2005). Evaluating 2-DNF formulas on ciphertexts. In Theory of cryptography (pp. 325-341). Springer Berlin Heidelberg.

Brakerski, Z., & Vaikuntanathan, V. (2014). Efficient fully homomorphic encryption from (standard) LWE. *SIAM Journal on Computing*, *43*(2), 831–871. doi:10.1137/120868669

Coron, J. S., Mandal, A., Naccache, D., & Tibouchi, M. (2011). Fully Homomorphic Encryption over the Integers with Shorter Public Keys. In Crypto (Vol. 6841, pp. 487-504). Academic Press. doi:10.1007/978-3-642-22792-9_28

Cramer, R., Damgård, I., & Nielsen, J. (2001). Multiparty computation from threshold homomorphic encryption. *Advances in cryptology—EUROCRYPT 2001*, 280-300.

Damgård, I., & Jurik, M. (2003). A length-flexible threshold cryptosystem with applications. In *Information Security and Privacy* (pp. 350–364). Springer Berlin Heidelberg. doi:10.1007/3-540-45067-X_30

Diffie, W., & Hellman, M. (1976). New directions in cryptography. *IEEE Transactions on Information Theory*, *22*(6), 644–654. doi:10.1109/TIT.1976.1055638

ElGamal, T. (1984). A public key cryptosystem and a signature scheme based on discrete logarithms. In *Advances in cryptology* (pp. 10–18). Springer Berlin Heidelberg.

Erkin, Z., Veugen, T., Toft, T., & Lagendijk, R. L. (2012). Generating private recommendations efficiently using homomorphic encryption and data packing. *IEEE Transactions on Information Forensics and Security*, *7*(3), 1053–1066. doi:10.1109/TIFS.2012.2190726

Gentry, C. (2009). *A fully homomorphic encryption scheme* (Doctoral dissertation). Stanford University.

Gentry, C., Sahai, A., & Waters, B. (2013). Homomorphic encryption from learning with errors: Conceptually-simpler, asymptotically-faster, attribute-based. In *Advances in Cryptology–CRYPTO 2013* (pp. 75–92). Berlin: Springer. doi:10.1007/978-3-642-40041-4_5

Gupta, D. S., & Biswas, G. P. (2015a). Securing Voice Call Transmission over Cellular Communication. *Procedia Computer Science*, *57*, 752–756. doi:10.1016/j.procs.2015.07.469

Gupta, D. S., & Biswas, G. P. (2015b). Secure Computation on Cloud Storage: A Homomorphic Approach. *Journal of Cases on Information Technology*, *17*(3), 22–29. doi:10.4018/JCIT.2015070103

Gupta, D. S., & Biswas, G. P. (2015c). Identity-Based/Attribute-Based cryptosystem using threshold value without Shamir's Secret Sharing. In *Signal Processing, Computing and Control (ISPCC), 2015 International Conference on* (pp. 307-311). IEEE.

Gupta, D. S., & Biswas, G. P. (2016). Cryptanalysis of Wang et al.'s lattice-based key exchange protocol. *Perspectives on Science*, *8*, 228–230. doi:10.1016/j.pisc.2016.04.034

Gupta, D. S., & Biswas, G. P. (2017a). A Secure Cloud Storage using ECC-Based Homomorphic Encryption. *International Journal of Information Security and Privacy*, *11*(3), 54–62. doi:10.4018/IJISP.2017070105

Gupta, D. S., & Biswas, G. P. (2017b). On Securing Bi-and Tri-partite Session Key Agreement Protocol Using IBE Framework. *Wireless Personal Communications*, 1–20. doi:10.1007/s11277-017-4399-5

Gupta, D. S., & Biswas, G. P. (2017c). An ECC-based Authenticated Group Key Exchange Protocol in IBE framework. *International Journal of Communication Systems*, e3363. doi:10.1002/dac.3363

Ibtihal, M., & Hassan, N. (2017). Homomorphic Encryption as a Service for Outsourced Images in Mobile Cloud Computing Environment. *International Journal of Cloud Applications and Computing*, *7*(2), 27–40. doi:10.4018/IJCAC.2017040103

Koblitz, N. (1987). Elliptic curve cryptosystems. *Mathematics of Computation*, *48*(177), 203–209. doi:10.1090/S0025-5718-1987-0866109-5

Kocabas, O., & Soyata, T. (2015). Towards privacy-preserving medical cloud computing using homomorphic encryption. In Enabling Real-Time Mobile Cloud Computing through Emerging Technologies (pp. 213-246). IGI Global. doi:10.4018/978-1-4666-8662-5.ch007

López-Alt, A., Tromer, E., & Vaikuntanathan, V. (2012). On-the-fly multiparty computation on the cloud via multikey fully homomorphic encryption. In *Proceedings of the forty-fourth annual ACM symposium on Theory of computing* (pp. 1219-1234). ACM. doi:10.1145/2213977.2214086

Melchor, C. A., Castagnos, G., & Gaborit, P. (2008). Lattice-based homomorphic encryption of vector spaces. In *Information Theory, 2008. ISIT 2008. IEEE International Symposium on* (pp. 1858-1862). IEEE. doi:10.1109/ISIT.2008.4595310

Miller, V. S. (1985). Use of elliptic curves in cryptography. In Advances in Cryptology—CRYPTO'85 Proceedings (pp. 417-426). Springer Berlin Heidelberg.

Munir, K., & Mohammed, L. A. (2012). Policy for Mobile Devices to Access Grid Infrastructure. In Grid and Cloud Computing: Concepts, Methodologies, Tools and Applications (pp. 1593-1608). IGI Global. doi:10.4018/978-1-4666-0879-5.ch704

Naccache, D., & Stern, J. (1998). A new public key cryptosystem based on higher residues. In *Proceedings of the 5th ACM conference on Computer and communications security* (pp. 59-66). ACM. doi:10.1145/288090.288106

Nobelis, N., Boudaoud, K., Delettre, C., & Riveill, M. (2012). Designing Security Properties-Centric Communication Protocols using a Component-Based Approach. *International Journal of Distributed Systems and Technologies*, *3*(1), 1–16. doi:10.4018/jdst.2012010101

Okamoto, T., & Uchiyama, S. (1998). A new public-key cryptosystem as secure as factoring. In Advances in Cryptology—EUROCRYPT'98 (pp. 308-318). Springer Berlin Heidelberg. doi:10.1007/BFb0054135

Paillier, P. (1999). Public-key cryptosystems based on composite degree residuosity classes. In Advances in cryptology—EUROCRYPT'99 (pp. 223-238). Springer Berlin Heidelberg doi:10.1007/3-540-48910-X_16

Rivest, R. L., Adleman, L., & Dertouzos, M. L. (1978). On data banks and privacy homomorphisms. *Foundations of Secure Computation, 4*(11), 169-180.

Sander, T., Young, A., & Yung, M. (1999). Non-interactive cryptocomputing for NC 1. In *Foundations of Computer Science, 1999. 40th Annual Symposium on* (pp. 554-566). IEEE.

Van Dijk, M., Gentry, C., Halevi, S., & Vaikuntanathan, V. (2010). Fully homomorphic encryption over the integers. In *Annual International Conference on the Theory and Applications of Cryptographic Techniques* (pp. 24-43). Springer Berlin Heidelberg.

Wei, Z. (2016). A pairing-based homomorphic encryption scheme for multi-user settings. *International Journal of Technology and Human Interaction, 12*(2), 72–82. doi:10.4018/IJTHI.2016040106

KEY TERMS AND DEFINITIONS

Bilinear Mapping: A function that transfers elements of a group to an element of a different group.

Cloud Computing: The practice of using a network of remote servers hosted on the internet to store, manage, and process data, rather than a local server or a personal computer.

Communication: Means of exchanging the information.

Cryptography: The science and art of writing secrets. The tools used are encryption and decryption.

Homomorphism: A transformation of one set into another that preserves in the second set the relations between elements of the first.

Protocol: A set of rules governing the exchange or transmission of data between devices.

Security: A degree of protection and resistance to the threats.

Section 4

On Higher Education Applications

This section of *Technology Management in Organizational and Societal Contexts* speaks to information technology applications in higher education. The scope of these six chapters speaks to tremendous potential for Information technology to transform and enhance higher education. Some of these chapters speak of the exploit of a particular technology while others speak to how education methods have been enhanced through information technology.

Coming from an engineering school in Portugal, Marina Duarte, Andresa Baptista, and Gustavo Pinto provide the first chapter. In their work, they employ Quick Response (QR) codes to enhance engineering laboratory work. Unknown until 1994, the authors use QR codes to connect students to short video segments and enhance hands on instruction for their students. In the second chapter, Paliktzoglou and Suhonens' work on microblogging shows an innovative use of technology to facilitate collaboration in a computer science program. Although students often use Twitter for microblogs, their work focuses on Edmodo, a specialized microblogging tool intended for use in education.

Antonio Sarasa's chapter speaks to an application for managing language competency examinations in university settings. Their system supports the needs of administrators, teachers and students. Özen, Kartal and Emre demonstrate via a case study an innovative use of Google Analytics in managing a learning management system (LMS). It demonstrates the power of web analytics in a higher education setting.

The final two chapters are more focused on learning. Manchado-Pérez, López-Forniés, and Berges-Muro begin this section with a chapter on innovation in learning through problem based learning. Their approach could easily be applied to the information technology field. In their work, they seek to adapt methodologies from a variety of disciplines to product engineering. In the final chapter of this book, Pinto-Llorente, Sánchez-Gómez, and García-Peñalvo utilize a qualitative research approach to better understand what characteristics students need to be successful in blended (that is, combined on-line and face-to-face learning) environments.

Chapter 6
Learning in the Laboratory:
Accessing Videos With Quick Response Codes

Marina Duarte
Polytechnic of Porto, Portugal

Andresa Baptista
Polytechnic of Porto, Portugal

Gustavo Pinto
Polytechnic of Porto, Portugal

ABSTRACT

Using QR codes to access videos in engineering laboratory classes might be a successful way of building a bridge from concrete to digital content. With QR codes placed on an apparatus, students know exactly which video to watch, allowing them to view the videos while performing the experiment or at home when writing the report. Low-cost videos do not require expensive equipment and software, and keeping them short assures a minimum download time for use with smartphones and tablets. The aim of this chapter is to evaluate the importance undergraduate engineering students attribute to these videos and their reaction to the possibility of accessing them with QR codes scanned by a smartphone or a tablet, using access statistics and video viewings to support the findings. Results show students attributed some importance to the videos, and that the QR codes are very helpful as means to quickly and easily access the videos.

DOI: 10.4018/978-1-5225-5279-6.ch006

INTRODUCTION

The aim of this chapter is to report the findings of research that evaluates the importance students attribute to low-cost, educational, short videos that explain the operating procedure of laboratory apparatuses and the students' reaction to the possibility of accessing videos with QR codes by using a smartphone or a tablet. For a better understanding and support of the results, access statistics and video viewings will also be used.

The rest of this chapter is structured as follows. Following this introduction, the second and third sections address the state of the art regarding the use of videos and QR codes in higher education. In the fourth section, solutions and recommendations are described and presented, namely the case study, the methods used in this research and the results. The final section presents a discussion of results and conclusions.

Background

Active learning in engineering is impossible without laboratories, and the recognition of their importance by engineering teachers dates back to 1980 (Fishenden & Markland, 2006). From the earliest days of engineering education, instructional laboratories have been an essential part of study programs (Feisel & Rosa, 2005; Pandermarakis, Sotiropoulou, Passa, & Mitsopoulos, 2012; Surgenor & Firth, 2011). This is expected because the overall goal of engineering education is the preparation of students to practice engineering and because students' understanding of a domain can be enhanced when they engage in laboratory experiments (Litzinger, Lattuca, Hadgraft, & Newstetter, 2011). In addition, many studies have emphasized the role of laboratory education in students' motivation, which is particularly important considering 'the continuous drop in student numbers taking engineering and science courses' (Abdulwahed, Nagy, & Blanchard, 2008, p. 2).

Krivickas and Krivickas (2007) argue that engineering education is inconceivable without laboratory instruction, but the modernised laboratory presents a challenge to the academic staff in developing new and more effective instruction and facing the 'many disadvantages such as constraints on time, resources, maintenance, expensive equipment, and safety hazards' (Abdulwahed et al., 2008, p. 2).

Undoubtedly, the massification of higher education is a challenge as laboratories requiring human and material resources depend directly on the number of students. But if, as Feisel and Rosa (2005) mention, this works against a quality laboratory experience, the introduction of Information and Communications Technology (ICT) works for it.

The broad use of ICT, which can be seen as an obstacle to conventional hands-on classes in the laboratory, really is an opportunity because almost every student has the technological means to access the World Wide Web, having technology that 'walks in' classes (Gradel & Edson, 2013). Students are used to accessing information with a click or a slide, and laboratories must keep up to date and not be restricted to the physical space of the laboratory itself.

The great majority of higher education students are what Prensky (2001) describes as digital natives because of their familiarity with and reliance on ICT. They live immersed in technology, 'surrounded by and using computers, videogames, digital music players, video cams, cell phones, and all the other toys and tools of the digital age' (Prensky, 2001, p. 1). Bennett, Maton, and Kervin (2008, p. 776) point out that 'they are held to be active experiential learners, proficient in multitasking, and dependent on communications technologies for accessing information and for interacting with others', but this might not be entirely true, as students' everyday technology practices may not be directly applicable to academic tasks. In their literature review, Halupa and Caldwell (2015) highlight the difficulty in implementing new methods and technologies in the classroom and the resistance of students to learning in a new way. Rogado et al. (2015) also report the difficulties students have in incorporating new technology in their learning.

It therefore is important that ICT is used in the laboratory with moderation and in an integrated way, blending it with hands-on experimental activities. Resorting to new and current devices and applications that students use in their everyday life and recognize as good practices (Ricoy & Couto, 2014) might be a good way to capture their interest and increase motivation, and by doing so, improve learning.

This purpose is well served by low-cost, educational, short videos that can be used to explain operational procedures, show what is expected to happen during a certain experiment, and allow students to watch them over and over again while performing the actual experiment. As these videos can be made available in a platform like YouTube, it is important that students know exactly what video to watch when in the laboratory. That is why the possibility of accessing videos with Quick Response (QR) codes is essential to building what Gradel and Edson (2013, p. 62) designate as a 'bridge from concrete to digital content'.

VIDEOS AND QR CODES

Teaching and Learning With Videos

Although the use of educational videos has been widely employed in past decades (Cruse, 2006), recently interest in their use has increased incrementally because of platforms such as YouTube (Chan, 2010; Giannakos, 2013; Meseguer-Martinez, Ros-Galvez, & Rosa-Garcia, 2017), video-based learning systems such as Khan Academy and edX, new for-profit companies such as Coursera and Udacity (Giannakos, 2013) and new teaching/learning models such as flipped learning (Fulton, 2014; Observatory of Educational Innovation of the Tecnológico de Monterrey, 2014). Moradi, Liu and Luchies (2016, p. 2) conclude that 'interactive instructional videos are an essential tool in student-centred curriculum'.

According to Caspi, Gorsky and Privman (2005), educational videos can be divided into three categories: demonstration (or procedural) videos, narrative videos and lecture session videos. In the laboratory, demonstration videos usually are used, as they are really good tools that allow and improve autonomous learning and are becoming much more effective than methodologies based on more traditional methods. Therefore, this methodology enables faculty, especially in technology-related areas, to develop new teaching and learning strategies, adding a new dimension to the teaching material, increasing learners' motivation and enhancing their learning experience (Cruse, 2006).

Being able to watch a demonstration of procedures in a video, while also hearing an explanation and reading the actual procedure, is more effective than just reading about them because of the opportunity to visualize the interaction with the apparatus. The presence of different stimuli (image, sound and text), are what makes it appropriate for more learning styles or preferences (Alqahtani, Al-Jewair, AL-Moammar, Albarakati, & ALkofide, 2015). When motion is involved, video is the more adequate educational resource (Telg, 1999). In a study conducted by Goset and Espinoza (2014, p. 93), 'the results showed a clear preference for the use of the demonstrative type of videos, specifically for information transmission'. The benefit of using videos is the fact that:

The procedures display, reinforced with text or narrative creates a common language and shared meanings. The complexity of the oral language to explain technical activities or situations is remedied by the visual language, giving the possibility to create a meaningful and relevant common conceptual basis. (Goset & Espinoza, 2014, p. 93)

Videos have the advantage of using

... audiovisual language that is synthetic and integral. It is entirely synthetic, since it merges audio and visual to create a new communication. It is not an addition, but a merging of sound and moving image, which allows the brain to integrate, at the same time, the information that it understands and the information that visual and acoustic memories have preserved, which give the whole its meaning. (Cloutier, 1975, p. 136)

Videos also offer the possibility to match learning pace with students' own needs (Caspi et al., 2005; Chan, 2010), and the increasing use of smartphones and tablets (Giannakos, 2013) make it possible to watch videos while operating the laboratory apparatuses. Furthermore, students can watch the videos at home as many times as needed, taking notes, pausing, and moving backward and forward according to their own needs (De la Fuente Sánchez, Hernández Solís, & Pra Martos, 2013). This makes learning more flexible (Montoya & Hernández, 2016), 'and allows students to review technical procedures before, during, or after the laboratory sessions' (Alqahtani et al., 2015, p. 1).

On the down side, students can experience some technical problems (Montoya & Hernández, 2016), and instructors need to have the knowledge, design skills and creativity to create the videos in addition to having the appropriate hardware and software. Although this endeavour may be time consuming, a good video can be used for years and easily can be made available on YouTube, which is free, user friendly and mobile friendly (Chan, 2010). Nowadays, the 'easy creation, distribution and instantaneous uploading and downloading of digital media is the norm' (Bramhall, Radley, & Metcalf, 2008, p. 2). It is not necessary to be an expert to make a good quality video as 'the fast development of new technologies and the decrease of costs related to video recording, editing and production, make this tool available to faculty, allowing self audiovisual creation, according to the own needs and by means of economic tools' (Fernandez et al., 2011, p. 3).

Using QR Codes

The video's purpose and audience determine how to present it and where to host it (Bolorizadeh, Brannen, Gibbs, & Mack, 2012). In the laboratory, it is essential for students to know which videos to watch and to be able to access them not only while operating the equipment and apparatuses, but also whenever it is needed. So, what is necessary is a way to add virtual information to the real world because by doing so, 'field and lab work may be enhanced by connections to existing content' (Gradel & Edson, 2013, p. 62). Bolorizadeh et al. (2012, p. 375) found that 'students are willing to use video resources as long as the videos can be found easily, the videos are short and present the desired information quickly, and the information is accurate'. Because nowadays there is a widespread use of smartphones (with cameras and internet connection), every one of these users has the means to read codes that store not only a lot of information, but also can provide a link to all kinds of resources online. From the codes available, the one that serves this purpose better is the QR code.

QR codes are two-dimensional matrix codes, easy to use and to create, that can be freely generated online. They were designed in Japan by Denso Wave Inc. in 1994 to meet the increasing market demands of information storage capacity, character types, and print size limitations of standard bar codes (Denso Wave Incorporated, n.d.-a), and they have been widely adapted for advertising purposes. Compared with other codes, like the barcodes, that are very popular because they can be scanned very quickly and accurately provide information with multiple features (Mobile-QR-Codes.org, n.d.-a), the QR codes have the advantage of storing more information- up to some hundred times more than the barcodes (Mobile-QR-Codes.org, n.d.-b)- while taking up a smaller printing area. These features of the QR codes, allows the delivery of information with speed and convenience, eliminating the need to manually key text when accessing documents, videos or an internet website (Bonifácio, 2012; Law & So, 2010; Lombardo, Morrow, & Le Ber, 2012). Because of their general use, like accessing promotions and additional information for consumers, purchasing a variety of items, initiating phone calls and sending short messages or emails, it is likely that students will be familiarized with them.

To read QR codes, a mobile device with a camera and a scanning application is needed. If the content being accessed is online, a connection to the internet also is required. When using QR codes, users' scanning history is saved in their QR readers, just as with internet browsers. This redundancy may hold a special valence for educational purposes because, once scanned, resources accessed by QR codes become immediately 'clickable' in the students' mobile devices (Gradel & Edson, 2013).

There are two kinds of QR codes: static and dynamic. Dynamic QR codes are preferable because it is possible to update the content without changing the address since the codes are an online service that redirects the user to the new content. Even if the content changes, the URL remains the same and reprinting of the code is not necessary. The online management of a dynamic QR code also makes it possible to retrieve statistics from the code's usage. This is not the case for static QR codes that point to a certain address. If the information is changed (for example, a new version of a video is uploaded to YouTube), it is likely that a new address is created, which means that a new static QR code is needed.

Nevertheless, the use of a free, online, dynamic QR code service is not without risks, as Lombardo et al. (2012, p. 22) point out, because 'relying on free services, based somewhere out on the Internet, creates the risk that the service can increase the pricing, or disappear altogether'. So a paid service might be a good investment.

In education, the use of QR codes is still in its infancy (Law & So, 2010), but with 'usage across both P-12 and higher education settings' (Gradel & Edson, 2013, p. 50). According to Law and So,

The low technical barrier of creating and reading QR codes allows innovative educators to incorporate them into their educational endeavours. The operations

to retrieve or store QR codes are incredibly simple and quick, and with mobile devices, make them the ideal educational tools for teaching and learning. (Law & So, 2010, p. 85)

Many research studies focus on the usefulness of QR codes because 'any new technology needs to be perceived as being useful in order for it to be accepted and assimilated into people's daily routines' (Shin, Jung, & Chang, 2012, p. 1417), pointing out its attractiveness for students (Ali, 2017; Durak, Ozkeskin, & Ataizi, 2016). The settings for much of this research are nursing practice (Jamu, Lowi-jones, & Mitchell, 2016), undergraduate classrooms, libraries, and museums (Traser et al., 2015).

SOLUTIONS AND RECOMMENDATIONS

Context of the Case Study

The current study was conducted in the School of Engineering of Polytechnic of Porto, in Portugal. Founded in 1852, the school has more than 6,000 students and 500 faculty and staff members, organized in departments that are responsible for 12 Bachelor of Science and 11 Master of Science degrees. As a polytechnic school, the emphasis is on students learning how to apply knowledge to become agents of global progress. This tradition of applied knowledge and research is grounded in a large number of laboratories and technical staff, enabling students to have a hands-on approach to learning.

The research took place in the Heat and Fluids Laboratory (H&F Lab). This lab is one of the laboratories of the Department of Mechanical Engineering, which is attended mainly by students earning a Bachelor of Science degree in mechanical engineering or civil engineering.

The H&F Lab has 339 square meters, with several apparatuses and instruments in hydraulics, fluid mechanics, thermodynamics, heat transfer, turbomachinery and refrigeration (see Figure 1). To help students to better use the equipment and understand underlying concepts, the laboratory has made available manuals and written operational procedures.

Videos in the H&F Lab

In the H&F Lab there are written operational procedures for all the apparatuses and experiments. Although these procedures are well illustrated, videos make many details more understandable because they present motion. Using written procedures as a

Figure 1. Heat and Fluids Laboratory (H&F Lab)

starting point, the authors planned to create two low-cost, educational videos: one regarding an operational procedure that is needed in several apparatuses (video 1) and another featuring an operational procedure only used in a particular apparatus (video 2). The educational purpose of the videos was to complement a teacher's live explanation of the operational procedures, namely by including (a) a brief explanation of the main features of the apparatus, (b) safety instructions, (c) how to operate the apparatus and its instrumentation and (d) what to measure and how to measure it. The videos were not associated with any particular course or programme.

The low-cost option relates to the ease with which this methodology can be applied in other contexts. Fernandez et al. (2011, p. 3) define low-cost educational video 'as a short demonstration stream video which has a very specific goal and has been created in a very short period of time, with few resources and that can be combined or embedded within other materials of a course'. The creation of these kind of videos implies, according to Simo et al. (2010, p. 2939), 'finding ways to produce and distribute audiovisual teaching equipment minimizing the costs(...) both in terms of physical resources and economic, as in actual time spent by faculty in the design and creation of audiovisual material'.

The choice was made to create mini-videos, defined as 'short videos that constitute a technological educational resource to convey particular information that will help to consolidate a certain learning' (De la Fuente Sánchez et al., 2013, p. 180). Hsin and Cigas (2013) recommend using short videos as a way to address several issues presented by long videos such as the difficulty in maintaining students' attention and the problem of a lengthy file download. Some authors (De la Fuente Sánchez et al., 2013; Hsin & Cigas, 2013; Sexton, 2006) report the use of videos under five minutes.

For the videos in the H&F Lab, the intention was for students to be able to download them with smartphones or tablets, which would be easier with very short videos. Because of that, the published videos were 1:51 minutes and 1:38 minutes long, for video 1 and video 2 respectively. Although this implied that the action taking place on the video would happen at a relatively fast pace, this possible limitation could be overcome by students stopping the video or rewinding it.

In the making of these short videos, planning was very important to ensure that the educational purpose of the videos was met. Resorting to the plan-shoot/capture-edit-deliver methodology (Bolorizadeh et al., 2012), the making of the videos comprised the following steps.

First was the development of the plan: the team chose two written operational procedures as the subject for the videos. These procedures were used to write the storyboard and the script, and the location was checked for light and sound conditions. The authors installed free or common license software for the editing of film, images (photos) and audio; the software included a video editor, an audio editor, an image and photo editor and a presentations editor. Free versions of software tend to include only a portion of the features available in the paid version of the software, which often is enough as long as the essential features are maintained. This simplifies its use and does not require advanced levels of computing.

The second step was shoot/capture. The team checked all equipment, using a standard digital camera to capture film (with audio) and photos, and made image and audio tests in the location, capturing film and audio using different plans (different angles and views), and different zoom (zoom in and zoom out). Finally, the image and audio quality was checked before finishing the shooting session. At this stage, .avi files were used for the recording, because of their high quality and ease with which they can be imported into editing software (Malaga & Koppel, 2016).

The third stage was the editing. The team downloaded all film and photos. Photos were edited for better quality (contrast, brightness and colour). All relevant film scenes were selected and still images and film were combined, in order to match the script and storyboard. The team included some text whenever it improved comprehension of the video content. This led to the first version of the video. Using this video, the script was adjusted, and the timing of narration was rehearsed to assure a correct diction and pronunciation of all the words. The team recorded the narration while watching the video, to ensure synchronization. The narrator was a person of the team and not a computer, for a more natural result. The audio file of the narration was cleaned of all background noises, and the tone of the voice was slightly improved (bass and treble). The team added this narration to the video, maintaining the location's audio whenever it was needed. Next, the team obtained feedback and included the suggestions in the final version, adding finishing elements before having experts in this scientific/technical field validate the content and accuracy of the information conveyed by the final version of the videos.

The final stage was delivering the videos, considering file size, portability and distribution. According to Malaga and Koppel (2016, p.11) 'the two main file formats that are appropriate for online teaching seem to be WMV and MPEG-4'. We chose MPEG-4 because it can play natively on newer computers and mobile devices (packs of codecs can be downloaded for older computers). The authors chose to host the videos on the YouTube platform (currently, the streaming video platform with the most visits, with a billion users; YouTube for Press, n.d.) and selected the appropriate quality/size for publishing the video. Links and other necessary access features for the video were created, and tested with several smartphones and tablets.

QR Codes in the H&F Lab

To overcome the difficulty of accessing the virtual world in the real world environment of the H&F Lab, the videos were linked to QR codes. The authors made a thorough search of free editors (online and offline) of dynamic QR codes and found a free online website. This website also provides access statistics and has no limit for the number of codes generated. These dynamic QR codes are black and white, black representing foreground squares and white representing background squares. The number of modules is 29 x 29 (version 3), which allows the storage of 32–127 characters, depending on the type of information (Denso Wave Incorporated, n.d.-b). This was enough for each video link.

QR codes include redundant information with the purpose of making corrections possible. So, even if damaged, this redundant data ensures the readability of QR codes. The correction level of data in the free dynamic QR codes generated online is 30% (The Marketing Bureau UK Ltd, n.d.), and the three pattern-detection squares on the corners allows the scanning of the code in any direction (omnidirectional).

The team conducted tests with several smartphones and tablets in a multitude of circumstances to ensure that these technical features were appropriate and that the codes worked in less than perfect conditions. Because there are a number of free apps available to scan QR codes, we downloaded and tested several, choosing the ones with higher ratings. All the apps were adequate.

The QR codes were printed (testing the appropriate size) and made available on the apparatus (see Figure 2 and Figure 3) and on information boards in the H&F Lab. All the students participating in the study were told how they could access

Figure 2. Student accessing a QR code with the smartphone

the videos using their smartphones and tablets and how to save this information for further use. Almost all the students had the technological means to access the videos, which was expected because in Portugal 46.4% of cell phones bought in 2012 were smartphones. For people 15–24 years of age, this value was 60% above average. These numbers represent an increase in smartphone use of 80% since May 2012 (Gaspar, 2014), and the more recent information shows that the number of smartphones users, in Portugal, continues to grow (GrupoMarktest, 2016), reaching in 2017 a 65% penetration (Wikipedia, 2017). Because most common uses amongst consumers of smartphones under 25 years old are accessing social networks (83%), finding information (67%), and weekly watching videos online (75%) (Fernandes, 2017), resorting to smartphones as educational tools will not require a different use.

Method

The participants were 41 students (36.6% female and 63.4% male) studying to earn a bachelor of science degree in civil engineering in the School of Engineering of Polytechnic of Porto. This was a convenience sample that represented 57% of the students attending the H&F Lab (72 students). Students attended the laboratorial classes in October and November of 2014, and had a deadline of two weeks to write

Figure 3. QR code of Video 1 (general procedure used with several apparatuses) and Video 2 (specific procedure used in one apparatus)

a report on each one of the different experiments done in the H&F Lab. Students worked in groups of three to four students, which means that there were 20 to 22 different groups.

The participants ranged in age from 19 to 66 (M = 27.59, SD = 9.439); 34 participants (82.9%) were Portuguese and seven (17.1%) were different nationalities. In addition, 32 (78.0%) were second-year students and seven (17.1%) were third-year students (4.9% were missing values), and 29 (70.7%) attended classes during the day, while nine (22.0%) did so at night (7.3% were missing values).

All participants collaborated voluntarily with this research.

To evaluate the importance of educational videos as pedagogical resources, the instrument used was an adapted version of the VINCERE questionnaire (FEUP-LEA, n.d.). Responses on the survey were measured using a 7-point Likert scale, where 1 is unimportant and 7 is full importance. The items are aggregated in three dimensions: ease and motivation, appropriateness to the contents and objectives and role in the evaluation. Three of the original 12 items were not used because they did not apply, and a new item was included because of the type of videos used. This adapted questionnaire had 10 items for a maximum of 70 points.

Some additional questions were included, namely regarding audio and image quality of the videos, and the importance of accessing them with QR codes. Participants also were asked to comment on the strengths and weakness of the videos.

A socio-demographic and academic questionnaire was used to characterise the participants, including such items as age, nationality, gender, course of studies, curricular year and class attendance schedule.

The procedures adopted for data collection and analysis began by approaching the participants during classes (with the permission and cooperation of their teachers) in December 2014 and asking them to complete a paper-and-pen questionnaire.

Participation was voluntary. The data were collected after the participants had attended practical classes in the H&F Lab.

The data collected was prepared and checked for errors. Then, the validity and reliability of the VINCERE questionnaire was assessed (with SPSS 21.0). To measure the validity of the scale, we evaluated the adequacy of data for exploratory factor analysis using the Kaiser-Meyer-Oklin criteria and Bartlett's test of sphericity. The sample size was adequate with a Kaiser-Meyer-Oklin value of 0.773 and Bartlett's test of sphericity ($\chi 2$ (21) = 117.412, p = 0.000) was statistically significant as required. These results allowed exploratory factor analysis with the principal components analysis extraction method and varimax rotation. Factors with an eigenvalue greater than one were retained. The indicator of the scale's reliability was internal consistency, and Cronbach's alpha was performed.

Content analysis was used for the participants' comments on the strengths and weaknesses of the videos, using emergent categories.

Results

The 10-item VINCERE questionnaire first was subjected to exploratory factor analysis (principal components analysis extraction method; varimax rotation). Three items (items 1, 3 and 5) were dropped, and two factors were identified. Factor 1 (appropriateness to the contents and objectives) was comprised of items 2, 4, 7 and 9 and explained 44.7% of the variance. Factor 2 (ease and motivation) was comprised of items 6, 8 and 10 and explained the remaining 30.6% of the variance. Table 1 presents the factor loadings.

Communalities ranged from 0.471 to 0.838. The alpha levels were good, being 0.830 for the total scale, 0.864 for Factor 1 (appropriateness to the contents and objectives) and 0.763 for Factor 2 (ease and motivation).

Table 1. Results of the Factor Analysis

Scale Items	Factor Loading	
	Factor 1	**Factor 2**
2	0.887	---
4	0.876	---
7	0.830	---
9	0.867	---
6	---	0.636
8	---	0.876
10	---	0.914

Table 2 includes descriptive statistics such as means, standard deviations and standard errors for the importance students attributed to the videos (total of VINCERE scale; maximum 49 points) and for its factors (maximum 28 points for Factor 1 and 21 points for Factor 2). The overall score was 57%, with 36% for Factor 1 and 21% for Factor 2.

The item that scored lower on average (M = 3.41) was related to not using other information sources because of the videos. The item that scored higher on average (M = 4.81) was related to the video being a clear example of what was intended to demonstrate.

On average, the audio and video quality was considered good; no participant found it bad while others considered it very good.

The videos were viewed 1.54 times on average (most participants reported watching them once; 25% of participants watched them two or more times to a maximum of five times). When questioned about the importance of being able to access the videos with QR codes in the laboratory, the average score was 4.53, with the 75% quartile scoring 6.75 (the highest of all items).

Participants' comments on the strengths of the videos included the quickness of being able to access information about the experimental procedure, the ease in accessing information and the clarity of the explanation given. Regarding the weaknesses, participants mentioned some difficulty in the first viewing, the videos being too fast and the explanation given not being enough. Participants also referred to the unavailability of more videos as a weakness.

Some confusion seemed to exist, with students thinking the videos were related to a particular course or teacher and not understanding why the videos were not available as resources for the courses. They had trouble understanding that it was the H&F Lab that created the videos and made them available, regardless of courses and study programmes.

Additional information was also obtained with statistics from YouTube. Video 1 (general procedure used in several apparatuses) was mainly viewed on a computer or smartphone (53% and 44%, respectively; 3% used a tablet). Views of this video spanned from October 2014, when students were in the laboratory, to January 2015, when the classes of this course ended. The majority of these views occurred in the

Table 2. VINCERE Scale Descriptive Statistics

Factors	N	M	SD	SE
Appropriateness to the contents and objectives	32	17.69	4.915	0.869
Ease and motivation	32	10.41	3.934	0.695
Total VINCERE: Importance of the videos	32	28.09	7.346	1.299

period that students were doing this procedure in the H&F Lab (October to November, as shown in Figure 4). Video 1 had a total of 100 views, with 25% having at least one minute (half the duration of video 1) and 50% having only the initial 30 seconds (see Figure 4) that correspond to the portion of the video in which the description of the apparatus is done. Although students had to hand-out to the teacher the report of the laboratorial coursework, two weeks after being in the laboratory, some students did it much later, which would explain some views in January. Because a long time had passed, these students probably relied on the view to write their reports, which explains why these visualizations are several minutes long.

Video 2 (specific procedure used in one apparatus) was mainly viewed on a computer or smartphone (57% and 36%, respectively; 7% used a tablet). Views of this video spanned from November 2014 to December 2014 (as shown in Figure 5), when students were in the laboratory doing this procedure. Video 2 had a total of 14 views, with 30% being of the full video and 20% having only the initial 30 seconds (see Figure 5).

CONCLUSION

Laboratories are of paramount importance for engineering education, but traditional hands-on classes face several challenges, which also can be seen as opportunities. Having students connected to the digital world, bringing to the classroom their own devices, widens the educational possibilities. Although we could argue that the teacher is totally dependent on students having smartphones with internet connection, in order to resort to this educational resources, it is also true that with the use of smartphones growing globally every year (Statista, 2017), this issue is becoming less and less of a problem by the day. This indicates that not only is it more and

Figure 4. Viewing pattern of video 1

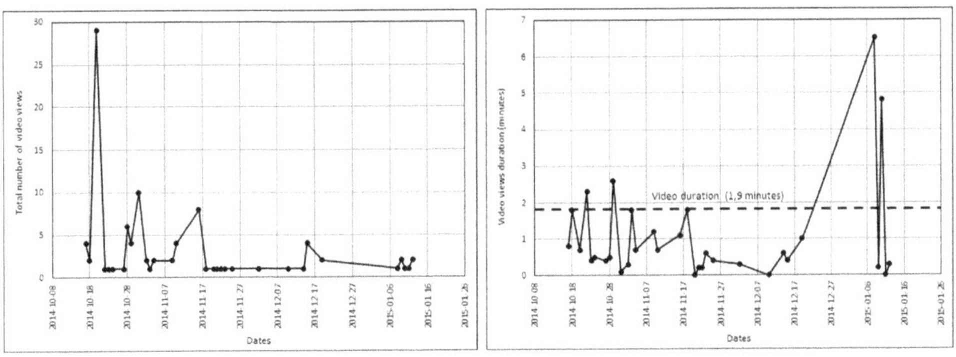

Figure 5. Viewing pattern of video 2

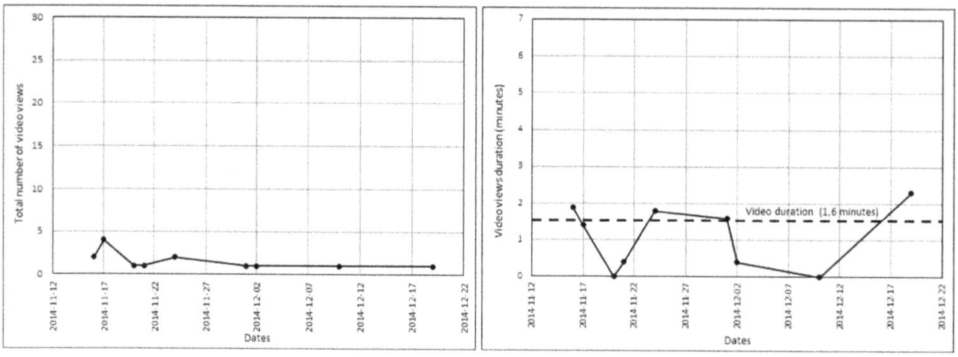

more likely that students have the resources to use ICT in classes, but also that they are used and motivated to resort to them in a variety of contexts.

With that in mind, two low-cost, educational short videos that explained operational procedures were created and made available in the H&F Lab through the use of QR codes. The aim of the research was to evaluate the importance students attribute to these instructional videos and their reaction to the possibility of accessing them with QR codes.

As the videos were created with free software by the laboratory staff, not multimedia experts, it was important to assess the quality of the final product. Students found them good or very good, showing that is possible, with modest means and some creativity, to make appropriate educational short videos.

The fact that these videos present very technical procedures and explanations might account for the fact that ease and motivation scored lower than appropriateness to the contents and objectives.

Students attributed only some importance to the videos (the overall score of the VINCERE questionnaire was 57%), which was expected, as only two very short videos were available.

The videos complemented other educational resources provided by the course teachers, and it was never intended to use them as substitutes. This explains why the item that was related to not using other information sources because of the videos scored lower on average.

The clarity of the explanation provided, which is fundamental in instructional videos of operational procedures, was rewarded by the students with the highest score and the demand for more videos. Other positive aspects were the quickness and ease of accessing information, showing that the QR codes are a helpful solution in the laboratory.

Most of the views, for both videos, were done during classes, but views pattern show that video 1 (related with several apparatus) was useful for students preparing the report after the deadline.

In conclusion, students attribute some importance to the videos, the videos used both in classes and also later on, and the QR codes in the laboratory were well received by the students who considered them very helpful as a means to access the videos.

ACKNOWLEDGMENT

This research received no specific grant from any funding agency in the public, commercial, or not-for-profit sectors. The authors thank Dr Ana Mouraz for sharing the Portuguese version of the VINCERE questionnaire and allowing its use in this research.

REFERENCES

Abdulwahed, M., Nagy, Z. K., & Blanchard, R. (2008). The Tri Lab, a Novel View of Laboratory Education. In *Proceedings of the EE2008 - International conference on innovation, good practice and research in engineering education*. Loughborough, UK: HEA Engineering Subject Centre.

Ali, N. (2017). Pre-service Teachers ' Perception of Quick Response (QR) Code integration in Classroom Activities. *TOJET: The Turkish Online Journal of Educational Technology*, *16*(1), 93–100.

Alqahtani, N. D., Al-Jewair, T., AL-Moammar, K., Albarakati, S. F., & ALkofide, E. A. (2015). Live demonstration versus procedural video: A comparison of two methods for teaching an orthodontic laboratory procedure. *BMC Medical Education*, *15*(1), 199. doi:10.1186/s12909-015-0479-y PMID:26537393

Bennett, S., Maton, K., & Kervin, L. (2008). The 'digital natives' debate: A critical review of the evidence. *British Journal of Educational Technology*, *39*(5), 775–786. doi:10.1111/j.1467-8535.2007.00793.x

Bolorizadeh, A., Brannen, M., Gibbs, R., & Mack, T. (2012). Making Instruction Mobile. *The Reference Librarian*, *53*(4), 373–383. doi:10.1080/02763877.2012.707488

Bonifácio, V. D. B. (2012). QR-Coded Audio Periodic Table of the Elements: A Mobile-Learning Tool. *Journal of Chemical Education*, *89*(4), 552–554. doi:10.1021/ed200541e

Bramhall, M., Radley, K., & Metcalf, J. (2008). Users as producers: students using video to develop learner autonomy. In *Proceedings of the EE2008 - International conference on innovation, good practice and research in engineering education.* Loughborough, UK: HEA Engineering Subject Centre.

Caspi, A., Gorsky, P., & Privman, M. (2005). Viewing comprehension: Students' learning preferences and strategies when studying from video. *Instructional Science*, *33*(1), 31–47. doi:10.1007/s11251-004-2576-x

Chan, Y. M. (2010). Video instructions as support for beyond classroom learning. *Procedia: Social and Behavioral Sciences*, *9*, 1313–1318. doi:10.1016/j.sbspro.2010.12.326

Cloutier, J. (1975). *A era de Emerec ou a comunicação audio-scripto-visual na era dos self-media* [The age of EMEREC or the audio-script-visual communication in the era of self-media]. Ministério da Educação e Investigação Científica - Instituto de Tecnologia Educativa.

Cruse, E. (2006). *Using Educational Video in the Classroom: Theory, Research and Practice.* Retrieved 6 June 2017, from http://www.libraryvideo.com/articles/article26.asp

De la Fuente Sánchez, D., Hernández Solís, M., & Pra Martos, I. (2013). El mini video como recurso didáctico na aprendizagem de materias cuantitativas [Mini videos as didactic resources in quantitative subjects]. *Revista Iberoamericana de Educación a Distancia*, *16*(2), 177–192. doi:10.5944/ried.16.2.9911

Denso Wave Incorporated. (n.d.a). *Answers to your questions about the QR code*Retrieved 14 November 2017, from http://www.qrcode.com/en/about/version.html

Denso Wave Incorporated. (n.d.b). *Information capacity and versions of the QR code*. Retrieved 15 March 2016, from http://www.qrcode.com/en/

Durak, G., Ozkeskin, E. E., & Ataizi, M. (2016). QR codes in education and communication. *Turkish Online Journal of Distance Education*, *17*(1), 42–58. doi:10.17718/tojde.89156

Feisel, L. D., & Rosa, A. J. (2005). The Role of the Laboratory in Undergraduate Engineering Education. *Journal of Engineering Education, 94*(1), 121–130. doi:10.1002/j.2168-9830.2005.tb00833.x

Fernandes, J. (2017). *Estudo da Google indica que adopção de smartphones triplica em Portugal* [Google report indicates that smartphones use in Portugal triples]. Retrieved 6 July 2017, from http://wintech.pt/53-noticias/google/23040-estudo-da-google-indica-que-adopcao-de-smartphones-triplica-em-portugal

Fernandez, V., Simo, P., Ines, A., Maria, A.-S., Nuria, S., Amante, B., & Federico, G. et al. (2011). 'Low-Cost Educational Videos' for Engineering Students: A New Concept Based on Video Stream and Youtube Channels. *International Journal of Engineering Education, 27*(3), 1–10.

FEUP-LEA. (n.d.). *Vincere: Ciências e Engenharia em português* [Vincere: Science and Engineering in Portuguese]. Retrieved 15 March 2016, from https://paginas.fe.up.pt/~lea/projvincere/

Fishenden, C., & Markland, E. (2006). A systems approach to elementary laboratory instruction. *European Journal of Engineering Education, 4*(3-4), 293–302. doi:10.1080/03043797800040306

Fulton, K. P. (2014). *Time for Learning: Top 10 Reasons Why Flipping the Classroom Can Change Education.* Corwin Press. doi:10.4135/9781483332772

Gaspar, A. (2014). Metade da população portuguesa já utiliza smartphones [Half of the Portuguese population uses smartphones]. *Jornal de Notícias.* Retrieved from http://www.jn.pt/PaginaInicial/Tecnologia/Interior.aspx?content_id=4114579

Giannakos, M. N. (2013). Exploring the video-based learning research: A review of the literature. *British Journal of Educational Technology, 44*(6), E191–E195. doi:10.1111/bjet.12070

Goset, J., & Espinoza, P. (2014). Perception of the use of teaching videos in preclinical training of Chilean dentistry students. *Journal Oral Of Research, 3*(2), 90–94. doi:10.17126/joralres.2014.022

Gradel, K., & Edson, A. J. (2013). QR Codes in Higher Ed: Fad or Functional Tool? Journal of Educational Technology Systems, 41(1), 45–67. doi:10.2190/ET.41.1.e

GrupoMarktest. (2016). *Smartphone continua a crescer em Portugal* [Smartphone continues to grow in Portugal]. Retrieved 5 July 2017, from http://www.marktest.com/wap/a/n/id~2046.aspx

Halupa, C. M., & Caldwell, B. W. (2015). A Comparison of Two Engineering Statics Courses: Traditional Lecture-based and Lecture-based with Online Supplemental Video. *International Journal of Higher Education*, *4*(1), 232–240. doi:10.5430/ijhe.v4n1p232

Hsin, W.-J., & Cigas, J. (2013). Short videos improve student learning in online education. *Journal of Computing Sciences in Colleges*, *28*(5), 253–259.

Jamu, J. T., Lowi-Jones, H., & Mitchell, C. (2016). Just in time? Using QR codes for multi-professional learning in clinical practice. [doi:10.1016/j.nepr.2016.03.007 PMID:27428702]. *Nurse Education in Practice*, *19*, 107–112.

Krivickas, R. V., & Krivickas, J. (2007). Laboratory instruction in engineering education. *Global Journal of Engineering. Education*, *11*(2), 191–196.

Law, C., & So, S. (2010). QR codes in education. *Journal of Educational Technology Development and Exchange*, *3*(1), 85–100. doi:10.18785/jetde.0301.07

Litzinger, T., Lattuca, L. R., Hadgraft, R., & Newstetter, W. (2011). Engineering Education and the Development of Expertise. *Journal of Engineering Education*, *100*(1), 123–150. doi:10.1002/j.2168-9830.2011.tb00006.x

Lombardo, N. T., Morrow, A., & Le Ber, J. (2012). Rethinking mobile delivery: Using Quick Response codes to access information at the point of need. *Medical Reference Services Quarterly*, *31*(1), 14–24. doi:10.1080/02763869.2012.641817 PMID:22289092

Malaga, R. A., & Koppel, N. B. (2016). A Comparison Of Video Formats For Online Teaching. *Contemporary Issues in Education Research*, *10*(1), 7–12. doi:10.19030/cier.v10i1.9876

Meseguer-Martinez, A., Ros-Galvez, A., & Rosa-Garcia, A. (2017). Satisfaction with online teaching videos: A quantitative approach. *Innovations in Education and Teaching International*, *54*(1), 62–67. doi:10.1080/14703297.2016.1143859

Mobile-QR-Codes.org. (n.d.a). *History of QR Codes*. Retrieved 5 July 2017, from http://www.mobile-qr-codes.org/history-of-qr-codes.html

Mobile-QR-Codes.org. (n.d.b). *How Are QR Codes Better Than Barcodes*. Retrieved 5 July 2017, from http://www.mobile-qr-codes.org/qr-codes-vs-barcodes.html

Montoya, M. S. R., & Hernández, D. (2016). Inverted Learning Environments with Technology, Innovation and Flexibility. *Journal of Information Technology Research*, *9*(1), 18–33. doi:10.4018/JITR.2016010102

Moradi, M., Liu, L., & Luchies, C. (2016). Identifying Important Aspects in Developing Interactive Instructional Videos. In *ASEE Midwest Section Conference*. Retrieved from https://conferences.k-state.edu/aseemidwest2016/files/2016/06/Paper-Moradi-Liu-Luchies-29kd4x4.pdf

Observatory of Educational Innovation of the Tecnológico de Monterrey. (2014). *Flipped Learning. Monterrey*. Retrieved from http://observatorio.itesm.mx/edutrendsaprendizajeinvertido

Pandermarakis, Z. G., Sotiropoulou, A. B., Passa, D. S., & Mitsopoulos, G. D. (2012). The Role of Engineering Educational Laboratories at a Thesis Level. *SEFI (European Society for Engineering Education) 40th Annual Conference Proceedings*.

Prensky, M. (2001). Digital Natives, Digital Immigrants Part 1. *On the Horizon*, *9*(5), 1–6. doi:10.1108/10748120110424816

Ricoy, M. C., & Couto, M. (2014). As boas práticas com TIC e a utilidade atribuída pelos alunos recém-integrados à universidade [Good practices with ICT and the use attributed by higher education freshmen]. *Educação e Pesquisa*, *40*(4), 897–912. doi:10.1590/S1517-97022014005000005

Rogado, A. B. G., Quintana, A. M. V., Conde, M. J. R., Martín, I. R., Abad, F. M., Miguelañez, S. O., & Diego, P. B. (2015, March 11). Adquisición de la competencia de prevención de riesgos en el laboratorio mediante tecnología móvil. *Revista del Congrés Internacional de Docència Universitària i Innovació* (CIDUI). Retrieved from http://www.cidui.org/revistacidui/index.php/cidui/article/view/663

Sexton, R. L. (2006). Using Short Movie and Television Clips in the Economics Principles Class. *Journal of Economic Education*, *37*(4), 406–417. doi:10.3200/JECE.37.4.406-417

Shin, D.-H., Jung, J., & Chang, B.-H. (2012). The psychology behind QR codes: User experience perspective. *Computers in Human Behavior*, *28*(4), 1417–1426. doi:10.1016/j.chb.2012.03.004

Simo, P., Fernandez, V., Algaba, I., Salan, N., Enache, M., Albareda-Sambola, M., & Rajadell, M. et al. (2010). Video stream and teaching channels: Quantitative analysis of the use of low-cost educational videos on the web. *Procedia: Social and Behavioral Sciences*, *2*(2), 2937–2941. doi:10.1016/j.sbspro.2010.03.444

Statista. (2017). *Smartphone users worldwide 2014-2020*. Retrieved 6 July 2017, from https://www.statista.com/statistics/330695/number-of-smartphone-users-worldwide/

Surgenor, B., & Firth, K. (2011). The role of the laboratory in design engineering education. *Proceedings of the 2nd Canadian Engineering Education Association*. doi:10.24908/pceea.v0i0.3848

Telg, R. (1999). *Producing an Educational Video*. AEC343/WC024. Retrieved from http://edis.ifas.ufl.edu/wc024

The Marketing Bureau UK Ltd. (n.d.). *FAQs*. Retrieved 15 March 2016, from http://www.freeqrcodetracker.com/pages/faqs

Traser, C. J., Hoffman, L. A., Seifert, M. F., & Wilson, A. B. (2015). Investigating the use of quick response codes in the gross anatomy laboratory. *Anatomical Sciences Education*, *8*(5), 421–429. doi:10.1002/ase.1499 PMID:25288343

Wikipedia. (2017). List of countries by smartphone penetration. Retrieved 14 November 2017, from https://en.wikipedia.org/wiki/List_of_countries_by_smartphone_penetration

YouTube for Press. (n. d.). Retrieved 14 November 2017, from https://www.youtube.com/yt/press/statistics.html

KEY TERMS AND DEFINITIONS

Digital Native: A person who grows up surrounded by technology using it in everyday tasks.

Dynamic QR Code: A QR code with an online resource that can be changed without the need of changing the QR code.

Educational Video: A video that is used with an educational purpose.

Low-Cost Educational Video: A video made with inexpensive equipment and software.

Mini-Video: An educational video of short duration that conveys a particular information.

QR Code: Two-dimensional matrix codes that can be freely generated online and are scanned by a camera.

Static QR Code: A QR code with an online resource that cannot be changed.

Chapter 7
Microblogging Case Study in Higher Education:
Edmodo in Finland

Vasileios Paliktzoglou
University of Eastern Finland, Finland

Jarkko Suhonen
University of Eastern Finland, Finland

ABSTRACT

Recent research indicates that even though social media networking sites are commonly used in higher education, very little empirical evidence is available concerning the impact of social media use on student learning and engagement. In this chapter, the experience of using Edmodo is analysed as learning aid to support group work in comparison with the level of familiarity, engagement, and frequency of use of social media technologies among university-level computer science students in Finland. The specific focus of the chapter is to examine the reception of the students towards the Edmodo platform. The data was collected through a social media familiarity questionnaire, Edmodo experience questionnaire, and interviews. The main findings are that the cohort was not very familiar with social media at the beginning of the course. This chapter provides experimental evidence that microblogging social networking sites and, more specifically, Edmodo can be used as an educational tool to help engage students more in the use of social media networking sites.

DOI: 10.4018/978-1-5225-5279-6.ch007

INTRODUCTION AND AIMS

The academic world has been experiencing a rapid growth in educational technology and more specifically social media tools, which have the potential to constructively complement traditional education and even replace it in the cases of distance/online learning. Social media tools are used in many institutions for educational purposes in numerous, innovative ways (Conole & Alevizou, 2010; Doyle, Sammon, & Neville, 2016; Paliktzoglou, Giousmpasoglou, & Marinakou, 2016; Oyelere, Paliktzoglou, & Suhonen, 2016) even to the extent of such tools being utilized in traditional face-to-face classrooms (Redecke, Ala-Mutka, Bacigalupo, Ferrari, & Punie, 2009 ; Bower, Hedberg, & Kuswara, 2010). Of particular note is that the revised Bloom's taxonomy of educational objectives (Anderson & Krathwohl, 2001) has been extended to include relevant terminology describing the learning process through the use of social media tools (Churches, 2007). This inclusion indicates the adoption of such tools in modern education. From the very beginning of the internet, educators introduced communicating with others through interfaces such as chat rooms, Internet forums, message boards, web communities and blogs (Albion, 2008).

Similar to other communication tools, social media tools have rules, conventions and practices which users have to adhere to in order to be accepted as legitimate participants in these online communities. As argued by Jacobs (2008) social media tools also have some potential pitfalls to negotiate, such as the unintended consequences of publicly posting sensitive personal information, confusion over privacy settings and contact with people one may not know.

According to Safran, Guetl, and Helic (2007) social media tools are part of a wider online phenomena, also in the field of education and learning, which enables self-expression, communication and more versatile user interaction online. Additionally Chatti, Jarke and Frosch-Wilke (2007) propose that in the era of culture centralized collaboration, one-size-fits-all, top-down, static and knowledge push models of traditional learning initiatives need to be changed to a more open, dynamic, emergent, social, personalized and knowledge pull model for learning. In recent research it has been argued that social media concepts open new windows for more efficient learning and have the potential to overcome many of the drawbacks of traditional learning models (Chatti et al., 2007; Maloney, 2007).

The motivation for this study stems from the researchers' opinion that there is a lack of empirical studies on the specific use of Edmodo (a microblogging social networking site) as an instructional tool, particularly in higher education. In this research, the use of microblogging is based on an adoption of social media driven learning frameworks by Chatti et al. (2007), which stresses the use of knowledge networking and community building to leverage, sustain and share knowledge in a collaborative way. Echoing Chatti's assertion, this research similarly places emphasis on penetrating classrooms boundaries to involve students, partners, stakeholders and different types of frequently overlapping, formal and informal communities. The microblogging project encourages a participatory culture where students believe that their contributions matter and feel some degree of social connection with one another (Jenkins, Purushotma, Clinton, Weigel, & Robison, 2006). Furthermore, microblogging sets the stage for learning as analogous to innovative processes of inquiry, where something new is created and original knowledge is either substantially enriched or significantly transformed during the process (Paavola, Lipponen, & Hakkarainen, 2002).

According to the authors' understanding most of the existing studies related to microblogging in educational settings have focused mainly on the use of Twitter in teaching and learning (Fei Gao, Tian Luo and Ke Zhang 2012; Lowe & Laffey, 2011). Therefore, we argue that our study on the use of Edmodo provides new perspectives to the scholarly discussion related to the use of microblogging in education. The study has two main aims, first, to investigate the familiarity of social media among Computer Science student and second, to analyze the experiences of using one specific social media networking site, Edmodo.

A cohort of Computer Science students participating to an online course at University of Eastern Finland participated in the study. The students used Edmodo in their group work related to the one of the topics of the course, Social Media. In the study, we collected data regarding students' views and experiences before and after the experience of using Edmodo during the course.

Their feedbacks on the use of microblogging were gathered through familiarity and experiences questionnaires and interviews. This paper will discuss the experiences and challenges faced by the researcher in designing a teaching and learning project integrating the use of microblogging. Also, there will be the inclusion of key reflections by the researcher on how microblogging can evolve from a purely recreational use to an authentic instructional use in higher education context of teaching and learning. Finally, the potential implications and recommendations for future implementations will be discussed.

MICROBLOGGING AS A SOCIAL NETWORKING SITE IN EDUCATION

Role and Benefits of Social Networking in Education

With integration into the daily lives of millions of people around the world, there has been a rapid growth in the use of social networking sites in education (Downs, 2005). Social interaction and collaboration work plays an important role in online community applications, which must be also considered in the context of learning (Safran, Doerr, Sherman, Doerr, Flaxman, & Winkler, 2007). In the contemporary media and knowledge intensive era of collaboration culture centralized, one-size-fits-all, top-down, static and knowledge push models of traditional learning initiatives need to be changed to a more open, dynamic, emergent, social, personalized and knowledge pull model for learning (Chatti, Jarke, & Frosch-Wilke, 2007; Doyle, Sammon, & Neville, 2016)). Social media concepts are opening new windows for more efficient learning and have the prospective to overcome many of the drawbacks of traditional learning model (Chatti et al., 2007). Social networking sites offer learners new and multiple ways of communication via the internet, whether through computers or mobile devices. Learners are able to create their own online pages or profiles and to construct and display an online network of contacts. They can then communicate on a one-to-one basis or private chat, or in a more public way such as a comment posted for public viewing.

The educational benefits of using social networking sites in education are many and various (Maloney, 2007). According to Selwyn (2009) social networking may: *"Benefit learners by allowing them to enter new networks of collaborative learning, based around interests and affinities not catered for in their immediate educational environment."* Ozkan and McKenzie (2008) argue that social networking skills are the necessity of a 21st century approach to teaching. Moreover, Greenhow (2009) brings out various benefits of using social networking sites in educational context, such as allowing exploration beyond the classroom settings, fostering creativity thought self-expression, exploring various viewpoints and making more informed decisions.

The Role of Microblogging Among Social Media Technologies

Microblogging is one of the social media technologies with applications in many domains, including education. The features of a microblogging platform vary from sending and receiving messages via the web, SMS, instant messaging clients, and by third party applications. As such microblogging is in many ways similar to blogging. When founded in 1999, blogs created a wave of online publishing for web users to write together, post their own thoughts, information and links (Blood, 2004). A key distinguishable difference between a blog and a microblog, is that the latter requires posts be made with a limited number of text characters as well as smaller audio and video files. According to McFedries (2007): *"A microblog can be seen as a weblog that is restricted to 140 characters per post but is enhanced with social networking facilities"*.

Microblogging naturally supports the use of mobile technologies, as it facilitates much more frequent updates than a blog, owing to participation with a concise reduction of text characters. According to Java, Song, Finin, and Tseng (2007), the shorter posting feature lowers users' requirement of time and thought investment for content generation. They note "*on average, a prolific blogger may update her blog once every few days; while a microblogger may post several updates in a single day*" (p.2). The success of microblogging in comparison to blogs is the ease of use and lightweight technology in order to follow up on the conversation and provide immediate input (Ebner, Lienhardt, Rohs, & Meyer, 2010). Moreover, as mentioned by Hsu & Ching (2011), microblogging is among the social media technologies that encourages conversation, and facilitates collective knowledge and content creation. In microblogging, users may explore, follow, reply or forward each other's posts, allowing interactions and collaborations to take place among people from virtually any corner of the world (Java et al., 2007).

In a survey of bloggers, Nardi, Schiano, Gumbrecht, and Swartz, (2006) describe different motivations for "why we blog". Their findings indicate that blogs are used as a tool to share daily experiences, opinions and commentary. Based on their interviews, they also describe how bloggers form communities online that may support different social groups in real world. Lento, Welser, Gu, & Smith (2006) examined the importance of social relationship in determining if users would remain active in a blogging tool called Wallop. A user's retention and interest in blogging could be predicted by the comments received and continued relationship with other active members of the community. Users who are invited by people with whom they share pre-existing social relationships tend to stay longer and active in the network. Moreover, certain communities were found to have a greater retention rate due to existence of such relationships. Mutual awareness in a social network has been found effective in discovering communities (Nardi et al., 2006)

Educational Impact of Microblogging

Although interactions via microblogging are often informal or sometimes playful (Dunlap & Lowenthal, 2009), the potential of microblogging as a tool of promoting learning has been identified by several scholars (Borau Ullrich, Feng, & Shen, 2009; Ebner & Schiefner, 2008; Gaonkar, Li, Choudhury, & Cox, 2008; Ramsden, 2008). In a critical literature review analysis conducted by Fei Gao et al., (2012), they suggest that microblogging has a potential to encourage participation, engagement, reflective thinking as well as collaborative learning under different learning settings. Microblogging can also be used to promote interaction, exchange of ideas and sharing of resources among students and the instructor (Click & Petit, 2010; Ebner & Maurer, 2008). Additionally, Hsu & Ching (2011) argue that microblogging in combination with "learning by doing" approach promotes knowledge construction and community building processes. A microblog can help students to take creative risks, demonstrate critical thinking and make sophisticated use of language and design elements. In addition, students also obtain creative, critical, communicative and collaborative skills that may be useful in scholarly and professional perspective.

Existing studies have examined, for instance, how instructors have used microblogging for extended communication beyond the classroom, such as posting learning materials (Lowe & Laffey, 2011) or announcing events and assignments (Perifanou, 2009). In Lowe and Laffey's (2011) study, the instructor used Twitter in a marketing class to post updates on recent marketing events, contemporary marketing issues and examples of key concepts. Students were able to access and exchange ideas about the updated information and materials in a timely manner.

study conducted by Ross, Terras, Warwick, & Welsh (2011) showed that microblogging can be used to enable interactions between audience and speakers in a live event or to encourage virtual participations from people worldwide. Microblogging made it easy for the audience to ask questions, have discussions, share resources and create shared comments on learning materials (Ebner, 2009). Such immediate participation provides a means for the audience to actively interact with the content and also allows presenters to respond dynamically to audience's reactions. Another study illustrated that microblogging is beneficial for documenting ongoing processes and just-in-time thoughts (Wright, 2010). In the study, eight graduate students in teacher education were asked to tweet three times each workday in response to their experiences of teaching and a list of questions such as "what do my students say about their learning right now?" and "what do I need to overcome or solve?" The activity encouraged students to share and reflect upon their teaching experiences when students were "walking in corridors," "in cars at the end of the teaching day" or "during lunch breaks" (p. 261) as the thoughts occurred. As a result, learning took place in dots of actions that were composed of small, discrete moments rather than in a linear and sequential manner (Ihanainen & Moravec, 2011).

Challenges of Using Microblogging in Education

Many studies discussed earlier identified several major challenges when microblogging was integrated in educational settings. The first However, as with any emergence of social media tools in education, its use does bring to light important considerations for educators. For example, educators must address a series of instructional and pedagogical challenges; like adapting to the need to communicate within a constraint of 140 character increments, managing a constant stream of comments, most of which are peripheral to the topics being discussed and even unethical behavior by students and potential distraction in a face-to-face classroom (Corbeil & Corbeil, 2011). Another challenge is related participants' unfamiliarity with microblogging. Though microblogging was gaining popularity, according to the Pew Internet Project surveys (Lenhart, Purcell, Smith & Zickuhr, 2010), only 8% of 12- to 17-year-old Internet users used Twitter, and only 19% of adult Internet users used Twitter or similar services to post updates or view others' updates. Studies suggested that many learners were not familiar with Twitter and found it difficult or even intimidating to use (Agherdien, 2011). The reluctance to learn or use the technology limited the scope of interactivity afforded by microblogging (Lowe & Laffey, 2011). As a result, some researchers suggested finding out creative ways to convince students of Twitter's benefits or establishing rewards to encourage its use (Rinaldo, Tapp, & Laverie, 2011). Furthermore, despite of its affordance in facilitating communication, microblogging can sometimes lead to "an unwieldy information flow, known as information overload" (Ebner et al., 2010, p. 98). The noise information posted online can be distractive (Holotescu & Grosseck, 2009), and some students may feel it is a waste of time reading posts containing less useful information (Rinaldo et al., 2011).

Some of the recent studies on Edmodo related to secondary education, (Picardo, 2011; Oyelere, Paliktzoglou, & Suhonen, 2016).) identified that Edmodo promotes collaboration between students and they are we able to formulate and deliver answers, and respond to one another in turn. Moreover, Jarc (2010) reports that student's progress can be easily tracked. The discussion threads in Edomdo encourages dialogue and collaboration in the sense that it is almost like having an interview with the students. Finally, he mentions that Edmodo provides a safe collaborative environment and reduces the amount of paper used in classroom.

Despite the prolific emergence of the use of microblogging in various educational contexts, much empirical research is yet to surface. In fact, studies focusing on the primary use of microblogging in higher education with an emphasis on the perspective of the students are almost non-existent; particularly investigating microblogging as a platform of social networking website where the use of other social media tools is seamlessly integrated. This research study aims to discuss the experiences, challenges and provide key recommendations for future similar implementation in an institute of higher education.

EDMODO MICROBLOGGING TOOL FOR EDUCATION

Edmodo is described on its official site as:

...a secure, social learning platform for teachers, students, schools and districts. We provide a safe and easy way for your class to connect and collaborate, share content and access homework, grades and school notices. Our goal is to help educators harness the power of social media to customize the classroom for each and every learner.

Edmodo is a social learning platform that provides educators' means for connecting, collaborations, sharing content and accessing homework, grades and school notices. Edmodo uses the micro-blogging model with main focus on education, and it is accessible from any mobile device or smart phone. Edmodo is a free online microblogging platform that focuses on learning and offers a safe and easy way to connect, collaborate and share ideas, files, events and assignments among instructors and students. Edmodo uses the microblogging model, but it has been especially designed for educational purposes. Basically Edmodo, as a micro blogging social learning platform, is accessible from any mobile device or smart phone. We evaluated a variety of microblogging platforms (including Twitter, Jaiku, Cirip.ro Postcros and Tumbl) and we selected Edmodo for the extensive functionality that it provides which includes the incorporation and use of other social media tools.

Edmodo has an easy to use, straightforward interface and features which uses the microblogging model and allows instructors and students to access the tool anytime, anywhere. Instructors can post messages and notes, discuss classroom topics, give assignments and grade class work, share content and materials and network and exchange ideas with their students and other instructors. Finally, Edmodo offers a class calendar and polls feature, a specific feature to store and share files, and the ability to integrate (RSS) streams.

As previously discussed, researchers have conducted vast amounts of research focused on micro blogging in education. Much attention has been afforded to Twitter and the domain of higher education whilst Edmodo has not received as much consideration. However, there have been a few studies related to the use of Edmodo in secondary education which have showed that Edmodo promotes collaboration between students (Picardo, 2008). Furthermore, Edmodo supports students to formulate and deliver answers and respond to one another in turn.

Additionally, a study by Jarc (2010) showed that students' progress can be easily tracked using a discussion thread and that this encourages dialogue and collaboration as it is almost like having an interview with the students. Jarc (2010) argues that Edmodo provides a safe collaborative environment and reduces the amount of paper used in the classroom. The above mentioned studies indicate that Edmodo can easily be adopted in the context of secondary education. However, very little research relates to Edmodo's potential in higher education settings.

Main Features of Edmodo

The easy to use and straightforward interface and features of Edmodo allows instructors and students to access their learning anytime, anywhere. Instructors post messages and notes, discuss classroom topics, give assignments and grade class work, share content and materials, and network and exchange ideas with their students but also with other instructors. Edmodo offers a class calendar and polls feature, specific feature to store and share files and the ability to integrate (RSS) stream.

Edmodo provides certain features that can facilitate multiple simultaneous discussions where the instructor can create collaborative "groups". Moreover, Edmodo assigns to each of the created groups a unique code that after that is circulated by the instructor to the class members. This feature of Edmodo provides at the same time autonomy and a secure environment where no one without access may view or participate in the discussion learning space. Additionally, with Edmodo students within their group have the ability to also create their own small groups" in order to facilitate more specific tasks. Edmodo allows instructors to post assignment reminders, build an event calendar and communicate to one specific or all the groups at once. Furthermore, the participants/students can also collaborate posting their ideas, sharing links, videos, and images. The accessibility feature of Edmodo in combination with the easy to use interface makes Edmodo an effective and promising learning environment.

Figure 1. Main features of Edmodo

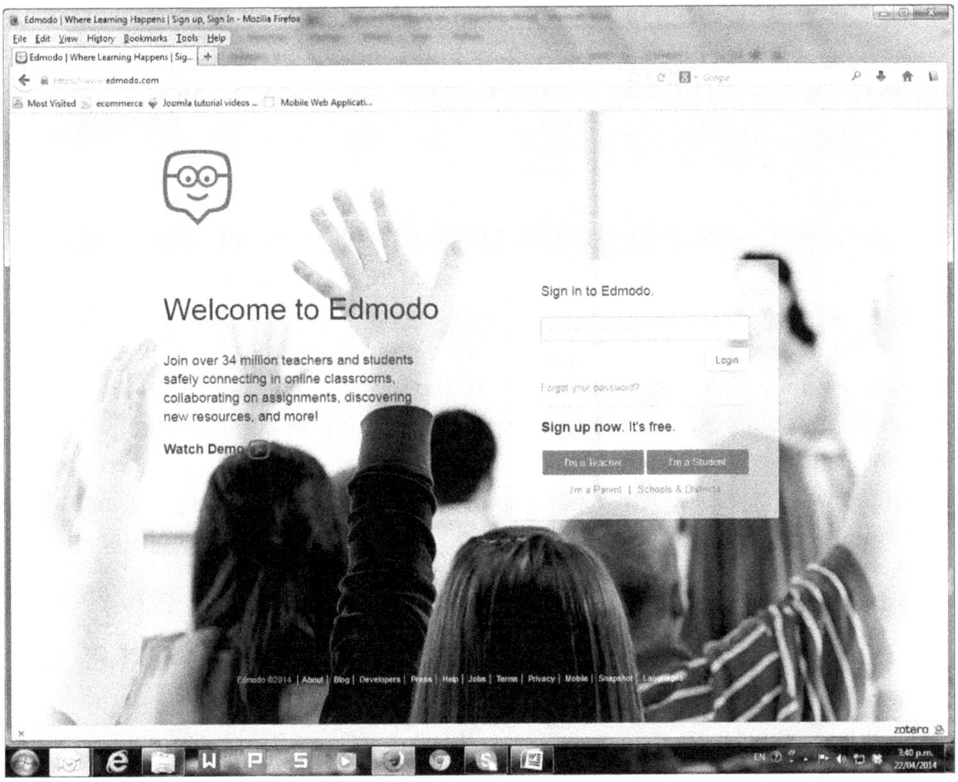

The Role of Edmodo in the Study

In our experiment, the Edmodo tool was introduced as a platform to support group work in one of the course modules during the Technologies for Development and Education online course at School of Computing, University of Eastern Finland. Edmodo was used alongside Moodle to support specifically the collaborative work of the students during the specific module of the course that focused on the topic of social media in education. In the beginning of the module, the students had a general introduction to Social media technology and more specifically Edmodo where they were provided instructions how to use the tool trough set of practical tasks. Then in groups of five students they start working with Edmodo and becoming familiar with Social media trough their assignments.

The students were informed that lately there has been a strong drive in developing and enhancing social media technology for e/m-learning purposes. Moreover, it was discussed that social media technologies could meet many of those needs, but also brings along new challenges. During the course module students became acquainted/trained with the social media technologies and they needed to reflect and discuss upon the potential of social media technologies used for e/m-learning purposes. Subsequently, they were given a short training in order to learn how to use the Edmodo system though video tutorials and practical tasks. Then student had to create their groups (five) according to the interest in working the following outlined assignment topics: 1) Social bookmarking and RSS in CS/HE, 2) The use of Blogs in CS/HE, 3) The use of Wikis – Microblogging in CS/HE, 4) Podcasts in CS/HE, 5) Social software (networking) – Virtual words in CS/HE.

All the lecture material in the form of articles, books, podcast and RSS was provide in Edmodo. It is worth mentioning that the instructor was constantly supporting, stimulating and facilitating collaboration on a day to day basis. Part of the learning strategy was to motivate student to collaborate, share resources and discuss their work in Edmodo. As part of the collaboration process, each of the students had to post at least three posts per week related to the topic of their group, and hen reply to posts from colleagues. The active participation in Edmodo was a requirement for students to pass the course module. The role of the instructor was to constantly trigger the work of the student groups in the form of microblogging posts including videos, links to books articles etc. In addition, the group forum was constantly feed with several related source using the RSS feature in Edmodo.

The nationality demographics are given in the pie chart of Figure 2.

RESEARCH DESIGN AND METHODOLOGY

Research data was collected from students enrolled in the Technologies for Development and Education online course at School of Computing, University of Eastern Finland. The student cohort was multi-ethnic (n=23, 7 female, 35% and 16 male, 65%), including five undergraduate (bachelor), 14 postgraduate (Master) and four PhD students. The majority of the students were aged 20-30 (14 participants, 70%), 7 over 30 (17%) and only one under 20(13%). There were two sources of data collection: questionnaires at the beginning and end of the experiment which provided students' perceptions on social media networking sites (Social Media Familiarity Questionnaire) including the use of Edmodo and interview session with the participants.

The first questionnaire activity was a part of the first week's "task to do" tasks (the students were given list of task that needed to be completed by the end of each week). The enrolled students in the course had an average response time for completing the questionnaire of seven to twelve minutes. The students were not given any prior information related to social networking sites, since we sought to discover if the students had any prior knowledge. The survey was split into two parts: the first part covered demographic information about the participants (the participants' name, e-mail address, age range, level of studies), the second part covered student engagement as user, the context and frequency of use; and with a list of technologies associated with the term Social media. The questions were prepared with Likert scale where the students could rate each item on a 1-to-5 response scale. The student engagement questions in Likert scale varied from "very familiar (frequent user)" to "not known", in the context of use the questions on the Likert scale varied from "Personal learning/studying (as tool for studying/learning" to "Other" and in the frequency of use on the Likert scale varied from "daily" to "never", where all the questions were compulsory to be answered.

The second (post-module) questionnaire activity was also a part of the last week's "task to do" tasks. This questionnaire was similar to the Social networking site familiarity questionnaire, but the specific focus was on the Edmodo tool. The enrolled students in the course had an average response time for completing the questionnaire five to ten minutes. The survey was again designed in two parts: the first part covered demographic information about the participants (the participants' name, e-mail address, age range, level of studies) the second part was focusing on the familiarity with the Edmodo tool. The questions were planned using a Likert 1-to-5 response scale. At the end of the survey one open ended questions was used where students were able to add comments or suggestions upon the future improvement on how to use Edmodo. The student frequency of use of Edmodo questions in Likert scale varied from "daily" to "never", in the engagement with Edmodo questions on the Likert scale varied from "completely agree" to "completely disagree" and in the motivation of use the students could from "module structure" to "other". Appendix 1 includes the questions used in the post-module questionnaire. At the end of the given above questionnaires there was a clear explanation about the nature of the research and there was a question asking from the participants their official approval to proceed with the assurance that all data will be treated confidentially.

Finally, semi-structured interviews were made in the end of the course so that information about the topics of the study could be obtained without the researchers being constrained by the questions. The interviews provided an understanding of the participants' experiences in using the microblogging tool, Edmodo, during their Social Media course module. A list of preliminary interview questions were used to elicit participants' initial responses before the in-depth interviews were conducted. The respondents had the possibility to give explanation about the central factors in the topics in their opinion and they are allowed to speak freely during the interview. While all informants were asked a standardized set of questions, informal interactions resulted in further researchers' probing and questioning. Questions ranged from surface level information, such as how the participants like the microblogging tool Edmodo, to more intuitive questions, such as, what is the participants' perceptions on the use of Edmodo for educational purposes. Appendix 1 lists the questions used in the interview. At the beginning of the interview there was a clear explanation about the nature of the research and the participants were asked for their official approval to proceed with the assurance that all data will be treated confidentially. Analysis of data gained from the interviews "entails classifying, comparing, weighing, and combining material from interviews to extract the meaning and implications, to reveal patterns, or to stitch together descriptions of events into coherent narratives" (Rubin & Rubin, 2005, p. 201). Each informant's interview was audio recorded then transcribed to cross reference for key themes of identity creation and consumer influence.

RESULTS

Social Media Familiarity Questionnaire

In the first questionnaire, the familiarity and frequency of use for the following social media tools was assessed: Blog,Wiki, Podcast (Audio & Video),Tag cloud, Social Bookmarking, Social Networks, Chat, Micro-blogging, Media sharing, Virtual worlds, Mashups, m-Learning. The results are given in figures #3 & #4 for familiarity and frequency respectively. Out of all the tools, students were mostly familiar with chat, social networks and media sharing platforms as they used these three most frequently. They were less familiar (aware of the application) with Mashups and tag cloud technologies. Prior to the start of experiments, most of the students were not familiar with Edmodo, which was the social media networking site of focus for this study.

Context of Use

In the context of use part, the same tools than in previous section was assessed. The results are presented in figure #5. The highest frequencies were encountered for Social Networks and Chat in the context of recreational use. Most of the students had not used Edmodo before.

The Edmodo Experience Questionnaire

The participants filled out the Edmodo experience questionnaire after they have been working with the tool during the course. The Edmodo experience questionnaire was similar to the Social networking site familiarity questionnaire, but the specific focus

Figure 2. Familiarity with Social media tools as users

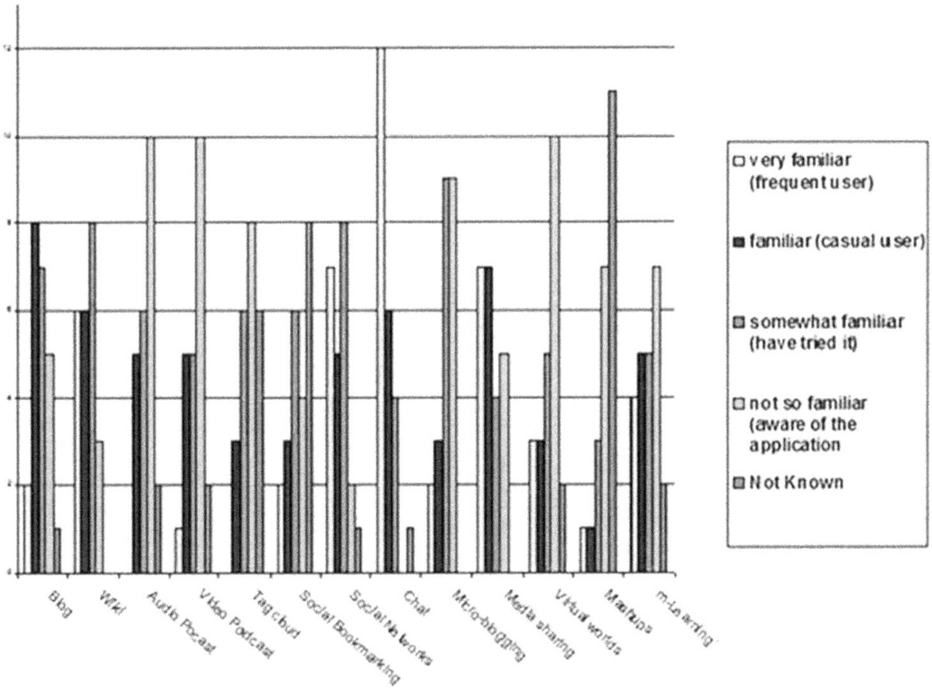

Figure 3. Frequency of use of Social media tools

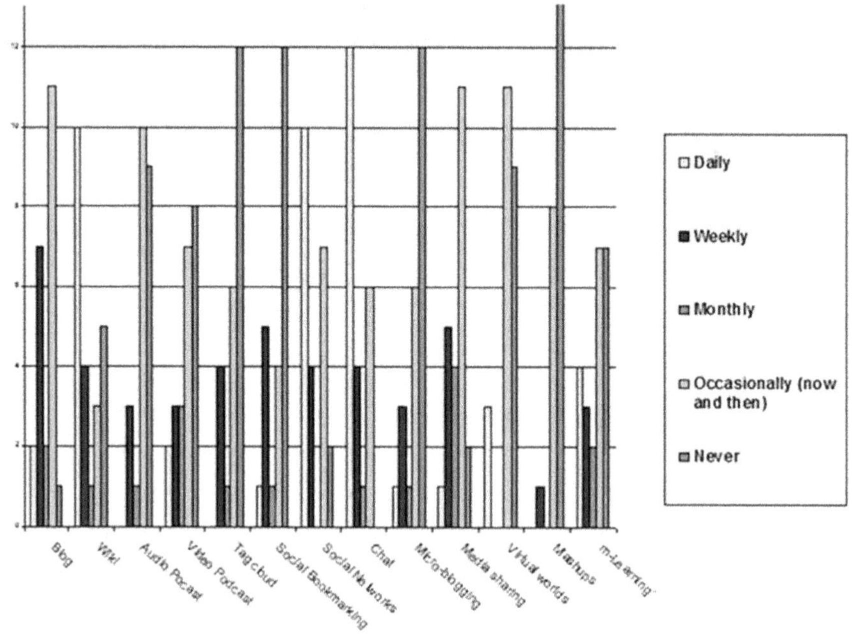

Figure 4. Context of use of Social media tools

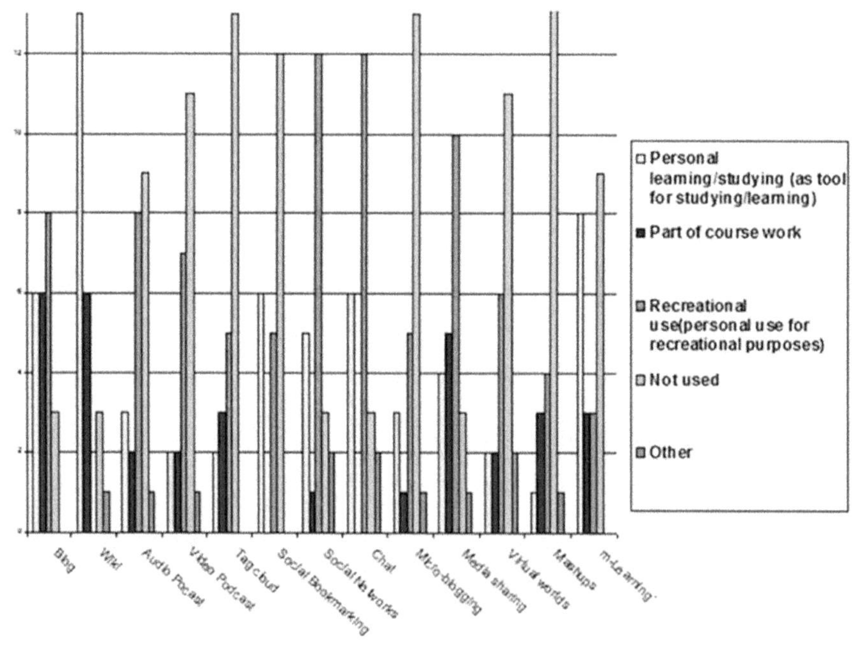

was on the Edmodo tool as opposed to the full range of the social media networking sites assessed at the beginning of the module.

The results of the experience questionnaire are presented in Figures 5, 6 and 7. The results show a positive impact of Edmodo. The "completely agree" response occurred with the highest frequency. The students were producing content and reading in Edmodo "daily" with the highest frequency. They mentioned that the "module structure" with the highest frequency motivated them using Edmodo. Finally, most of the students when asked to comment or suggest improvement in the use of Edmodo mentioned that this is great tool form learning and they would definitely suggest using this tool in other courses.

Interview

The interview was analysed in two stages. Analysis of data from transcripts requires an open mind and immersion in the text (Morrison, Haley, Bartel Sheehan, & Taylor, 2002). Stage one consisted of the preparation of transcripts, the findings of themes, concepts and events and coding these in order to retrieve information on specific ideas (Polonsky & Waller, 2005). The process of coding the themes required the researcher to engage with the data and fully understand the meanings and insights that relates to the data. The second phase used the data to formulate a description of the setting (Rubin & Rubin, 2005) and interpret how these themes and insights relate to the research questions (Polonsky & Waller, 2005). Here we present the results of the analysis according to the important themes that emerged from the interviews.

Figure 5. Impact of Edmodo

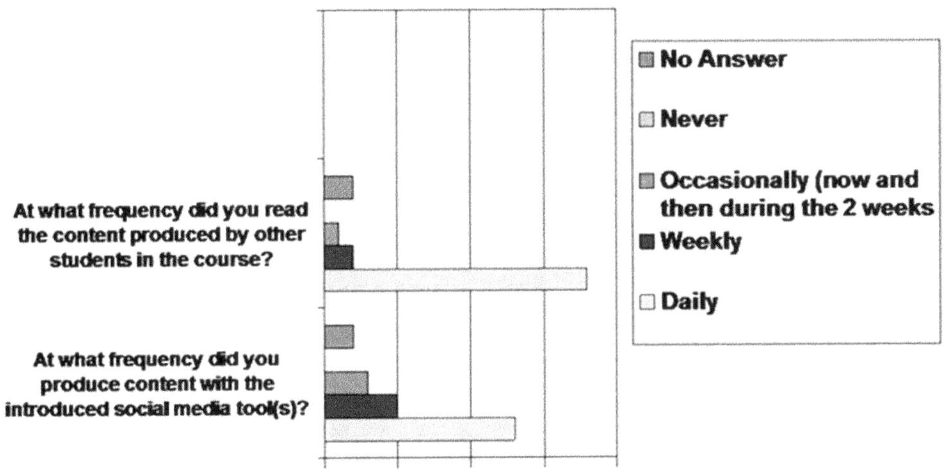

Figure 6. Impact of Edmodo (Frequency)

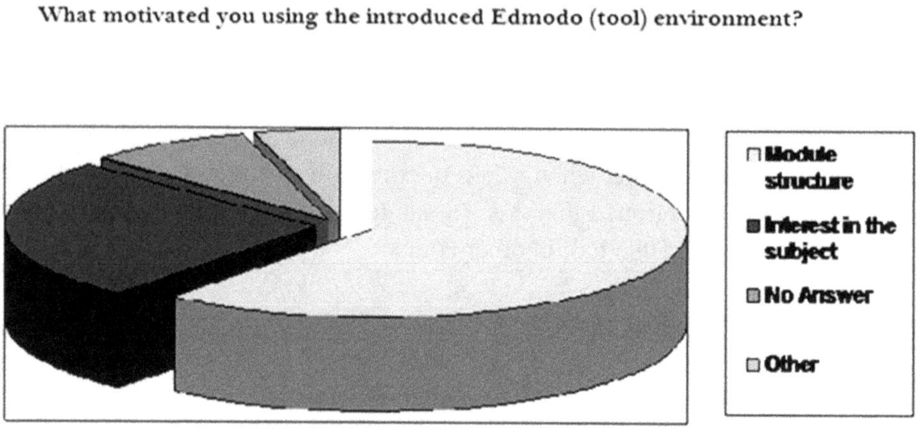

Figure 7. Impact of Edmodo (Motivation)

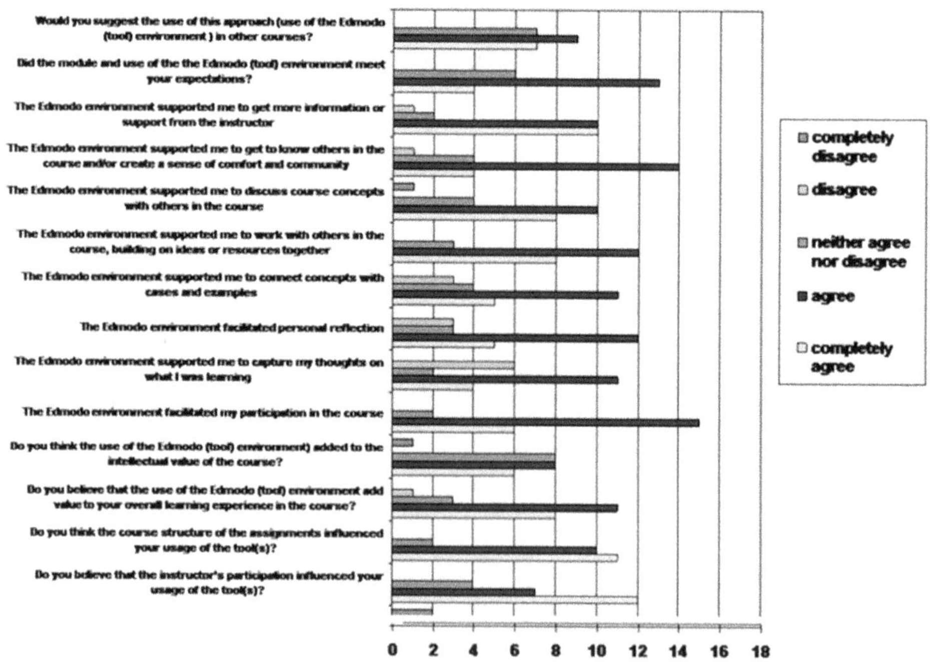

Collaboration and Feedback

Collaboration among students was easy in many cases. Students felt confident to work collaboratively. Overall, the students felt that they received useful feedback from fellow students and the Instructor through forum posts. One of the students stated that: *"When people see and provide constructive feedback I can understand and improve upon my mistakes"*. Furthermore, emphasizing on the positive collaborative learning opportunities a student commented, *"Edmodo is more for discussing and sharing between students and teacher. So in my opinion, Edmodo is very useful for working in groups."*

Motivation and Familiarization

Many student participants stated that they need time to get used to Edmodo. Such familiarity for any social networking tool introduced in an instructional course is expected. Although, the exact time span needed for students to be comfortable with its use is varied. Students commented, *"It's easy after having practiced with it"*, *"Need more time to practice with it"*, and *"It is time consuming"*.

Nevertheless, they remained motivated to be able to use the tool effectively for their learning needs in the course. It is noteworthy that the learning curve for Edmodo's integration in the student participant's course was not that steep. This is due to the fact that there are similar features in Edmodo as in other learning management systems and microblogging tools.

Technical Issues

Some students faced technical problems with the Edmodo platform that can be frequently encountered with many online tools, problems, browser-specific problems, problems with the invitations to become contributors to Edomodo. Some of these problems stem from user frustrations with the limitations of the features available in Edmodo. For example, a student participant shared, *"File upload feature in Edmodo needs to be upgraded. I have to re-upload another file with a different name. It is not easy to keep the document version and very easy to confuse others who are not the document's owner if the document was marked as shared."* Some students experimented extensively with the functionality of Edomodo (e.g. using widgets for embedding other Social media tools) and, hence, faced other types of more elaborate technical problems. More issues will be discussed in a later section on negative reception.

Comparison With Other Educational Technology Tools

Students had also used the Moodle Learning Management System as part of the course. As such, it was natural for the student participants to compare the use of Edmodo with the Moodle during the interviews. A student mentioned that a disadvantage of the Moodle is that it is not so intuitive, which indicates that the student involvement in the use of Edmodo is appreciated by parts of the cohort. On the other hand, some students found it simpler to understand and work within the Moodle environment.

A student participant did however note that the use of the two different tools would suit different learning tasks or learning activities. He said, *"They suit for different task. Edmodo is better to be used to organize project works, as it suits better for collaborative tasks (own areas for groups, post and comment communication)."* This is synonymous with another student participant's comment that *"Edmodo seemed to be better for group tasks."*

Other areas of comparison between the two tools include the general interface. It is interesting that with the simplest function that is made available in a tool, users tend to prefer a tool more than the other. A participant shared, *"Moodle has a clearer user interface when it comes to sharing material as the icons are smaller and bunched up together. Also the discussion is separated into its own easy to find section."*

Noting the student participants' increased familiarity and expertise in using both the tools, Moodle and Edmodo, it is noteworthy to mention some of their feedback during the interviews turned into interesting suggestions. This was particularly true with the integration of both tools into one system that is deemed to be more efficient and meaningful for their learning. One participant said, *"It would be good if Edmodo can be integrated into Moodle or other LMS, or, Edmodo can add LMS features into it."* This is commendable as the student participants are able to 'think out of the box' and imagine a better system or tool that would effectively address their concerns, issues they face and cater to their learning needs.

Positive and Negative Aspects of Edmodo

There were some negative remarks by students mentioning that it was not easy to navigate in Edmodo. *"When you log on you were not able to know where to go to. "*

Drawing from the interview data, the student participants raised some drawbacks of using Edmodo; in particular, its social media tools usability did not reach user satisfaction. A student participant shared, *"Sometimes, it was difficult to notice new messages, because the page focus did not move to the new message."*

However, on a positive note, some students did express that Edmodo is indeed helpful for simple tasks. Many students stated that they understood the purpose of Edmodo and mentioned the following keywords during the relevant part of the interview discussion *"to collaborate"*, *"knowledge sharing"*, *"improves your technical skills"*. Almost all students stated that they like Edmodo even if they had some issues adjusting with it (technical, steep learning curve etc). Some admitted that they did not like it at the beginning but changed their views as the course progressed. Overall, students commented that they enjoyed using online tools for study or personal pleasure.

Further discussions on the student participants' negative and positive receptions on the use of Edmodo led to recommendations for its improvement. Some of these points proved useful for future educators to consider when integrating Edmodo's use in their course work in higher education.

A participant suggested,

...a collaborative writing and chat feature so that it would bring more value to the system because then you could really collaborate and make progress inside the system easily with other members.

Such suggestions came up with increased awareness among the students on the various affordances of social networking tools that are available. It is noteworthy that students are able to perceive the most functional and meaningful tools that they can leverage on in order to maximize their learning.

Views on Educational Impact

Collaborative learning can be benefited by social media tools within microblogging sites like Edmodo, because of the influential characteristics for the field of collaborative learning. Students used them because those tools allow them to feel connected to their tutors and other students of the class. One interview participant said, *"Using Edmodo as a group work tool works. Edmodo offers a good environment for online group projects."* While another commented, *"It is very interactive. Teachers can provide better support through the platform and offer good materials for the students to use in studying. Therefore both will benefit of using Edmodo."*

On the other hand, social network tools remove the physical distance among people and generally bring them together. From the educational point of view, the collaboration and information sharing of social media tools in Edmodo is a plus because many students has the opportunity to use the documents whenever they need. One respondent talked about this point, *"I feel that I benefit from the use of Edmodo. With the system I could discuss, share important resources and comment on other people's progress."*

Furthermore, another student highlighted the fact that both educators and students are able to contribute and build upon one another's contribution. She said, *"The best of Edmodo is putting students and teachers at the same communication level, breaking the barrier between them so joining ideas and thoughts is at the first place."*

Instructor's Remarks

Working with Edmodo was interactive and enjoyable though a significant amount of time and effort is required on the part of the course instructor to moderate Edmodo, especially when there are more than three groups working on diverse subjects. While using groups is indeed useful and promotes collaboration and the exchange of ideas, the researchers have certain reservations as to how much effort the instructors will need to put into the facilitation of big cohorts/groups where there can be more than 30 contributors. The fact that there are no clear instructions as to the taxonomy in Edmodo could cause problems in following the flow of both course teaching and learning. On the other hand, as students and instructors become more experienced, less instructor input is anticipated, and the above mentioned issues could become less pertinent. Additionally, as argued in the literature, cultural differences play an important role in the acceptance of Edmodo. Students in our case were not skeptical when using Edmodo as a leaching tool (Cheung et al., 2011). On the contrary, it was observed that the reception was more than welcome. Moreover we would suggest the following improvements in order that Edmodo increases its usefulness

1. Create a dedicated add-on feature to improve the learning process. This could be implemented by improving Edmodo's taxonomy and relating all the material (links, sources, conversations) to the outlined "problem" of the course.
2. Provide a common "space" for all the groups in Edmodo, where all the students and the instructors could communicate and share ideas. At present the groups work in isolation.
3. Add an embedded chat feature that will provide a synchronous communication among students and the instructors. At present there is not such feature available in Edmodo.

Having implemented the features suggested above, the authors argue that Edmodo will become more widespread in the higher education domain and that this will lead to a better reception of the tool.

DISCISSION AND CONCLUSION

The motivation for this study stems from the lack of empirical studies on the specific use of Edmodo as a microblogging social networking site as an instructional tool, particularly in Higher Education. There are studies focusing on the use of Twitter in teaching and learning (Fei Gao, Tian Luo and Ke Zhang, 2012), however much is yet to be investigated on the use of Edmodo, often a much 'ignored' microblogging tool.

Additionally we believe that the level of familiarity of students on the use of social media in teaching and learning is crucial to ensure its effective and meaningful integration in a given course. Therefore, the level of familiarity of the students on the use of social media is crucial (Doyle, Sammon, & Neville 2016 ; Paliktzoglou,, Giousmpasoglou, & Marinakou, 2016). Our results, which are comparable to previous experiments (Oyelere, Paliktzoglou, & Suhonen 2016), (Raitman, Augar, & Zhou, 2005) and (Safran, 2007) focus specifically on Computer Science students from University of Eastern, demonstrates that students are aware of social media but less familiar with mashups and tag cloud technologies. From the student replies it is evident that social media are used most frequently for recreational use. In terms of familiarity with social media, chat, social networks and media sharing platforms are among the most familiar. Considering the above facts we were able to use a certain approach toward our students in order to examine their acceptance and experience in using Edmodo. Our findings indicate that encouraging the use of Edmodo for educationally relevant purposes has a positive impact. From out finding it is evident that Edmodo promotes collaboration among students and sets the learner confident to work collaboratively receiving useful feedback from fellow students and the instructor through forum posts.

The learning curve for Edmodo's integration in the student participant's course is not steep due to the fact that there are similar features in Edmodo as in other learning management systems and microblogging tools. However from our experience some technical drawbacks were also encountered with the Edmodo platform, frequently encountered with many online tools. These drawbacks varied from browser-specific problems to user frustrations with the limitations of the features available in the Edmodo platform and other types of more complex technical problems. Moreover, on a positive note, the receptions of Edmodo was positive and there is a clear indication that it can be used as online tools for study or personal pleasure where collaboration and information sharing are plus points. From the educational point of view, Edmodo allows collaborators to feel connected with the class and removes the physical distance among people and generally brings them together. It is noteworthy to highlights the fact that in Edmodo both educators and students are able to contribute and build upon one another's contribution. Additionally, students were active participants and collaborators in Edmodo producing and reading content on a daily basic. The level of acceptance of Edmodo was high and students believe that it is great tool for learning. Moreover, the authors believe, as the students mentioned, that the structure of the module helped in the easy adoption of Edmodo. Edmodo is positively suggested by students to be used also in other courses. However, the researchers had the feeling that even though Edmodo is a valuable educational tool, the integration of two crucial educational features, a chat for instant and synchronous communication and collaborative writing feature in the platform would improve the functionality and "maturity" of the platform. Finally, it is noteworthy that some of the above mentioned points might be proved useful for future educators to consider when integrating Edmodo's use in their course work in higher education.

This study provides experimental evidence that using Edmodo can increase student engagement, and thus, Edmodo as a microblogging social networking site can be used as an educational tool. Moreover, we provided evidence to suggest that students and the instructor were both highly engaged in the learning process through communication and collaboration. In summary, we have focused on the investigation of the overall familiarity of students with social media and further focused on the use of the Edmodo as a learning and collaboration tool in Higher Education. Additionally, we presented the finding from the students' perceptions, engagement and experiences. The emphasis was to support the students as collaborators in becoming confident and self-directed learners. The familiarity with social media was less than anticipated, with the exception of social networks and chat which were also associated with the highest frequency of use. Moreover, the student positively suggested the use of Edmodo in other courses. It is worth mentioning that one specific limitation of our experiment was that the participants were restricted to the specific setting of Computer Science students from University of Eastern Finland. It would be advisable to conduct similar studies with a broader set of participants in order to compare them to this setting. However, from our results it is evident that social media and more specifically Edmodo can support social-constructivist models of pedagogy and that Edmodo as a social media tool has the potential to play an important role in the future of learning environments. Finally, as there is continuing growth in the use of social media in Higher Education, it is hoped that this study will motivate further controlled studies of Edmodo in order to evaluate how emerging technologies can be more efficiently utilized and adopted in Higher Education.

REFERENCES

Agherdien, N. (2011). Twitter and Edulink: balancing passive consumption with knowledge creation. In P. Balcaen (Ed.), *Proceedings of the 6th international conference on e-learning* (pp. 489–492). Reading: Academic Conferences, Ltd.

Albion, P. R. (2008). Web 2.0 in Teacher Education: Two Imperatives for Action. *Computers in the Schools*, *25*(3-4), 181–198. doi:10.1080/07380560802368173

Anderson, L. W., Krathwohl, D. R., & Bloom, B. S. (2005). *A taxonomy for learning, teaching, and assessing*. Longman. Retrieved from http://www.lavoisier.fr/livre/notice.asp?ouvrage=1603118

Blood, R. (2004). How blogging software reshapes the online community. *Communications of the ACM*, *47*(12), 53–55. doi:10.1145/1035134.1035165

Borau, K., Ullrich, C., Feng, J., & Shen, R. (2009). Microblogging for language learning: Using twitter to train communicative and cultural competence. In *Advances in Web Based Learning–ICWL 2009* (pp. 78–87). Springer. doi:10.1007/978-3-642-03426-8_10

Bower, M., Hedberg, J. G., & Kuswara, A. (2010). A framework for Web 2.0 learning design. *Educational Media International*, *47*(3), 177–198. doi:10.1080/09523987.2010.518811

Chatti, M. A., Jarke, M., & Frosch-Wilke, D. (2007). The future of e-learning: A shift to knowledge networking and social software. *International Journal of Knowledge and Learning*, *3*(4), 404–420. doi:10.1504/IJKL.2007.016702

Churches, A. (2007). Educational Origami. *Bloom's and ICT Tools*. Retrieved from http://edorigami.wikispaces.com/Bloom%27s+and+ICT+tools

Click, A., & Petit, J. (2010). Social networking and Web 2.0 in information literacy. *The International Information & Library Review*, *42*(2), 137–142. doi:10.1080/10572317.2010.10762855

Conole, G., & Alevizou, P. (2010). *A literature review of the use of Web 2.0 tools in Higher Education.* A report commissioned by the Higher Education Academy. Retrieved from http://www.jisctechdis.ac.uk/assets/EvidenceNet/Conole_Alevizou_2010.pdf

Corbeil, J. R., & Corbeil, M. E. (2011). The birth of a social networking phenomenon. *Cutting-edge Technologies in Higher Education*, *1*, 13–32. doi:10.1108/S2044-9968(2011)0000001004

Downes, S. (2005). Feature: E-learning 2.0. *Elearn magazine, 2005*(10), 1.

Doyle, C., Sammon, D., & Neville, K. (2016). A design science research (DSR) case study: building an evaluation framework for social media enabled collaborative learning environments (SMECLEs). *Journal of Decision Systems, 25*(sup1), 125-144.

Duffy, P. D., & Bruns, A. (2006). *The use of blogs, wikis and RSS in education: A conversation of possibilities*. Retrieved from http://eprints.qut.edu.au/5398

Dunlap, J. C., & Lowenthal, P. R. (2009). Tweeting the night away: Using Twitter to enhance social presence. *Journal of Information Systems Education*, *20*(2), 129–135.

Ebner, M. (2009). Interactive lecturing by integrating mobile devices and micro-blogging in higher education. *CIT. Journal of Computing and Information Technology*, *17*(4), 371–381. doi:10.2498/cit.1001382

Ebner, M., Lienhardt, C., Rohs, M., & Meyer, I. (2010). Microblogs in Higher Education–A chance to facilitate informal and process-oriented learning? *Computers & Education*, *55*(1), 92–100. doi:10.1016/j.compedu.2009.12.006

Ebner, M., & Maurer, H. (2008). Can Microblogs and Weblogs change traditional scientific writing? In *World Conference on E-Learning in Corporate* (Vol. 2008, pp. 768–776). Government, Healthcare, and Higher Education. Retrieved from http://www.editlib.org/p/29699

Ebner, M., & Schiefner, M. (2008). Microblogging-more than fun. *Proceedings of IADIS mobile learning conference*, 155-159.

Edmodo – The Secure, Social Learning Network for Teachers and Students. (n.d.). Retrieved May 8, 2013, from http://trendingeducation.com/?p=69

Edmodo: What students think. (n.d.). *Box of Tricks*. Retrieved May 8, 2013, from http://www.boxoftricks.net/2008/09/edmodo-what-students-think/

Ellison, N. B. (2007). Social network sites: Definition, history, and scholarship. *Journal of Computer-Mediated Communication*, *13*(1), 210–230. doi:10.1111/j.1083-6101.2007.00393.x

Gao, F., Luo, T., & Zhang, K. (2012a). Tweeting for learning: A critical analysis of research on microblogging in education published in 2008–2011. *British Journal of Educational Technology*, *43*(5), 783–801. doi:10.1111/j.1467-8535.2012.01357.x

Gao, F., Luo, T., & Zhang, K. (2012b). Tweeting for learning: A critical analysis of research on microblogging in education published in 2008–2011. *British Journal of Educational Technology*, *43*(5), 783–801. doi:10.1111/j.1467-8535.2012.01357.x

Gaonkar, S., Li, J., Choudhury, R. R., Cox, L., & Schmidt, A. (2008). Micro-blog: sharing and querying content through mobile phones and social participation. In *Proceedings of the 6th international conference on Mobile systems, applications, and services* (pp. 174–186). Retrieved from http://dl.acm.org/citation.cfm?id=1378620

Greenhow, C., Robelia, B., & Hughes, J. E. (2009). Response to comments: Research on learning and teaching with Web 2.0: Bridging conversations. *Educational Researcher, 38*(4), 280–283. doi:10.3102/0013189X09336675

Holotescu, C., & Grosseck, G. (2009). Using microblogging in education. Case Study: Cirip. ro. In *6th International Conference on e-Learning Applications*. Retrieved from http://www.cblt.soton.ac.uk/multimedia/PDFs%20Twitter/8551345-Using-microblogging-in-education-Case-Study-Ciripro.pdf

Hsu, Y.-C., & Ching, Y.-H. (2011). Microblogging for strengthening a virtual learning community in an online course. *Knowledge Management & E-Learning: An International Journal, 3*(4), 585–598.

Ihanainen, P., & Moravec, J. (2011). Pointillist, cyclical, and overlapping: Multidimensional facets of time in online education. *International Review of Research in Open and Distance Learning, 12*(7), 27–39. doi:10.19173/irrodl.v12i7.1023

Jacobs, S. (2008). *College Degree.com | The Facebook Classroom: 25 Facebook Apps That Are Perfect for Online Education*. Retrieved 11 June 2013, from http://www.collegedegree.com

Jarc, J. (2010). *Edmodo – a free, web 2.0 classroom management tool*. Retrieved 11 June 2013, from http://trendingeducation.com/?p=190

Java, A., Song, X., Finin, T., & Tseng, B. (2007). Why we twitter: understanding microblogging usage and communities. In *Proceedings of the 9th WebKDD and 1st SNA-KDD 2007 workshop on Web mining and social network analysis* (pp. 56–65). Retrieved from http://dl.acm.org/citation.cfm?id=1348556

Jenkins, H., Clinton, K., Purushotma, R., Robison, A. J., & Weigel, M. (2006). *Confronting the challenges of participatory culture: Media education for the 21st century. MacArthur Foundation*. Retrieved 10/29/2009 from http://www. digitallearning. macfound. org/atf/cf/%7B7E45C7E0-A3E0-4B89-AC9C-E807E1B0AE4E% 7D/JENKINS_WHITE_PAPER.pdf

Lenhart, A., Purcell, K., Smith, A., & Zickuhr, K. (2010). *Social media & mobile internet use among teens and young adults*. Pew Internet & American Life Project. Retrieved from http://web.pewinternet.org/~/media/Files/Reports/2010/PIP_Social_Media_and_Young_Adults_Report_Final_with_toplines.pdf

Lento, T., Welser, H. T., Gu, L., & Smith, M. (2006). *The ties that blog: Examining the relationship between social ties and continued participation in the wallop weblogging system. In 3rd Annual Workshop on the Weblogging Ecosystem* (p. 13). Retrieved from http://citeseerx.ist.psu.edu/viewdoc/download?doi=10.1.1.61.6907&rep=rep1&type=pdf

Lowe, B., & Laffey, D. (2011). Is Twitter for the birds? Using Twitter to enhance student learning in a marketing course. *Journal of Marketing Education, 33*(2), 183–192. doi:10.1177/0273475311410851

Maloney, E. (2007). What Web 2.0 can teach us about learning. *The Chronicle of Higher Education, 25*(18), B26.

McFedries, P. (2007). Technically speaking: All a-twitter. *Spectrum, IEEE, 44*(10), 84–84.

Morrison, M. A., Haley, E., Sheehan, K. B., & Taylor, R. E. (2011). *Using qualitative research in advertising: strategies, techniques, and applications*. SAGE Publications, Incorporated.

Nardi, B. A., Schiano, D. J., Gumbrecht, M., & Swartz, L. (2004). Why we blog. *Communications of the ACM*, *47*(12), 41–46. doi:10.1145/1035134.1035163

Oyelere, S. S., Paliktzoglou, V., & Suhonen, J. (2016). M-learning in Nigerian higher education: An experimental study with Edmodo. *International Journal of Social Media and Interactive Learning Environments*, *4*(1), 43–62. doi:10.1504/IJSMILE.2016.075055

Ozkan, B., & McKenzie, B. (2008). *Social networking tools for teacher education*. Society for Information Technology & Teacher Education International Conference; Retrieved from http://www.editlib.org/p/27640

Paavola, S., Lipponen, L., & Hakkarainen, K. (2004). Models of innovative knowledge communities and three metaphors of learning. *Review of Educational Research*, *74*(4), 557–576. doi:10.3102/00346543074004557

Paliktzoglou, V., Giousmpasoglou, C., & Marinakou, E. (2016). Assessment of Students' Familiarity, Adoption, and Use of Social Media in Bahrain. *International Journal of Technology Diffusion*, *7*(4), 82–90. doi:10.4018/IJTD.2016100105

Perifanou, M. A. (2009). Language micro-gaming: Fun and informal microblogging activities for language learning. In *Best Practices for the Knowledge Society. Knowledge, Learning, Development and Technology for All* (pp. 1–14). Springer. Retrieved from http://link.springer.com/chapter/10.1007/978-3-642-04757-2_1

Polonsky, M. J., & Waller, D. S. (2010). *Designing and managing a research project: A business student's guide*. Sage Publications, Incorporated. Retrieved from http://books.google.com/books?hl=en&lr=&id=D3aXAgQpkJUC&oi=fnd&pg=PR1&dq=Polonsky,+M.,+%26+Waller,+D.+(2005).+Designing+and+managing+a+research+project%3B+A+business+student%27s+guide+(1st+ed.).+Thousand+Oaks,+CA:+Sage+Publications.+&ots=1udeGVc2_8&sig=qcYfugQ0JdN_4fkHoGnIJe31mh0

Raitman, R., Augar, N., & Zhou, W. (2005). Employing wikis for online collaboration in the e-learning environment: Case study. In *Information Technology and Applications, 2005. ICITA 2005. Third International Conference on* (Vol. 2, pp. 142–146). Retrieved from http://ieeexplore.ieee.org/xpls/abs_all.jsp?arnumber=1488944

Ramsden, A. (2008). *How and why are people using Twitter: A small group study*. Retrieved from http://opus.bath.ac.uk/id/eprint/471

Redecker, C., Ala-Mutka, K., Bacigalupo, M., Ferrari, A., & Punie, Y. (2009). *Learning 2.0: The impact of Web 2.0 innovations on education and training in Europe*. Final Report. European Commission-Joint Research Center-Institute for Porspective Technological Studies, Seville. Retrieved from ftp://ftp.jrc.es/pub/EURdoc/EURdoc/JRC55629.pdf

Rinaldo, S. B., Tapp, S., & Laverie, D. A. (2011). Learning by Tweeting Using Twitter as a Pedagogical Tool. *Journal of Marketing Education*, *33*(2), 193–203. doi:10.1177/0273475311410852

Ross, C., Terras, M., Warwick, C., & Welsh, A. (2011). Enabled backchannel: Conference Twitter use by digital humanists. *The Journal of Documentation*, *67*(2), 214–237. doi:10.1108/00220411111109449

Rubin, H. J., & Rubin, I. S. (2011). *Qualitative interviewing: The art of hearing data*. SAGE Publications, Incorporated.

Safran, C., Guetl, C., & Helic, D. (2007). The Impact of Web 2.0 on Learning at a Technical University-A usage survey. In *World Conference on E-Learning in Corporate* (Vol. 2007, pp. 436–443). Government, Healthcare, and Higher Education. Retrieved from http://www.editlib.org/p/26364/

Safran, R. J., Doerr, V. A., Sherman, P. W., Doerr, E. D., Flaxman, S. M., & Winkler, D. W. (2007). Group breeding in vertebrates: Linking individual-and population-level approaches. *Evolutionary Ecology Research*, *9*(7), 1163.

Selwyn, N. (2009). Faceworking: Exploring students' education-related use of Facebook. *Learning, Media and Technology*, *34*(2), 157–174. doi:10.1080/17439880902923622

Wright, N. (2010). Twittering in teacher education: Reflecting on practicum experiences. *Open Learning*, *25*(3), 259–265. doi:10.1080/02680513.2010.512102

KEY TERMS AND DEFINITIONS

Collaborative Learning: An educational approach where students learn together and from each other.

Cooperative Learning: An educational approach that organizes activities in the course room into social learning and academic experiences.

E-Learning 2.0: An extension of traditional e-learning which focused on learning management systems (LMS) to focus more on learning as a network phenomenon.

Edmodo: An information technology company noted for microblogging applications used in education.

Social Media: Computer-mediated technology that facilitates the sharing of information, especially user-generated content (closely connected with Web 2.0).

Web 2.0: World wide websites that support user-generated content.

APPENDIX

Questions in the Post-Module Questionnaire

Table 1.

Q1: At what frequency did you produce content with the introduced social media tool(s)?
Q2: At what frequency did you read the content produced by other students in the course?
Q3: Do you believe that the instructor's participation influenced your usage of the tool(s)?
Q4: Do you think the course structure of the assignments influenced your usage of the tool(s)?
Q5: Do you believe that the use of the Edmodo (tool) environment adds value to your overall learning experience in the course?
Q6: Do you think the use of the Edmodo (tool) environment) added to the intellectual value of the course?
Q7: What motivated you using the introduced Edmodo (tool) environment?
Q8: Which of the following learning activities have been facilitated or enhanced by using the Edmodo (tool) environment?
Q9: The Edmodo environment supported me to capture my thoughts on what I was learning
Q10: The Edmodo environment facilitated personal reflection
Q11: The Edmodo environment supported me to connect concepts with cases and examples
Q12: The Edmodo environment supported me to work with others in the course, building on ideas or resources together?
Q13: The Edmodo environment supported me to discuss course concepts with others in the course
Q14: The Edmodo environment supported me to get to know others in the course and/or create a sense of comfort and community Q15: The Edmodo environment supported me to get more information or support from the instructor Q16: Did the module and use of the Edmodo (tool) environment meet your expectations? Q17: Would you suggest the use of this approach (use of the Edmodo (tool) environment in other courses? Q18: Do you have other comments or suggestions to help improve the use of the introduced Edmodo (tool) environment in courses?

Questions in The Semi-Structured Interview

1. What did you like in Edmodo?
2. What features did you like?
3. What was working well? What needs to be improved? Can please list three good aspects (or features/ functionalities) of Edmodo and three aspects (or features/functionalities) that needs improvement.
4. How would you compare Edmodo vs Moodle?

5. Please describe any problems that you may have faced using Edmodo.
6. What features you think should be added to Edmodo? Explain the benefits or added value of such features.
7. What is you opinion/perception of the use Edmodo for educational uses?
8. Do you think that students/ teachers could benefit of using Edmodo? Explain how.

Chapter 8
An App to Manage Grammar Level Tests in Language Schools

Antonio Sarasa
University Complutense Madrid (UCM), Spain

ABSTRACT

A typical task in language schools is to assign a level of skill to the students with respect to the language they are going to study. For this, the students are evaluated through a set of level tests that measure the student's competence. Normally, the assessment consists of an oral competency test, a written competency test, and a listening comprehension test. A characteristic of this type of test is the possibility of evaluating them remotely (usually they are tests with questions where every question has associated a set of possible answers where only one of the answers is correct) without the necessity of the presence of any evaluator. For this reason, it is possible to create computer applications that automate the evaluation and management of test results. This chapter presents a computer application that has been created with the aim of automating the process of management and evaluation of level tests for students of Spanish as a foreign language. The application has been tested at the University of Zaragoza.

INTRODUCTION

This article is contextualized in language schools that exist in universities. In this sense, many universities have language training services that offer students and university staff the opportunity to learn a language (note that this service is very

DOI: 10.4018/978-1-5225-5279-6.ch008

useful and used in the case of foreign students who visit a university and who must learn the native language of the country).

To perform in one of these language courses, students must be leveled by performing a set of tests (Kumaravadivelu, 1994) that allow the school to determine the level of linguistic competence that the student has with respect to the language to be studied. In general, three types of tests are performed (Finocchiaro, 1983). These are an oral comprehension test (McKay, 2002) that measures the ability to understand and be understood by a native person, a test of listening comprehension that measures the student's ability to understand conversations between native people (Rivers 1981) and a written comprehension test (Hadley, 1993) that measures the domain of vocabulary and grammatical constructs in language. Sometimes a fourth test of reading comprehension is also performed which is independent of the written comprehension test and consists of reading a text and then answering questions about the text read.

Usually the oral test (Lado, 1961) is performed in the presence of a teacher with whom a topic is discussed. In the auditory test (Pinto-Llorente, 2016), schools use a multimedia resource such as a video, a song or a recorded conversation that the student must listen to and after which questions are asked to see what he has understood about the audition. Finally, the written test (Nuttall, 1996) usually consists of exercises such as writing an essay on a proposed topic, rearranging the components of a sentence so that the sentence is correct or completing parts of a sentence that are empty with the more appropriate words (Pinto-Llorente, 2014).

Level tests can be automated through some computer applications. In this sense, there are several ways of automating them (Yang, 2009) that can be classified in: specific evaluation tools or general evaluation tools (Squires, 1997).

The specific evaluation tools (Voorheis, 2004) evaluate only one specific language. Its functionality is oriented to the particular characteristics of the language to be evaluated. In most cases, these tools are not free, being necessary to pay some kind of license to use them. The main advantage they present (Gottliebson, 2010) is the specificity with respect to the particularities of the language to be evaluated. However, they present a disadvantage, the impossibility of reusing them to evaluate other languages (Pinto-Llorente, 2015). Likewise, in many cases, they have little possibility of adaptation and configuration (usually it is software that cannot be modified by users, adaptations of their interfaces cannot be made, and they offer a very poor set of user management services).

The generic tools of evaluation (Bachmann, 2005) are not particularized to a specific language. Its main features are: a) They are flexible and adaptable to different contexts (Elbeck, 2014) b) In many cases, they allow one to establish the type of evaluation process (conditional navigation, free navigation ...), c) They allow to create banks of questions, d) They allow to create tests with different types of questions (Shohamy, 1988) such as: multiple choice, free text, jumbled words ... (Abello, 2008) and e) It is possible to set the way of how the questions must be shown (for example, random presentation of questions and answers ...). In this type of tools, there are both free tools and payment tools. However, there are more free tools than payment tools such as Google Drive forms (Ji, 2015), QuestionPro, thatquiz, testmoz, and others.

There are other alternatives to those discussed above. On the one hand are a set of online applications oriented to the development of tests questions and the management of the answers to the tests. Some examples of these applications include Socrative, Mentimeter, Thatquiz (Dervan, 2014), LearningPod, Propofs, GoConqr, questbase, Gnowledge, Testmoz, QuestionPro, Educaplay and ExeLearning.

Socrative is an application that allows one to perform tests, evaluations, activities, and the like and manage data by the teacher. Likewise, it has as the main advantage the ability to be used from different devices and not having to register to be used by students. Mentimeter (Hainzt, 2014) allows one to create surveys in a simple way without the need to make any record, just add questions and options of answer that users must to respond to. Finally, thatquiz (Yuuichi, 2006) allows one to generate exercises and see results immediately. For this, there are a set of pre-established exercises or one can create new exercises and manage any kind of content.

LearningPod (http://www.learningpod.com/) is an application that offers two types of services. On the one hand the package provides a bank of questions and questionnaires on various topics that can be used by students to assess the level of their knowledge. To facilitate the search, you have a search engine that filters by the type of test, the level of the student and the subject of the questions. On the other hand, it offers teachers the possibility to create new tests and add questions to the bank. questions. The application uses groups of questionnaires called "learning pod" as the unit of work. Each consists of a series of questions with multiple A, B, C, and D response options, where there is only one correct answer. Each time a question is answered, the system provides feedback regardless of whether or not the question is answered. It also presents a graph that shows statistics about the number of people who have tried to answer the questions, how many have succeeded and how many failed.

ProProfs Quiz School (https://www.proprofs.com/quiz-school/) is a tool that allows one to develop test online in a simple way. It offers the user a generator of different types of questions such as multiple-choice questions, open-ended questions, single-answer questions, true / false questions or fill-in questions, etc. Likewise, it allows one to configure characteristics of the questionnaires such as: choice of colors, insertion of images, backgrounds and the like. Other features present in the tool, refer to the possibility of analyzing the results, offer personalized feedback according to the results obtained by the student, offer help while the student answers the questionnaire, set a time limit, and other features. Finally, created questionnaires and questions can be tagged with metadata to be retrieved from an application browser, and one can add comments, share with other users, or embed them into blogs, web pages or educational platforms.

GoConqr (https://www.goconqr.com/en/quiz-maker/) is an application and an educational platform that offers services of creation, recovery and sharing of digital educational resources. With respect to the creation of resources, it allows the creation of 4 types of resources: tokens, mental maps, slides and tests. The tabs are oriented to facilitate the memorization of data of different types. The tool allows the user to create new tabs or to use tab patterns or tabs already published by other users. The tabs can contain text or images, and once created can be shared by email or social network. With regard to mind maps, maps can be created starting from a central idea, and deploying nodes to develop that idea. One can create maps from scratch or use maps already published and modify them. A map can contain images or text, and once created can be shared by email or social network. Another resource, are slides, which can contain text, images, links to web pages and videos. In addition, other resources generated with GoConqr such as tokens or mind maps can be included. To create them, you can create new slides or retrieve slides published by other users, and modify them. Once published, they can be shared by email or by a social network. Finally, you can create tests with different types of questions (multiple selection, true or false, drag and drop, etc.). Questions can include pictures. Once the questions are answered, it is possible to visualize the incorrect answers and offer feedback or explanations about the answers given or about how they can be improved. One can create from scratch or retrieve already published tests and modify them. The tests can be published to groups of users interested in a specific topic in a way that facilitates its reuse. Once published, they can be shared via email or social networks. As already mentioned, all the resources generated by the tool are published in a repository that can be consulted by any user. To do this, each resource is tagged with keywords and areas of knowledge. Likewise, from the tool, it is possible to visualize resources and create or subscribe to study groups in specialized topics. These groups receive information about any resource related to the topic of interest of the group. Finally, note that all the contents of the platform, can be downloaded locally to use without having an internet connection.

QuestBase(www.questbase.com) is a free application that allows one to create and manage evaluations, tests and exams online or printed. It provides a web editor that allows one to import questions from external files (Excel, .acc or created using wizards). In the questions one can insert text, images, and video. In particular QuestBase facilitates the inclusion of formulas (true / false, multiple options, single answer, and the like). On the other hand, questions can be sorted by topics or by different levels of difficulty, and can be configured to show randomize both the questions and possible answers, assign scores to each correct or incorrect question, associate a feedback that will be displayed when the question has been answered, or add instructions or clues to the question. After the questionnaires have been created, can be published online, or shared by email or on a web page. It is also possible to print the questionnaires in paper. The content may be public or private. Other possible options are: managing different versions of a questionnaire, issuing certificates to users who have passed a questionnaire, adding logos to the questionnaires, establishing security mechanisms on the questionnaires, privacy options, etc. Finally, it is possible to generate real-time reports of the answers given containing graphs and statistics, export them to other external programs for analysis of information, or export responses to formats such as Excel, Access, CSV or PDF.

Gnowledge (http://www.gnowledge.com/) is a free tool that offers services to create, publish, share or perform tests, exercises and evaluations. To do this, it is necessary to create a free account, from which the user can access three modules: "create", "share" and "learn". Through each module one can design and create questions and questionnaires using a form that allows creation step by step. The form indicates the characteristics of the questionnaire or question. One can configure features such as: randomly displaying questions and answers, the possibility of repeating the same tests, and others. All the generated resources are indexed and remain public for other users. In addition, the results and answers of the users are stored for analysis or consultation.

Testmoz (https://testmoz.com/) is a free application that allows one to create quizzes quickly and simply online. To do this, a document is generated to which a name is assigned, a set of usage options such as an input password or welcome message, the questions and answers are added, and published. Each test has associated a permanent URL that is controlled by password, so that you can control when and who will access the test. Also, the application allows you to generate reports that indicate who did the exam, when did (date and time), and the qualification you obtained. This application is suitable for short and quick tests, but it has the disadvantage that it does not allow the insertion of multimedia, it can only contain text.

QuestionPro(https://www.questionpro.com) is a web application that allows one to create and publish questionnaires. To create a questionnaire, one can make a new one or use pre-designed templates that can be modified. These templates are divided into categories. It is also possible to import a Word document that contains content. It is possible to configure aspects such as backgrounds, color, etc. Its main characteristics are: it supports different types of questions (true / false, multiple answer, etc.), one can distribute the questionnaires via QR code, email or insert it in a web page or a blog, it is possible to generate real-time reports with statistics and graphs, one can track users who have completed a questionnaire, and one can export responses to CSV or Exce format.

The Educaplay application (https://es.educaplay.com/) is a free tool that allows to create multimedia educational activities in different languages, choosing among 14 types of activities (soups of letters, crossword puzzles, completing texts, dialogues, dictations, relating, etc.). In addition, it has a repository with activities performed by other people that can be reused. It can be used from a given device using HTML5 technology, and is integrable within blogs or platforms such as Moodle. Likewise, all the activities created can be exported to be used without internet connection. ExeLearning (http://exelearning.net) is an open source tool that facilitates the creation of educational content without having to be an expert in HTML or XML. It is a cross-platform application that uses content trees, multimedia elements, interactive self-assessment activities ... facilitating the export of content generated in multiple formats: HTML, SCORM, IMS, and others. It allows one to create a basic navigation tree that will facilitate browsing, writing text and copying from other applications, including images, sounds and videos, embedded multimedia elements such as videos, presentations, texts or audios. Also, it can include activities carried out with other applications, and supports creating simple activities: test questions, true / false, white space. For this last item, the package has components called iDevices. Each idevice represents a different pedagogical or technical utility. There are 4 types of idevices include interactive activities that allow the student to interact directly (fill in holes, multiple choice questions, etc.), no interactive activities that propose activities that cannot be answered directly (reading activity, case study, etc.), presentation (free text, prior knowledge, etc.) and presentation of non-textual information (images and web pages). The main disadvantages of these tools are the impossibility of reusing content in other contexts other than the online domain. Indeed, their functionalities are limited and are unstabile. In many cases, the life cycle of these applications is very short because they are not economically profitable (Takagi,2006). In this sense, a user can lose all the work done because they are not able to export them to other systems.

On the other hand, the test generation tools provided by Learning Management Systems (LMS) such as Moodle (Gracia, 2010) can be used in comparable ways. These tools (Kakasevski, 2008) offer similar functionalities to the generic tools of creation of level tests (Ozkan, 2009) but present additional advantages. These include integration with other services within a learning environment (Garcia-Peñalvo, 2015) such as user management, communication tools or management of qualifications, etc. (Garcia-Holgado, 2013). The main disadvantage of this alternative in its impossibility of reuse. Normally, tests done in the LMS context are not exportable or usable outside the LMS. In this sense, it will be necessary to be an LMS registered user.

Finally, it is necessary to comment that this type of tool can be divided according to the format that is used to store and to represent the tests of questions: standardized or non-standardized. Non-standardized formats represent tests in a particular way. The problem is that they cannot be reused in different systems from which they were created. However, the standardized formats represent the tests in a standard way that allows them to be reused in other systems. Some examples of standardized formats (Kravtsov, 2008) are SCORM (Bohl, 2002), IMS QTI (Vogten, 2006) or IMS Simple Sequencing (Shen, 2004) to organize the Information, its deployment process and the coding of test questions.

This article describes an app for the creation and management of tests to level the linguistic competence of students of Spanish as a foreign language. This tool was created based on a real case at the University of Zaragoza. In this university, there is a Spanish as a foreign language school in which students and teachers can study. The titles obtained are not official. The management of the school and the institutional support offered by the university is less than in the case of regulated studies. For example, language school students do not receive an official e-mail account at the university, they cannot use the virtual campus services and the university does not keep track of their activities.

Normally, students participating in these studies are outside Spain, and they hope to be able to know the level of linguistic competence they have and the course they will study in the school before coming to Spain. Also, for reasons of organization and planning it would be interesting to know the number of students that will be in each level before starting the courses.

The manual process of leveling students consists of performing the leveling tests to students a few days before starting the course. The tests are evaluated, and according to the results obtained, students are assigned a level of linguistic competence. With this system, from the point of view of class planning, there are problems with regard to the reservation of classrooms, teacher assignments, activity planning, etc. Another problem of the system refers to the management of the information associated with the students: level of linguistic competence original or achieved, exams done, personal data, enrollments made, and the like. This information is not stored digitally (it is on paper), so that it cannot be consulted efficiently and quickly if a student has taken several courses in different years. Several items are not visible including what level he demonstrated, what grades he had, what teachers he had, and the like. This is so in many cases, because it is not possible to have access to this information and hence it is necessary to do new leveling).

In order to solve these problems, it is necessary to have a computer tool that allows students to assess their level of competence online remotely, manage the data associated with each student (such as personal data, examinations, enrolled courses, etc.), keep track of students' progress, and perform related tasks. This would allow an efficient management of the organization and planning of the courses. For example, the number of students at each level could be known with sufficient time to update plans.

Based on the needs, two software tools were created in two different phases. In the first phase, a tool was created based on Google Drive forms. Since the tool created had some limitations, in a second phase a web application with more functionality was created compared to the first. This application was developed in the context of a final grade project at the School of Computer Science at the Complutense University of Madrid. The following sections describe the work done. In the next section the requirements of the first tool are described; the following section presents the first tool based on Google Drive; the next section shows the limits and problems present in the first tool and motivates the need to create a new tool with additional requirements; the fifth section describes the second tool implemented. Finally, in the last section a set of conclusions and possible lines of future work are raised.

REQUIREMENTS

In order to obtain the functional requirements of the system, interviews were carried out with the people who work in the language school at the University of Zaragoza (teachers, students and administration staff). The objective was to know how the language school worked, and what the user needs were. Based on the information gathered from the interviews, a set of requirements was obtained:

- Access to the application must be done through the web without the need for specific software for its use.
- The application should be easy to use by anyone with knowledge of IT at the user level.
- The application must store records of all level tests made for each specific student.
- Any student can take a placement test without the requirement to be enrolled in any course.
- The application must provide the tools necessary to create level tests with questions of various types. These tools can only be used by a teacher, in no case by a student.
- Must be independent of university computer systems.
- Must be easy to operate and with few maintenance requirements.
- It should be possible to update the application with new features.

Taking into account the functional requirements obtained, two possible solutions were analyzed:

1. **Using the LMS of the University of Zaragoza:** The university has a Moodle system. This LMS provides a tool to create tests with questions of different types, allows the management of the results and the students who perform the tests and is easily configurable (Kakasevski, 2008). This solution has as main advantage the integration with the computer system of the university. With respect to students, this solution would allow students to have a user ID of the university, so that they could access other tools related to the educational process (university software, enrollment management, etc.). The main problem presented by this solution is the fact that the university considers these studies to be unregulated, and is not responsible for its management (they must be managed independently). In this sense, the university is not responsible for storing information about the students enrolled in the language school, maintaining a history of exams and courses, etc. In addition, from the technical point of view, this solution presents the drawback of the unnecessary functionality of Moodle for language students.
2. **Using a Free Online Tool:** In this sense, the possibility of using the Google forms (Rowe, 2013) was studied. This tool allows to create evaluation tests with different types of questions in a very simple way, allows to download test results in a Google Drive spreadsheet for analysis, and allows you to make small scripts to automate certain tasks (create reports, manipulate the spreadsheet, filter data, etc.). Likewise, Google forms offer the advantage of having universal access via the web and the simplicity of use for both the student and the teacher.

According to the analysis performed and the functional requirements obtained, it was decided that it was preferable to make a simple tool to use and that it were be easy to process and analyze the results. In this sense, the solution using Google Drive forms fits better than the solution based on using the services offered by an LMS (Garcia-Holgado, 2016).

In the next section, the application done on Google Drive is described.

THE FIRST APPLICATION

This section describes the application that has been developed using Google Forms. Thus, in the first section the functional requirements are presented and in the second section the application implementation is described.

Functional Requirements

Teachers at the language school designed the test questions on paper. The purpose of the test was to validate knowledge about vocabulary and grammatical constructions. In the test, there were questions of different types: multiple choice, filling in an incomplete sentence, understanding a text and writing short essays. The questions were grouped according to the levels of knowledge (De Europa, 2002) that a student may have (A, B, and C).

With respect to the presentation of the questions and their implementation, the following requirements were set:

- First, the student will fill out a form where it must indicate: personal data (e.g. name, email and age) and information about previous knowledge about the language they would like to study, the level of approximate language proficiency, objectives they want to achieve with the course, etc. All information is optional except personal data.
- Next, the student will take the test. The questions of the test will be grouped according to the levels of linguistic proficiency of the European framework: first, the questions concerning level A, then the questions concerning level B and finally the questions concerning level C. In the test, there will be some questions that are not mandatory.
- Regarding the qualification, correct answers will score 1 point, incorrect answers will discount 1 point and blank answers will not be considered (0 points).
- Regarding the duration of the test the maximum duration of the test should not exceed 60 minutes. If this duration were exceeded, the test will not be corrected and will be considered that the student has not passed the test.
- Once the multiple choice test questions are completed, student will fill incomplete sentence type questions. These types of questions are objective and it is possible to automate the correction if the correct answers are previously known.
- When the student has finished the above exercises, the student will receive a message in the email entered in the first form, indicating the level obtained. Also, the teacher will receive an email for each student who has performed the test. The email will be accompanied by a text document showing the questions and answers of each student. The correction and level obtained will appear in the questions that can be corrected automatically. In the rest of questions will not appear the level obtained because the teacher must correct them manually.

- Once the teacher has corrected and inserted the scores for the questions that must be corrected manually, then it will send a new email to the student with the final mark obtained in the level test.
- It is necessary to prevent students from taking the test more than once in the same temporal period, since the level test will be the same during that period of time, and therefore could have the answers of the questions from previous attempts.

Implementation

Google Drive is a free Google software package that offers services like spreadsheets, documents, forms, etc. In the case of Google forms, it is possible to design tests with questions of different types (multiple-choice questions, open-ended questions, and the like). Each form is online (it is possible to access them via an associated URL). In addition, one can establish different levels of sharing: private, shared with certain people or public for anyone. Each form is associated with a Google spreadsheet, so each time a person fills out a form, the responses entered are recorded in a row in the worksheet where each question on the form corresponds to a column in the worksheet. Each entry in the worksheet is identifiable because it has an associated time value, which indicates the time the row was inserted (and therefore the time it was sent by the user). The advantage of storing the information inserted in a form in a spreadsheet is the ability to display data graphically or apply some sort of statistical calculation to create reports. Also, Google spreadsheets support the ability to define scripts that will run when certain events occur. These scripts allow the manipulation of data in spreadsheets: read rows or certain fields, write new data, delete, query data from different spreadsheets, and the like. In order to encode scripts, each spreadsheet has an online programming console in which the programmer can make script code, and associate it with an event and a particular spreadsheet.

Taking into account the possibilities offered by Google Drive, it was used as follows to implement the functional requirements proposed (note that when the implementation was performed, there was some currently features of Google forms that it was not available):

- Two different Google forms were created, one for the introduction of personal data and the initial survey (see Figure 1) and another for the test (see Figure 2). The last Google form was structured on several pages in order to facilitate the reading and inserting of the answers for the student.
- In Google Drive, there is no way to associate the data entered in a form to a particular user, since it appears anonymous when inserted into the spreadsheet. To identify each user's data, and to associate the data entered into the first

Figure 1. Google form for personal data

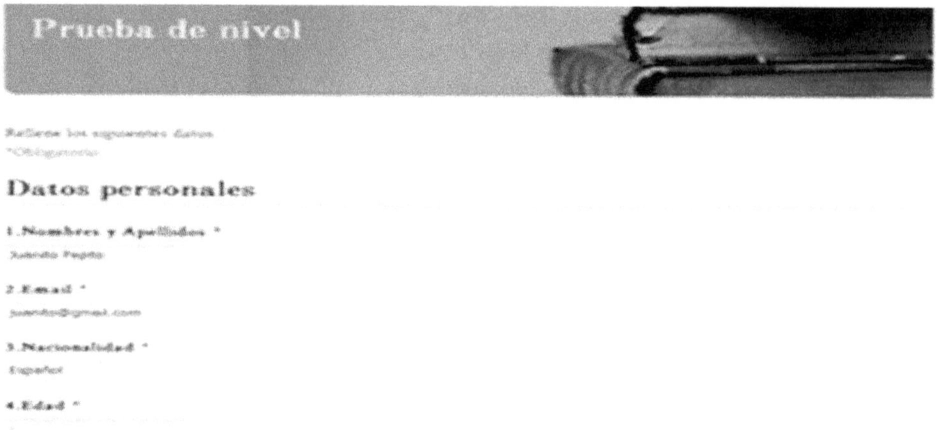

form and second form to the same user, it was decided to use the email of the student as a primary key. In this sense, it is mandatory for the student to fill in the email field in both forms. Otherwise, the system prevents the form from being submitted to the server.

- In order to sequence the performing of the two Google forms, the completion message of the first Google form was modified (see Figure 3) and a link was inserted on the second form (in which the level test was coded).
- Google Drive does not offer any mechanism for establishing a time limit to fill out a Google form and does not allow counting how long one has been invested in completing and submitting a form. In order to get the system requirement about duration of the test (<=60 minutes), the time stamp was used that is added to the data sent by users when it is inserted into the spreadsheet associated with the form. This time stamp indicates the time in milliseconds at which the user sent the form. So, to calculate the duration of the test, it calculates the difference between the time stamps of the sending of the first form containing personal data and the sending of the second form containing the level test. The amount obtained is in milliseconds and it is converted to minutes.
- Google Drive does not have any mechanism for the automatic correction of the answers of the questions that appear in a form and is not able to associate the records entered in a form to concrete users. In order to implement a process of automatic correction, which were able to distinguish the data entered by each user and assign a grade, it was necessary to code this process using Google scripts. In this sense, the following scripts were written:

Figure 2. Google form for the test

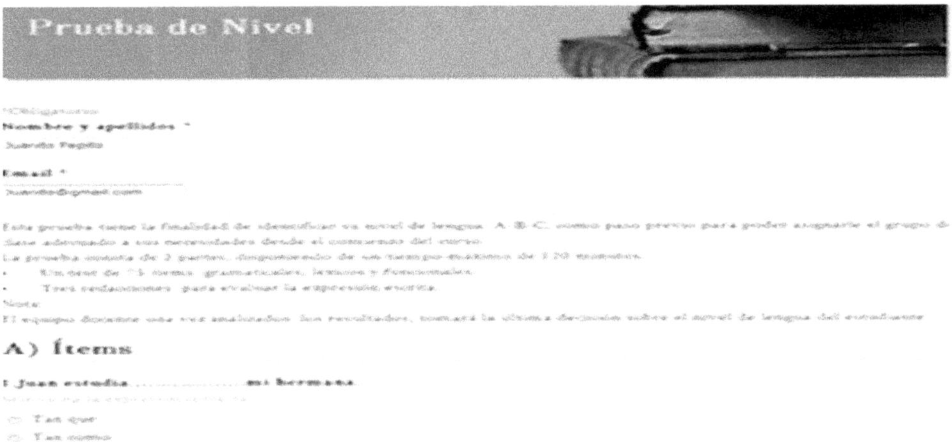

Figure 3. Modified completion message

- ○ Script with the goal to monitor that a user does not perform the level test more than once - Each time that new data are inserted into any spreadsheet associated with the forms, a search is performed into the spreadsheet using the email field from the new data inserted. If the search finds more than one row with the same email then it deletes the duplicate rows and keeps the oldest row.
- Script for the automatic correction of questions entered by a user - When a user submits a form to the server, the input data is inserted into a row on the spreadsheet associated with the form. A Google script has been associated with that event, which performs several actions. First, it evaluates the multiple-choice questions and those of the fill in an incomplete sentence

Figure 4. Email sent to the student

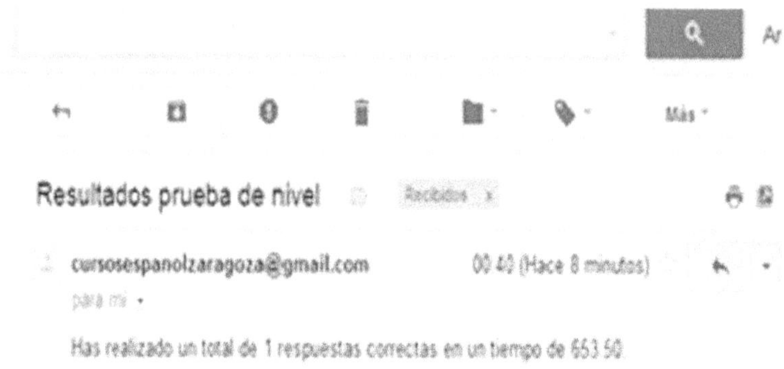

type. Second, it sends an email to the student who has inserted the data, using the mail address inserted in the first Google form (see Figure 4).

This generates a document in Google Drive that contains the questions and answers the student submitted and the grade obtained (it also includes the essays written by the student which must be evaluated directly by the teacher) (see Figure 5). Next, it sends an email to the teacher in which a link to this document appears (see Figure 6).

Finally, Figure 7 shows a screenshot of the code for the scripts.

THE LIMITS OF THE FIRST APPLICATION.

Using Google's forms-based application showed that it had some important limitations such as:

- Difficulty in managing the data of each student (tests, results, and enrollments, ...)
- Limitations in fixing and controlling the duration of a test.
- Difficulty managing different types of tests.
- Inability to manage different types of users (administrator, student or teacher).
- In this sense, it would be necessary to create a new application with additional requirements:
- Registration and management system for three types of users: students, teachers and administrators.
- Availability of a search feature of previous tests.

Figure 5. Document with answers inserted by student

Figure 6. Email sent to the teacher

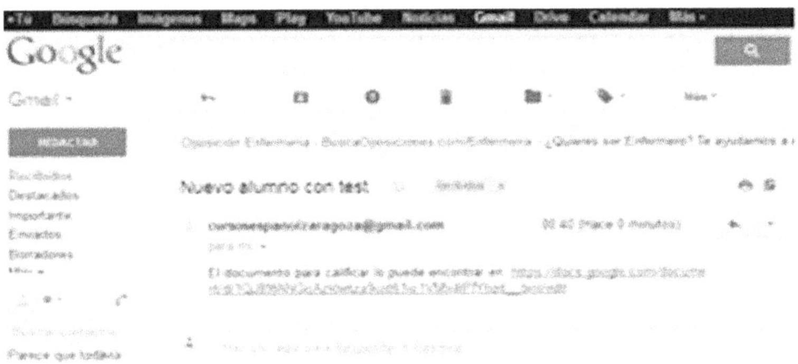

- Availability of tools oriented to teachers to create tests with questions of different types.
- Management of leveling tests directly by teachers.
- The administrator should be able to manage all system elements (users, tests, generated information, etc.).
- The application will offer two interfaces, a free access interface that will allow anyone to access the service information and a restricted access interface for registered users.

Based on the new requirements, the authors decided to create a new application using web technology.

THE SECOND APPLICATION

In this section, the authors describe the second application developed. The architecture of the application is described in the first subsection and in the second subsection the authors describe the functionality of the application.

Architecture of the Application

The design of the application was carried out taking into account the new requirements obtained. In this sense, a client-server architecture was used in which the browser acts as the client that communicates with a server and requests pages coded in PHP. These pages act as an interface with a MySQL database that stores all the system information: exams, teachers, and students. Functionality is sufficiently modularized so that it is easy to maintain and expand the system.

Functionality of the Applications

The application has a main page from which users can access its functionality. On this page is the public information for the language service. On the upper right of the page, there is a form where users can login or register. Figure 8 shows a screenshot of the main page.

Access to the application is via a userid and password. The userid is an email. Anyone can register in the application by clicking on the link on the home page called "Register". This link gives access to a form (see Figure 9) in which the user fills in a set of personal data: name, sex, date of birth, telephone number, zip code, city, nationality, and access data: email and password. Registration is not automatic, since the administrator of the application receives the request for registration and chooses whether to accept the request. Likewise, the application for teachers' registration is done from a different page, which is not accessible from the application. However, it uses a similar form and the registration is equally subject to acceptance by the administrator. In addition, a link allows the user to remember her access data in case it is forgotten. To do so, the user must click on the link: "Have you forgotten your access data?"

There are three types of users: administrator, teacher and student. The "student" user type has access to a page that has two fixed tabs called "history" and "profile" and another dynamic one. The tab "profile" gives access to a page where students can make changes to their personal and access data. The tab "history" links to a page in which all the tests that a student has performed are listed along with the answers and the grade obtained. Finally, the dynamic tab is displayed whenever there is a level test active so that the student can do it. Once the test has been performed, the

Figure 7. Main page of the application

Figure 8. Form of the registration

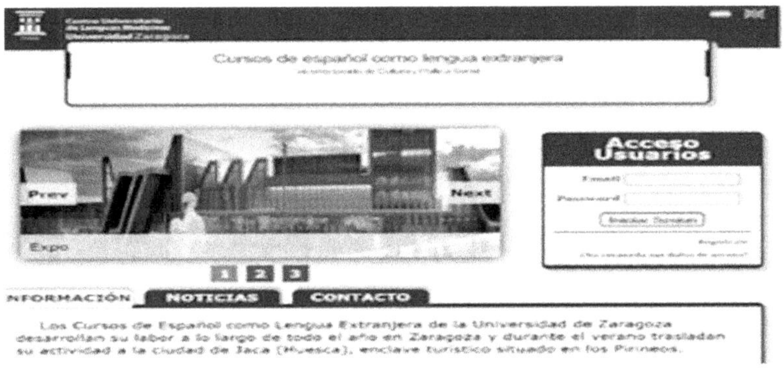

dynamic tab disappears. When the teacher corrects the test, the test will appear in the history list. Figure 10 shows a test done by a student that is listed on the history page.

The page associated with the "teacher" user type has the following tabs: students, exams, exams pending, statistics, history and profile. The "Students" (see Figure 11) tab gives access to a list of all students registered in the system and displays information about their nationality, date of birth and email. In addition, if the student's name is clicked on, a tab appears with expanded information about student and a list of all level tests that have been done to the date. Each test can be visited by clicking on the test.

The "Test" tab (see Figure 12) allows access to a list of level tests where there appears information about each exam: the academic year in which the test was used; type of test (annual, intensive or semi-annual); the number of questions of the test ; the duration of the test, the state of the test (shared if the test can be seen and modified by any teacher or private when the test can only be accessed and modified

Figure 9. Test from the history page

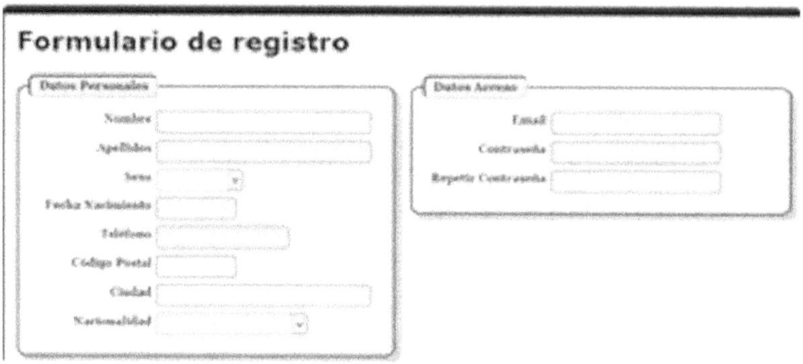

Figure 10. Student tab

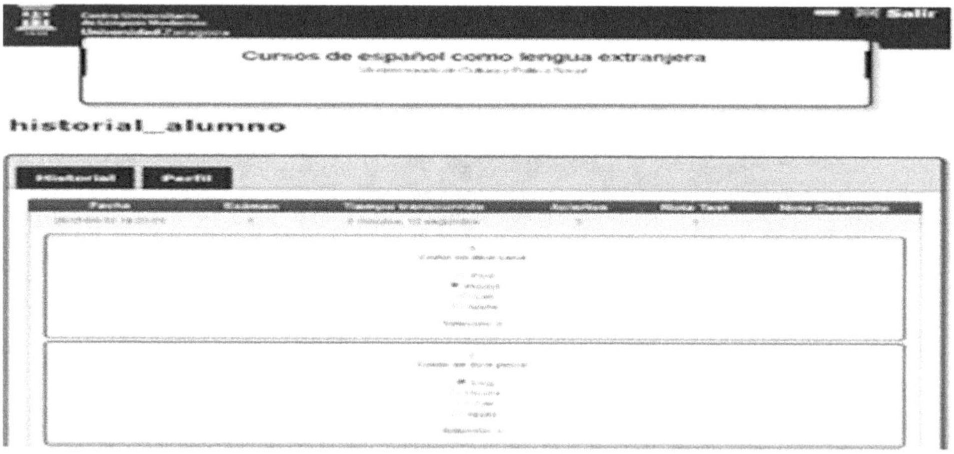

by the teacher who created the test); and the activation that indicates whether the test is active or not.

Because there can only be one active test at a time, it is necessary to click on the exam in order to disable or enable it. In addition, each test can be edited or removed by two icons that appear next to each test. The status field can be changed by clicking on it, so that if you have the value "shared" and it is clicked on, it passes to the "private" state, and vice-versa. Another editable field is the duration, so that if you click on that field then a window appears in which you can set the duration in minutes for the test. Finally, if you click on the name of the teacher or on the edit icon, one can access the contents of the test (see Figure 13), where questions can be edited, deleted or new questions added.

Figure 11. List of tests

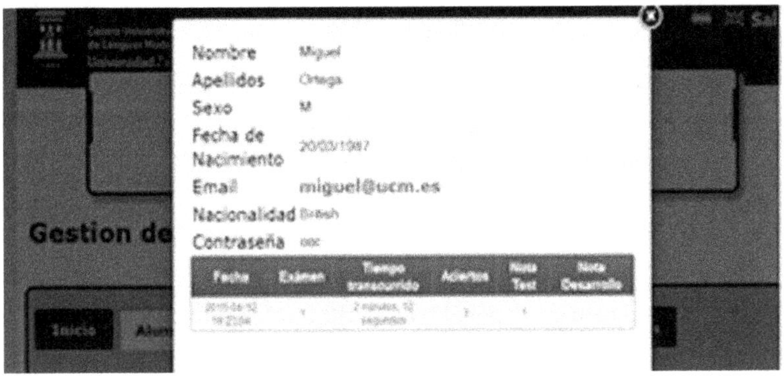

Figure 12. Test edition

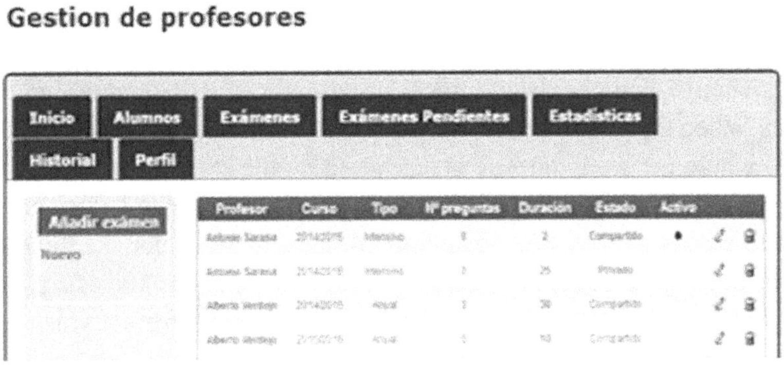

Figure 13. Inserting a new question

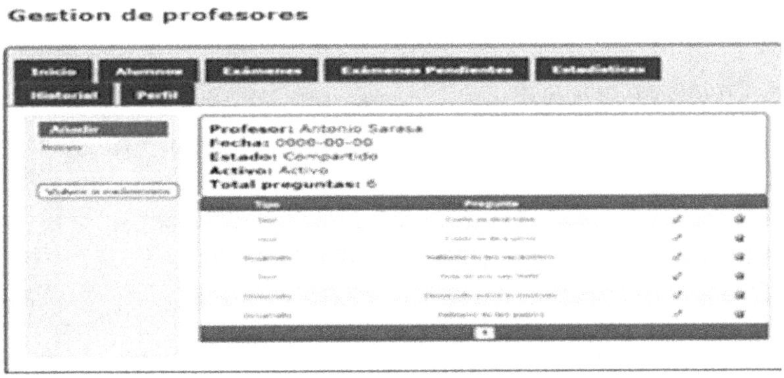

When a question is edited or created, a window appears (see Figure 14) in which the question and its type can be inserted. In the case of multiple answers, the possible answers and which one is correct can be inserted.

Finally, in the "Test" tab, one can carry out the creation of a new test. On the top left, there is a link called "Add new test." When this link is clicked on, a window (see Figure 15) appears where the duration of the test and the type of test must be inserted: semi-annual, annual or intensive.

When the link "New", is clicked on, the new empty test appears in the list of tests with the data entered by the teacher. Then the teacher can click on the test in order to edit it, and add the questions that the test will contain. In the "Pending tests" tab (see Figure 16) a list of tests appears that have been performed by registered students and which have not yet been reviewed by the teacher. For each test, the student's name, an internal identifier of the type of test performed, a score and two icons, one to evaluate the test and another to delete the test, are shown.

Tests may contain multiple-choice questions and / or essay questions. Multiple-choice questions are evaluated automatically without teacher intervention. However, the teacher should evaluate development questions manually and assign a note in the system. When the teacher clicks on the icon to evaluate, a window appears (see Figure 17) where the essay questions answered by the student appear, and next to each question a field where a score between 0-10 can be assigned. Once the teacher has assigned grades to each essay question, the system will calculate the final mark obtained by adding the scores from multiple-choice questions and essay questions. Next, the test disappears from the teacher's list of pending tests and is sent to the list of tests from the history page of the corresponding student.

In the "Historical" tab (see Figure 18), tests that have already been used once as level test (and therefore are inactive) are listed. Information is shown on who has

Figure 14. Creating a new test

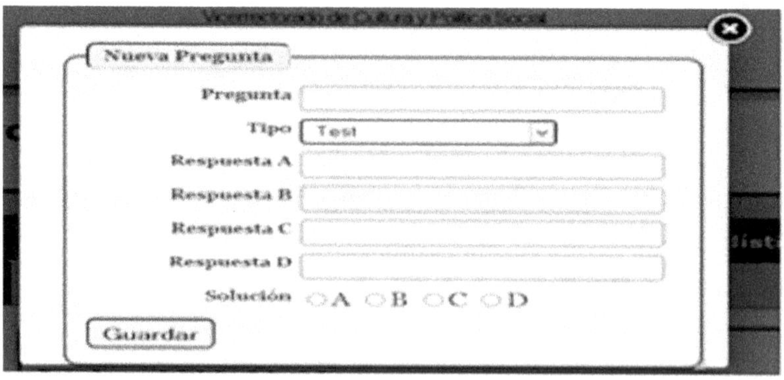

Figure 15. List of pending tests

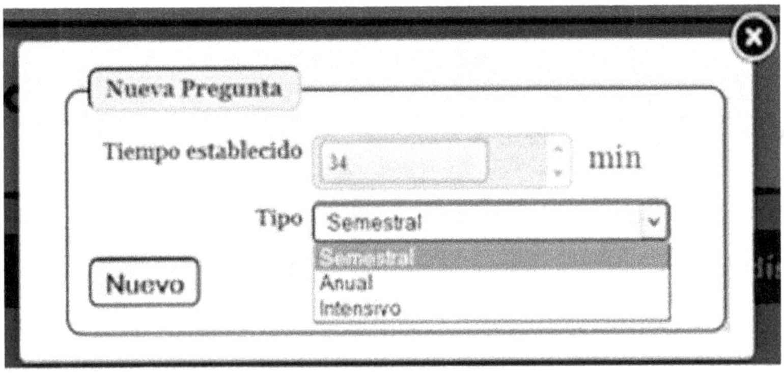

Figure 16. Correction of essay questions

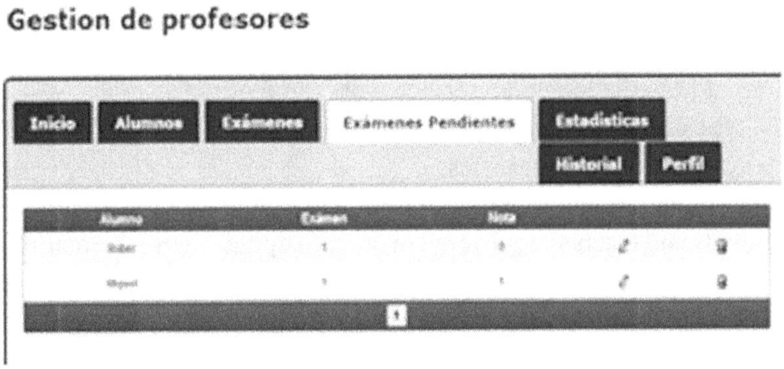

taken the test, the number of questions, the date of completion and the assigned duration. Besides, when the test is clicked on, it becomes visible.

The aim is that these tests cannot be re-used in the future because students could know its contents. However, teachers can see these tests.

Finally, the teacher has the “Statistics” tab that shows different measures of the results obtained by students in each exam, such as the mean, the highest and lowest grade obtained. Teachers also have the “Profile” tab in which teachers can modify their personal data and access data.

Note that the information displayed in the various tabs is common to all teachers so that all teachers can see the same students, the same students pending evaluation, and historical data. The only different information for each teacher appears in the “test” tab, where each teacher can view all shared tests, and private tests, but the private tests from other teachers cannot be seen. In addition, it is important to note that the activation and deactivation of a level test can be performed by any of the

Figure 17. Historical of tests

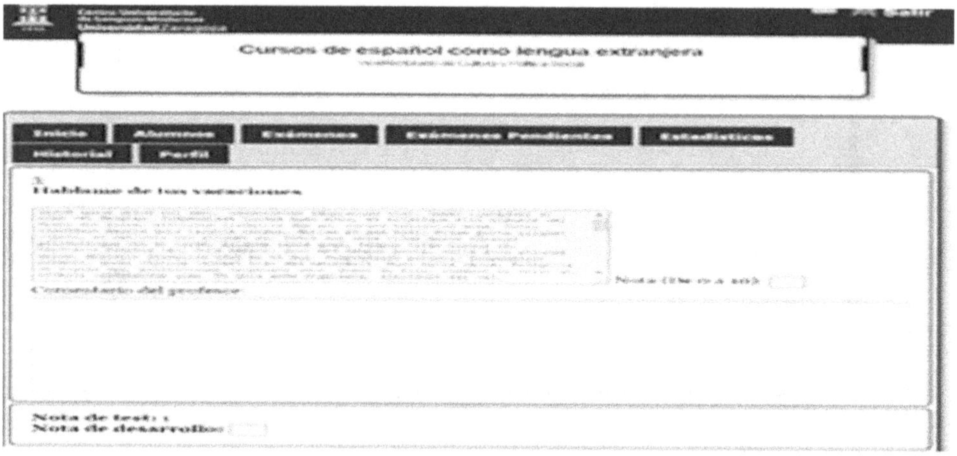

teachers. There is no user with more privileges than another. This behavior reproduces the structure of how the language service works in reality.

The last type of user is the administrator. It has four tabs (see Figure 19), called students, teachers, tests and statistics.

With the "Students" and "Teachers" tabs (see Figure 20), the administrator can manage students and teachers registered in order to update their information, remove them, activate them when they have applied to register in the system or temporarily disable them. In each of these tabs, there is a search box at the top of the page that facilitates finding a particular teacher or student.

Using the "Test" tab, the administrator can list the tests given by all teachers, which are shared, and view the content thereof. Although the administrator can see the active test, he cannot deactivate and activate another, since this is considered an exclusive feature for teachers.

Finally, the administrator has a set of links at the top right side of the main interface from which she can manage the images shown in the dynamic film that appears in the public part of the application, as well as the management and publication of news in the "News" tab (see Figure 20) in the public interface of the application. The administrator also has a link called "Edit Account" with which his access information can be updated.

Figure 18. Interface of the administrator user

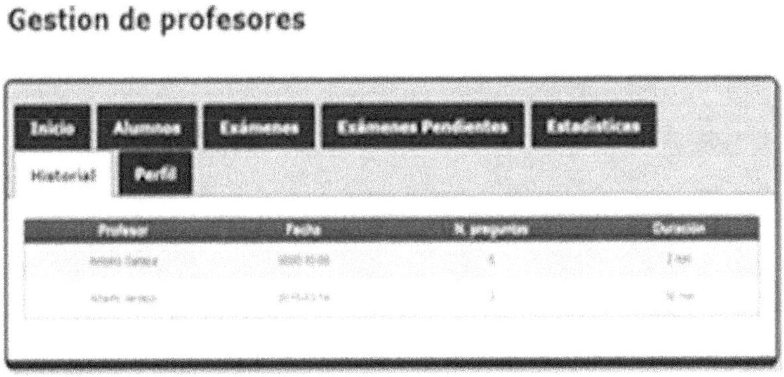

Figure 19. Tab for student management

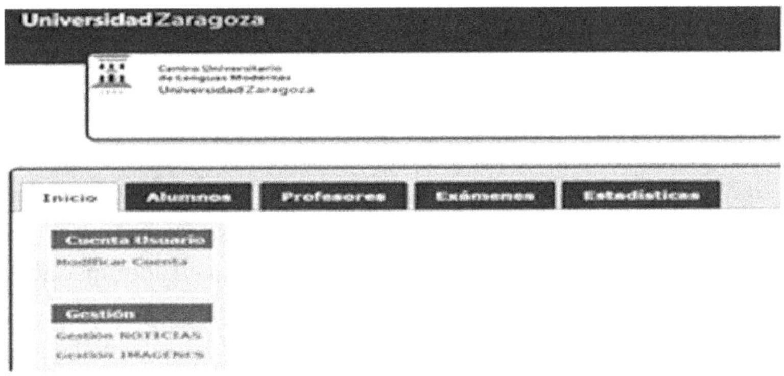

CONCLUSION AND FUTURE WORK

This article has described the problems that are presented in the language schools to manage the process of leveling the linguistic competency of students who are going to do a language course and to manage the information associated with their students. In this sense, different free or non-free products were analyzed. The analysis showed that a great variety of applications exist with very diverse functionalities. However, its use in specific cases presents problems such as difficulty of adaptation, complexity of functions, problems of extensibility of functionality, or the need to use licenses.

Also, the article has presented a real case in the context of a Spanish as a foreign language school at the University of Zaragoza. For this case two computer applications have been developed to create and manage tests to level the linguistic competency of students of Spanish as a foreign language. The first application described was implemented using Google Forms. This option presented some problems such as the configuration of the duration of the test time, the management of the users

Figure 20. News management

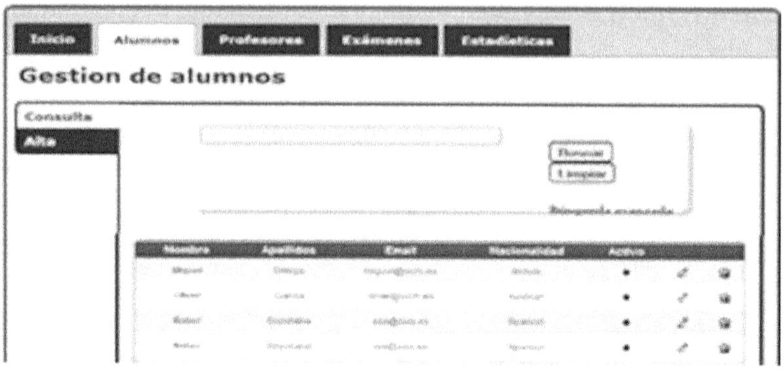

and the tests, and the maintenance of the information of the students. For these reasons, the authors decided to create a web application with greater functionality. In this second application, the authors automated some of the tasks that need to be performed to level the student's language competency, as well as it was addressed the problem of persistence and management of the data associated with the students. The web application is designed using a client-server architecture. From the point of view of usability, the application presents simple and short navigation paths, and interfaces adapted to each type of user (students, teachers and administrators). Also, the application is easily extensible with new features.

The application implements a set of basic services, which is why the main future line of work is to add new services to the user. These include exporting a student's information to standard formats (pdf, csv, ...), extending the type of questions or elements that can be included in a test (such as images, videos, rich text), adding a new tab in the application oriented to the student where to leave study material, adding communication tools between users (such as chat, Forums, ...), and adding a social network so that students can share experiences among them.

ACKNOWLEDGMENT

The authors would like to thank the facilities and the time devoted by the staff of the teaching service of Spanish as a foreign language at the University of Zaragoza, and especially its director, Vicente Lagüéns Gracia. In addition, the authors would like to thank students Evelín Otarola de la Cruz, Miguel Ortega Gómez and Oliver Garcia González, who carried out the coding of the second tool described in this paper. This work has been partially supported by the projects Santander-UCM GR3 / 14 (group 962022) and eLITE-CM S2015/HUM-3426

REFERENCES

Abelló Gamazo, A., Rodríguez González, M. E., Urpí Tubella, A., Burgués Illa, X., Casany Guerrero, M. J., Martín Escofet, C., & Quer Bosor, M. C. (2008). LEARN-SQL: automatic assesment of SQL Based on IMS QTI specification. In *8th IEEE International Conference on Advanced Learning Technologies* (pp. 592-593). IEEE.

Bachmann, S., & Moroney, W. F. (2005, September). A Consumer's Guide to Free Internet Questionnaire Development Tools. *Proceedings of the Human Factors and Ergonomics Society Annual Meeting*, *49*(24), 2085–2089. doi:10.1177/154193120504902404

Bohl, O., Scheuhase, J., Sengler, R., & Winand, U. (2002, December). The sharable content object reference model (SCORM)-a critical review. In Computers in education, 2002. proceedings. international conference on (pp. 950-951). IEEE.

De Europa, C. (2002). *Marco común europeo de referencia para las lenguas. Strasburgo: Consejo de Europa*. Ministerio de Educación, Cultura y Deporte/ Instituto Cervantes.

Dervan, P. (2014). Increasing in-class student engagement using Socrative (an online Student Response System). *AISHE-J: The All Ireland Journal of Teaching and Learning in Higher Education*, *6*(3).

Elbeck, M. (2014). Selecting a free web-hosted survey tool for student use. *E-Journal of Business Education & Scholarship of Teaching*, *8*(2), 54.

Finocchiaro, M., & Sako, S. (1983). Foreign Language Testing: A Practical Approach. Regents Publishing Company, Inc.

García-Holgado, A., & García-Peñalvo, F. J. (2013). The evolution of the technological ecosystems: An architectural proposal to enhancing learning processes. In F. J. García-Peñalvo (Ed.), *Proceedings of the First International Conference on Technological Ecosystems for Enhancing Multiculturality (TEEM'13)* (pp. 565-571). New York, NY: ACM. doi:10.1145/2536536.2536623

García-Holgado, A., & García-Peñalvo, F. J. (2016). Architectural pattern to improve the definition and implementation of eLearning ecosystems. *Science of Computer Programming*, *129*, 20–34. doi:10.1016/j.scico.2016.03.010

García-Peñalvo, F. J., Hernández-García, Á., Conde-González, M. Á., Fidalgo-Blanco, Á., Sein-Echaluce Lacleta, M. L., Alier-Forment, M., & Iglesias-Pradas, S. et al. (2015). Learning services-based technological ecosystems. In G. R. Alves, & M. C. Felgueiras (Eds.), *Proceedings of the Third International Conference on Technological Ecosystems for Enhancing Multiculturality (TEEM'15)* (pp. 467-472). New York: ACM. doi:10.1145/2808580.2808650

García-Peñalvo, F. J., & Seoane-Pardo, A. M. (2015). *Una revisión actualizada del concepto de eLearning*. Décimo Aniversario. *Education in the Knowledge Society, 16*(1), 119–144. doi:10.14201/eks2015161119144

Gottliebson, D., Layton, N., & Wilson, E. (2010). Comparative effectiveness report: Online survey tools. *Disability and Rehabilitation. Assistive Technology, 5*(6), 401–410. doi:10.3109/17483101003793404 PMID:20441440

Gracia, M. J. F., García, R. T. M., Papí, J. A. F., & Aycart, G. J. F. (2010). Pruebas objetivas en Moodle, Campus Virtual y sistema tradicional. Análisis comparativo de su gestión. In VIII Jornades de Xarxes d'Investigació en Docència Universitària [Recurso electrónico]: noves titulacions i canvi universitari= VIII Jornadas de Redes de Investigación en Docencia Universitaria: nuevas titulaciones y cambio universitario (pp. 1659-1673). Universidad de Alicante.

Hadley, A. O., & Reiken, E. (1993). Teaching Language in Context, and Teaching Language in Context--Workbook. Heinle & Heinle Publishers.

Haintz, C., Pichler, K., & Ebner, M. (2014). Developing a Web-Based Question-Driven Audience Response System Supporting BYOD. *J. UCS, 20*(1), 39–56.

Ji, Z., García Seoane, J., Vázquez Osorio, T., Flores Herráez, R., & Álvarez Vázquez, M. (2015). *Empleo de Google Drive en evaluación universitaria*. Academic Press.

Kakasevski, G., Mihajlov, M., Arsenovski, S., & Chungurski, S. (2008, June). Evaluating usability in learning management system Moodle. In *Information Technology Interfaces, 2008. ITI 2008. 30th International Conference on* (pp. 613-618). IEEE. doi:10.1109/ITI.2008.4588480

Kravtsov, H., & Kravtsov, D. (2008). Knowledge Control Model of Distance Learning System on IMS Standard. In Innovative Techniques in Instruction Technology, E-learning, E-assessment, and Education (pp. 195-198). Springer Netherlands. doi:10.1007/978-1-4020-8739-4_35

Kumaravadivelu, B. (1994). The postmethod condition: (E) merging strategies for second/foreign language teaching. *TESOL Quarterly, 28*(1), 27–48. doi:10.2307/3587197

Lado, R. (1961). *Language Testing: The Construction and Use of Foreign Language Tests. A Teacher's Book*. Academic Press.

McKay, S. L. (2002). *Teaching English as an international language: Rethinking goals and perspectives*. OUP.

Nuttall, C. (1996). *Teaching reading skills in a foreign language*. Heinemann.

Ozkan, S., & Koseler, R. (2009). Multi-dimensional students' evaluation of e-learning systems in the higher education context: An empirical investigation. *Computers & Education*, *53*(4), 1285–1296. doi:10.1016/j.compedu.2009.06.011

Pinto-Llorente, A., Sánchez-Gómez, M. C., & García-Peñalvo, F. J. (2015). Student Perception of the Use of a Blended-Learning Model to Improve Grammatical Competence. In G. R. Alves, & M. C. Felgueiras (Eds.), *Proceedings of the Third International Conference on Technological Ecosystems for Enhancing Multiculturality (TEEM'15)* (pp. 91-98). New York: ACM. doi:10.1145/2808580.2808595

Pinto Llorente, A. M., Sánchez Gómez, M. C., & García-Peñalvo, F. J. (2014). Assessing the effectiveness of a technological model to improve written skills in English in higher education. In F. J. García-Peñalvo (Ed.), *Proceedings of the Second International Conference on Technological Ecosystems for Enhancing Multiculturality (TEEM'14)* (pp. 69-74). New York: ACM. doi:10.1145/2669711.2669881

Pinto Llorente, A. M., Sánchez Gómez, M. C., & García-Peñalvo, F. J. (2014). Students' essential characteristics for learning English in a hypermedia modular model. In F. J. García-Peñalvo (Ed.), *Proceedings of the Second International Conference on Technological Ecosystems for Enhancing Multiculturality (TEEM'14)* (pp. 107-111). New York: ACM. doi:10.1145/2669711.2669887

Pinto Llorente, A. M., Sánchez Gómez, M. C., & García-Peñalvo, F. J. (2016). Assessing the Effectiveness of Interactive and Collaborative Resources to Improve Reading and Writing in English. *International Journal of Human Capital and Information Technology Professionals*, *7*(1), 66–85. doi:10.4018/IJHCITP.2016010105

Rivers, W. M. (1981). *Teaching foreign-language skills. University of Chicago Press.*

Rowe, M., Bozalek, V., & Frantz, J. (2013). Using Google Drive to facilitate a blended approach to authentic learning. *British Journal of Educational Technology*, *44*(4), 594–606. doi:10.1111/bjet.12063

Shen, L. P., & Shen, R. M. (2004). Learning content recommendation service based-on simple sequencing specification. In *Advances in Web-Based Learning–ICWL 2004* (pp. 363–370). Springer Berlin Heidelberg. doi:10.1007/978-3-540-27859-7_47

Shohamy, E. (1988). A proposed framework for testing the oral language of second/foreign language learners. *Studies in Second Language Acquisition*, *10*(02), 165–179. doi:10.1017/S0272263100007294

Takagi, M., & Teshigawara, Y. (2006). A WBT system enabling to create new or similar quizzes collaboratively by students. In *Proceedings of the 3rd International Conference on Educational Technology* (pp. 263-268). ACTA Press.

Travis, L. (2010). One of many free survey tools: Google docs. *Journal of Electronic Resources in Medical Libraries*, *7*(2), 105–114. doi:10.1080/15424065.2010.482902

Vogten, H., Martens, H., Nadolski, R., Tattersall, C., Van Rosmalen, P., & Koper, R. (2006, July). CopperCore Service Integration-Integrating IMS Learning Design and IMS Question and Test Interoperability. In *Advanced Learning Technologies, 2006. Sixth International Conference on* (pp. 378-382). IEEE. doi:10.1109/ICALT.2006.1652450

Voorheis, C. M., Singleton, H. M., Silvers, T. J., Snead, A. E., & Moroney, W. F. (2004, September). A consumer's guide to internet questionnaire development tools: II. *Proceedings of the Human Factors and Ergonomics Society Annual Meeting*, *48*(21), 2446–2449. doi:10.1177/154193120404802101

Yang, Q., Li, J. J., & Weiss, D. M. (2009). A survey of coverage-based testing tools. *The Computer Journal*, *52*(5), 589–597. doi:10.1093/comjnl/bxm021

Yuuichi, S., Toshihiro, K., Seisuke, Y., & Hiroshi, N. (2006, July). Web-based rapid authoring tool for LMS quiz creation. In *Information Technology Based Higher Education and Training, 2006. ITHET'06. 7th International Conference on* (pp. 617-620). IEEE. doi:10.1109/ITHET.2006.339676

KEY TERMS AND DEFINITIONS

E-Learning: It is a system of distance education that use of services offered by the internet.

Google Drive: It is a Google service that offers office tools.

Google Forms: It is a Google Drive tool to create question forms.

Language Schools: It is a school of education oriented to the teaching of foreign languages.

MySQL: It is a system of management of relational databases.

Php: It is a programming language that allows to create programs that run on a web server.

Test Service: It is a computer application whose purpose is to create and manage tests.

Web Application: It is a type of computer application that runs in a web environment.

Chapter 9
Analysis of a Learning Management System by Using Google Analytics:
A Case Study From Turkey

Zeki Özen
İstanbul University, Turkey

Elif Kartal
İstanbul University, Turkey

İlkim Ecem Emre
Marmara University, Turkey

ABSTRACT

Changes and developments in technology are also reflected in user behavior. Therefore, companies that monitor and analyze user behavior to gain competitive advantage will be one step ahead of others. Google Analytics is one of the most popular web analytics tools that perform this function. This chapter observes changes in user habits between 2011 and 2017 by using the Google Analytics data of Enocta Academic Education Platform (EAEP) that offers course content and learning management system (LMS) services for organizations and universities in Turkey. Thus, it advises to LMS providers in terms of improving their software considering changes in user habits. In addition, different approaches of improving e-learning services are discussed. This case study shows that Google Analytics, one of the web analytics tools, can provide useful information for e-learning platform or LMS providers in order to understand behaviors and needs of the visitors as well as their own strengths and weaknesses.

DOI: 10.4018/978-1-5225-5279-6.ch009

INTRODUCTION

Rapid developments in Information and Communication Technologies (ICTs) bring changes together. These changes nowadays affect almost every field of life. Today's society is therefore called as knowledge society where ICTs has an important place in daily life. Economic, social and cultural life is changing based on digital technologies and its effects of them can be obviously observed. One of the most important characteristics of knowledge society is seen as continuous learning or lifelong learning. Unlike physical workers in industry society, brain power and intellectual capabilities of workers are more important features in knowledge society (Yılmaz, 2012b). Since knowledge is produced, evaluated, transferred or processed by knowledge workers, these workers should be capable of handling all kind of processes related to knowledge. Education has a significant role for these workers in order *to "make the upcoming generation active participants in the society; teaching, learning as well as evaluation in the information digital society can be supported using relevant theories and models"*(Hošková-Mayerová & Rosická, 2015).

Like any other field, characteristics of learning have changed with recent technological advances. Traditional learning techniques fail to catch today's expectations and rapidly changing technological developments. Due to this reason, e-learning concept shows up as a concept that combines ICTs with learning concept. As the amount of the data produced by using e-learning platforms and with the emergence of ICT in almost every industry, analyzing e-learning data has gained in importance. So that tools like Google Analytics enables researchers to make efficient analyses which may have useful and meaningful results for learners' perspective as well as organizations.

In this chapter, the authors aim to explain how Learning Management System (LMS) providers can improve their e-learning services using Google Analytics based on a case study. In this scope, Google Analytics data of Enocta Academic Education Platform (EAEP), which offers course content and LMS services for organizations and universities in Turkey, is used to observe changes in user habits between 2011-2017. Thus, it is aimed to give advices to LMS providers in terms of improving their software considering changes in user habits. Therefore, in the following section, authors give definitions of e-learning concept, LMS and web analytics and after that literature summary about Google Analytics and Google Analytics analysis in e-learning.

Background

There are different definitions of the e-learning concept in the literature, which are mainly similar and emphasize the usage of web-based technologies or ICT in the learning processes. "Online learning", "Web-based learning", "computer based instruction", "Web-based training", and "virtual education" concepts can be gathered under the concept of e-learning which is a broad concept (Hewitt, 2015). E-learning also refers to "technology-enhanced learning" or "digital learning" (Wheeler, 2012). In general, the terms online learning and e-learning are often used interchangeably (Hewitt, 2015). E-learning is gaining in importance so both academic and commercial institutions are interested in it (Tîrziu & Vrabie, 2015). In the literature, there are different definitions of e-learning:

E-learning is learning at a distance that uses computer technology (usually the internet) (Henderson, 2003).

E-learning is essentially the network-enabled transfer of skills and knowledge. E-learning refers to using electronic applications and processes to learn. E-learning applications and processes include web-based learning, computer-based learning, virtual classrooms and digital collaboration (Mihalca, Uta, Andreescu, & Intorsureanu, 2008).

E-learning is the use of Internet technologies in order to create a rich learning environment which includes a large variety of instruction and information resources and solutions, and also to deliver this environment. And the goal of e-learning is to improve individual and organizational performance (Yılmaz, 2012a).

E-learning as a term refers to a variety of different forms of technology-supported learning, usually characterized as the application of knowledge, information and educational technology to link people to each other and / or with educational resources, for the purpose of education (formal or informal) (Ehlers & Hilera, 2012).

E-Learning describes a set of technology-mediated methods that can be applied to support student learning and can include elements of assessment, tutoring, and instruction (Wheeler, 2012).

E-learning (electronic-learning) refers broadly to the use of information and communication technologies for the electronic delivery of instructional content and the support of educational processes (Hewitt, 2015).

All these definitions emphasize the significant role of the technology in education for e-learning and boosting power of e-learning. E-learning concept enables training processes to be independent of time and space. Besides it enables cost reduction, which is related to physical classes, accommodation, travel, physical training tools etc. Not only for the companies but also for the academia e-learning concepts are getting more attention (Chang, 2016). Both industry and academia need to keep the skills of students and employees up-to-date in order to enable individuals to adapt to the requirements of knowledge society (Chang, 2016). New concepts have emerged with the technological advances such as "digital literacy" which indicates the need of the new skills to cope up with the "digitally driven world" (Wheeler, 2012).

E-learning environment provides different type of course contents such as text documents, videos, podcasts, images, audio, animation etc. Hewitt (2015) states that e-learning includes affluent multimedia content. This makes the course more understandable for learners. According to Arkorful & Abaidoo (2015) individuals can select which part of the course that they want to focus. They also indicate that self-pacing is one of the advantages of e-learning. At this point "e" part of e-learning needs little bit more technical care for efficient education environment. As a result of technological advances, many other different platforms blogs, wikis or new microblogging tools like Twitter or Tumblr have emerged, which also enable to share knowledge and such platforms enhance collaborative learning platforms besides e-learning platforms (Wheeler, 2012).

There are different e-learning platforms/environments. Learning Management System (LMS) is a tool used for systematic management of e-learning. Basically, an LMS provides course content for learners and also enables to track down and evaluate the whole learning process (Watson & Watson, 2007). There are open source LMS options such as Moodle, ATutor, Eliademy, Forma LMS, Dokeos, ILIAS, Opigno, OLAT etc. (Pappas, 2015a). Moodle and Blackboard are two of the popular e-learning environments that combine many of different tools (i.e. blogs, online forums, real-time chat, instant messaging) providing a combination of useful tools for instructors (Hewitt, 2015). In addition, there are commercial LMS options such as Adobe Captivate Prime, Oracle Learning Management, The Office 365 LMS, Docebo etc. Among all of these systems, Moodle is the most popular LMS with 73.8 million users in terms of actual users whereas Edmodo has the highest number of customer as 120,000 (Pappas, 2015b). According to Capterra's LMS research report (Medved, 2015), industries using LMS are education (21%), technology (12%), manufacturing (9%), healthcare (7%), consulting (7%), software development (4%), non-profit (3%), real estate (3%), event management (3%), government (2%) and others (29%). Report also indicates that assessment / testing tools, blended learning, student portal, SCORM compliance, and learning paths / tracks are the most used features (Medved, 2015). Live/video conferencing, integrated talent management, mobile learning, gamification and social learning are seen as the most desired features in an LMS (Medved, 2015). Companies either establish or outsource LMS services for e-learning. Users can reach course contents through a web site on an e-learning environment and there are many different factors that are directly or indirectly related to the improvement of these services such as web analytics, LMS log analyses etc.

Web analytics, includes gathering, measuring, tracking analyzing and reporting web usage data to investigate behavior of visitors (Hasan, Morris, & Probets, 2009). Web Analytics *"is the measurement, collection, analysis and reporting of Internet data for the purposes of understanding and optimizing Web usage"* (Web Analytics Association, 2008). It is defined as the process of analysing visitor behaviors and often is used as a part of customer relationship management (Rouse, 2005). Web analytics studies aim to analyze performance of the web site and optimize its web usage depending on these analyses (Tutorials Point, 2015). Life of web analytics can be divided into four parts: birth of web analytics (between 1993 and 1998), emerge of page tagging (between 1999 and 2004), growth of web analytics (between 2005 and 2009) and transition to digital analytics (between 2010 and present) (Dykes, 2014). Waisberg and Kaushik (2009) emphasize that web analytics is not an only technology to generate reports but also it is a process that suggests a skilled cycle to optimize a web site. Steps of the web analytics process are given below (Figure 1):

Step 1: Define goals (to find an answer for the question: "Why does the web site exist?")
Step 2: Build key performance indicators (KPIs) (to determine how close is the web site to its goals)
Step 3: Collect data (web logs, JavaScript tagging etc.)
Step 4: Analyse data (some basic metrics such as visits, bounce rate, page views, web site traffic, etc.)
Step 5: Implement changes

Many tools have been developed for web analytics analyses. Google Analytics, Yahoo Web Analytics, Crazy Egg, Compete, Google Website Optimizer, Optimizely are some examples of web analytics tools (Dubois, 2010). Hasan et al. (2009), states that there are two different methods, which are used for web analytics. First

Figure 1. Steps of the web analytics process (Waisberg & Kaushik, 2009)

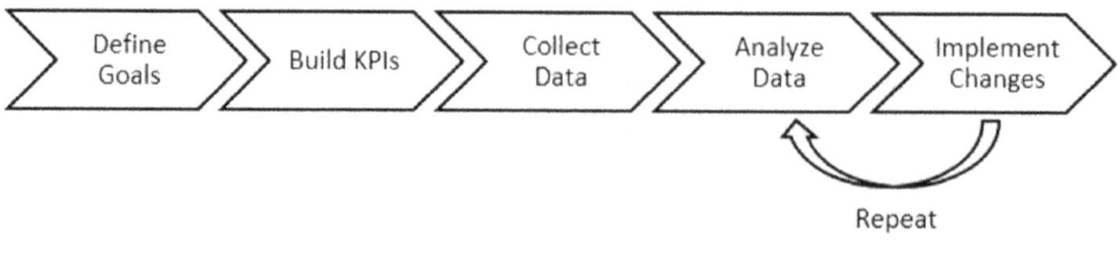

is analyzing the web server log files and the second one depends on tagging the pages, used by visitors.

Google Analytics, which is an important figure in web analytics, uses page-tagging method (Hasan et al., 2009). It is a free web analytics solution that provides webmasters with insightful information about how visitors find and interact with their web sites (Fang, 2007). Basically, it creates reports on visitor (audience), acquisition, behavior and conversion analysis (Tutorials Point, 2015):

- **Visitor Analysis:** Age, gender, the affinity reach and market segmentation, language, location, new and returning visitors, their frequency and engagement, browsers, operating systems, network, mobile device, benchmarking channels, locations, devices, flow of user activity details.
- **Acquisition Analysis:** Web site traffic details.
- **Behavior Analysis:** Web site content, speed, search and event details.
- **Conversion Analysis:** Different goals can be settled then the goal status can be examined. E-commerce tracking can be performed. Source of conversion can be reported.

A web admin is able to start to use Google Analytics by inserting a small piece of code (to the header parts of each web page or only to header part of a master page). It gives time series data (Plaza, 2009) so that it become possible to make comparisons by indicating different time periods. Google Analytics provide many benefits to the users and it is used by various organizations. Marketing campaign optimization, website usability improvement, target audience identification and budget allocation can be considered as the major business benefits of Google Analytics (Holloway, 2014). Customizable dashboard, conversion visualization, data export in various formats and advanced segmentation features are seen as the advantages of Google Analytics (Bekavac & Praničević, 2015).

Google Analytics is used in many web sites such as business, news, shopping, technology, entertainment, social and sports (marketingland, 2015). By the way, according to BuiltWith (2016); 66,1% of Quantcast's Top 10.000 sites (by traffic) and 8,1% of entire Internet is using Google Analytics. In addition, in the literature there are many papers about Google Analytics practices in different areas. Tourism is one of the sectors in which admins of the tourism web sites prefer Google Analytics to measure the performance of the web site. Time series data analysis, which is obtained by using Google Analytics, is performed by Plaza (2011) to optimize a tourism website for developing online marketing purpose. Web technologies are utilized for library services in education. Fang (2007) has used Google Analytics to improve content and design of a library web site. After making some changes on the library web site, the web usage is compared with initial status. According to the results, new visitors and returning visitors of the library web site have increased by 21% and 44% respectively. It is thought that ease of use is one of the important elements in order to make successful e-commerce (Hasan et al., 2009). For that reason, they have evaluated the usability of an e-commerce web site by using Google Analytics. Three different food composition websites are evaluated by (Pakkala, Presser, & Christensen, 2012) using visitor statistics. The impact of a CyberTraining project, which aims to provide an e-book about cyberbullying, is measured by Mc Guckin & Crowley (2012).

Companies such as Domino's, Lenovo, L'Oreal, Nest, Novartis and Panasonic have different success stories based on Google Analytics Solutions (Google Analytics Solutions, 2017). Fairmont, which is a luxury hotel company, benefit from Google Analytics to improve social media marketing (Google Analytics, 2013). The performance of Twitter campaign is monitored and better understanding of behavior of social media referred visitors is provided. American Cancer Society used Google Analytics to identify user segments which belongs to cancer.org website (Google Analytics, 2015b). Then, visitor behavior is examined based on scoring to find out if they have reached their goals. As a result of these efforts; donations rise 5.4%. AlfaStrakhovanie LLC, which is one of Russia's largest insurance companies, take the advantage of Google Analytics in terms of increasing their transaction rate and adjusting the prices for each visitor segments (Google Analytics, 2015a). Mumzworld (Google Analytics, 2015c), which is a shopping website and has products for mothers, babies and children, takes the advantage of Google Analytics some e-commerce issues such as increasing total purchases and total revenue from advertising channels over three-month period (Google Analytics, 2015c).

As stated in above Google Analytics is used in different sectors that includes LMS. Google Analytics provides useful information about behaviors of LMS users's. After integrating it to an LMS, it can be possible to understand some key factors such as (Edwards, 2013):

- Which devices are people using?
- Which browsers are learners using to take courses?
- When and where does training happen?
- How long does it take learners to complete a course?
- Where do users have problems?
- Which promotional campaigns work best?

Google Analytics offers various possibilities to e-learning content providers in order to evaluate e-learning activities (Pappas, 2016). On top of the list, there is the time that students spend online. Corrections can be made by observing the period of time that they spend online hereby course admins can determine if the time required for that course is adequate. Furthermore, average time per page can be considered to make the page (also course content, topics, etc.) more interesting for students or employees and make them spend more time on it. Also, Google Analytics provides device information which indicates the type of the used devices to access to the page. Furthermore, platform access information which indicates that whether access to the page is made using desktop PC, laptop, tablet or mobile devices, can be observed. These information help LMS providers to make page contents more user friendly by using responsive design tools. In this way, user experience can be improved and user satisfaction can be increased. Another advantage of Google Analytics for an e-learning platform is information about site traffic. Previous and following web sites and social media platforms of visitors can be analyzed, so advertisements can be given or routing can be done to that channels (xyleme.com, 2014). Google Analytics also reports which keywords are used by a visitor to access the e-learning platform. Visitors' demands can be predicted based on these keywords (xyleme.com, 2014). Google Analytics also provides time information that indicates access to a page. By determining the most visited time periods of the platform, technical tasks such as maintenance and updating of LMS platform can be more efficient. Thus, a problem can be solved easily in a short time or problem never occurs so that the platform is totally protected.

There are many studies in education field that are performed using Google Analytics. Özen, Bakioğlu, & Beden (2014) have examined the user habits of an academic education platform with Google Analytics data of EAEP for the year 2011. Their results have shown that it can be decided which improvements or developments to be made in LMS by analyzing user habits with Google Analytics. Wohlers & Jamieson (2014) have shared Google Analytics statistics of the Moodle LMS that belongs to Eastern Institute of Technology. Ellis (2012) has emphasized the importance of designing, developing, and evaluating the e-learning artefacts. Therefore, a case study, which is based on experiences at Massey University, the relation between analytics data and creating online course contents has explained. As a part of Rodriguez (2012)'s study, Google Analytics is used to report number of visits of new and returning visitors of a connectivist massive open online course. Yamamoto, Özen, & Beden (2012) has analysed web logs of Okan University in Turkey that is provided by EAEP. According to their studies, browsers that are mostly preferred by visitors, days that the system is mostly used and some other significant findings are obtained.

Before examining the changes in user habits in Turkey, the general situation in the world should be considered in terms of desktop PC, portable PC, tablet and smart phone usage. On one hand, according to International Data Corporation (IDC) data; global smartphone shipments from 2010 to 2016 increased 7 times, tablet shipments increased 13 times (statista, 2016a). On the other hand, global desktop

PC and portable PC shipments decreased (statista, 2016b). At the same time in 2016 mobile and tablet Internet usage exceeds desktop usage worldwide for the first time (StatCounter, 2016). Based on all these information, it can be said that habits of LMS users have changed in the last five years, too (Figure 2).

GOOGLE ANALYTICS ANALYSIS ON AN EDUCATION PLATFORM

If e-learning is considered as a type of online business, the following goals (Chan, 2008) should be kept in mind too: Selling online, building the community, automating the customer services, generating new leads and amplifying organization's voice to visitors. At this point, there is a need for a well-organized website and monitoring it to move towards all these goals.

In this study, Google Analytics data of EAEP, which is used by 19 universities in Turkey, is used to compare user habits and to observe the changes in user behaviors of this platform. Six different time periods are chosen by considering the possible course start and end time. Only last year ends on 1st August.

- From September 1, 2011 to August 31, 2012,
- From September 1, 2012 to August 31, 2013,
- From September 1, 2013 to August 31, 2014,
- From September 1, 2014 to August 31, 2015,

Figure 2. Global smartphone, tablet, portable PC and desktop PC shipment between 2010 and 2016 (in million units)

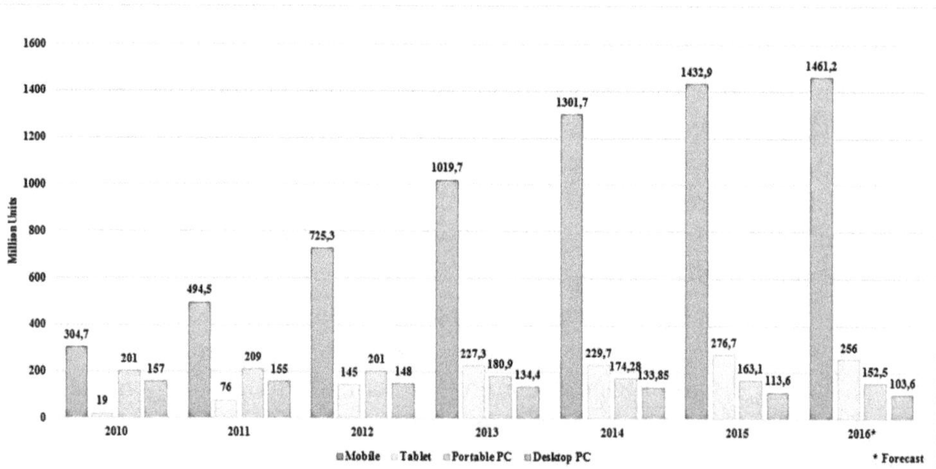

- From September 1, 2015 to August 31, 2016,
- From September 1, 2016 to August 1, 2017.

Figure 3 shows the visits which belongs to different time periods from 2011 to 2017 month by month. According to Figure 3, general overview indicates that almost each year the visits are higher than other months on November and on May. It is thought that exams of the courses are taken on these months, so that visit count peaked for two times in every year. Furthermore, visits are nearly doubled from 2011 to 2016. The highest visit count is gathered in November 2015. In addition, from 2015-2016 to 2016-2017 visits have decreased.

Table 1 indicates the comparison of general information (visits, unique visitors, pages/visit, average visit duration, page views, new visits, and bounce rate) from 2011 to 2017. The highest number of visits, unique visitors, and page views belongs to period 2015-2016. Although the number of visits have increased each year, there is a dramatic decline in 2016-2017. In addition, the highest average visit duration is obtained in 2013-2014 period and 2015-2016 period is nearly the same. Moreover, bounce rate has become higher and reached from 5,74 to 17,63 at the end of six years.

Figure 4 indicates the difference of daily visits between 2011-2012 and 2016-2017. As it is stated before, the peak points of visits can be seen in November and May in Figure 4. Although six years have elapsed since then, it was observed that there was a significant increase only between October and January, while the numbers of visits in the other months are almost the same.

Table 2 shows visits and bounce rates of the top 10 cities in 2016-2017. In this period Isparta, Antalya and Konya have higher bounce rates than other cities in the

Figure 3. Visits of different time periods from 2011 to 2017 month by month

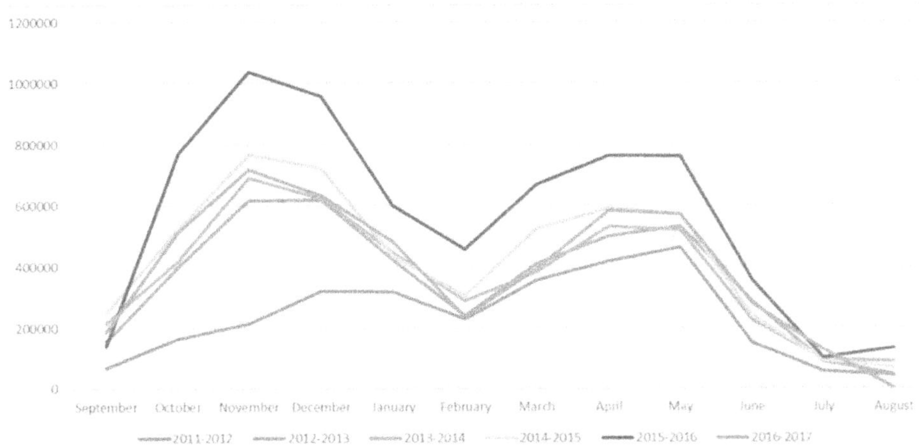

Table 1. Comparison of visit information between 2011-2017

Years	Visits	Unique Visitors	Pages / Visit	Average Visit Duration (sec)	Page Views	New Visits (%)	Bounce Rate (%)
2011-2012	2.791.239	631.370	14,3	00:08:46	39.926.199	22,43	5,74
2012-2013	4.298.265	1.053.175	14,22	00:08:55	61.112.511	24,17	8,24
2013-2014	4.516.451	1.000.804	14,53	00:09:05	65.615.693	21,78	6,93
2014-2015	5.075.862	1.119.502	12,8	00:08:47	64.954.508	21,68	10,05
2015-2016	6.735.786	1.317.124	12,22	00:09:02	82.306.328	19,28	11,46
2016-2017	4.714.628	878.537	10,55	00:08:03	49.762.476	18,26	17,63

list. Cities that have highest visit numbers are İstanbul, Ankara and İzmir, respectively. Volume of visits in terms of cities in 2016-2017 can be seen in Figure 5.

Change in usage for Chrome, Internet Explorer, Firefox and Safari can be seen in Figure 6. According to data of 2016-2017, Chrome is at top of the list (Visits: 3.494.214). Internet Explorer (Visits: 334.002), Firefox (Visits: 279.508) and Safari (Visits: 268.247) are following it in 2016-2017. It is clear that Chrome has the highest increase rate from 2011-2012 to 2016-2017. The usage of Internet Explorer has decreased dramatically while Firefox remains almost the same for all years. Safari usage has also increased slightly but still remains as the least preferred browser.

Operating system used by the visitors is also an important factor in the analyses. Table 3 shows the changes in operating system preferences between 2011-2012 and 2016-2017. Windows is at top of the list with the highest visit count each year but its usage has decreased in 2016-2017. Android also has the highest increase ratio, approximately 42 times. iOS change ratio is the second highest one. These results may indicate that the usage of e-learning platforms on mobile technologies have increased. iOS, iPhone and iPad usage results are combined in 2011-2012 since they are in the top 10 list of operating system usage.

Figure 4. The difference of daily visits between 2011-2012 and 2016-2017

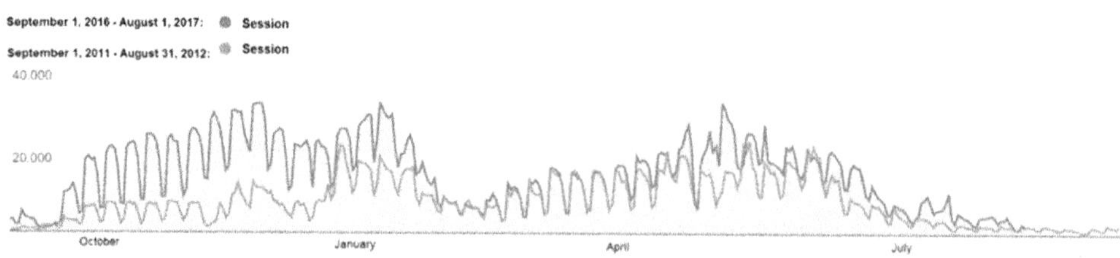

Table 2. Visits and bounce rates according to cities in 2016-2017

Cities	Visits	Visits (%)	Bounce Rate (%)
İstanbul	1.234.463	26,64	16,87%
Ankara	819.597	17,69	14,29%
İzmir	378.355	8,17	16,47%
Bolu	222.140	4,79	10,55%
Adana	212.032	4,58	21,45%
Manisa	169.265	3,65	11,20%
Isparta	159.029	3,43	41,46%
Konya	127.503	2,75	23,43%
Mersin	111.666	2,41	18,27%
Antalya	109.663	2,37	29,73%
Total	4.633.809	98,29	17,63%

Figure 5. Volume of visits in terms of cities in 2016-2017

In this study mobile device, information of the users is also examined. It can be seen in Table 4 that Apple and Samsung are sharing top two in 2011-2012 and 2016-2017. However, their rankings have changed. Usage of Android devices have increased. Ratio column shows the usage percentage of each device amongst others. Change ratio is calculated by dividing usage amount of 2016-2017 to 2011-2012.

Figure 6. Change in usage for Chrome, Internet Explorer, Firefox and Safari

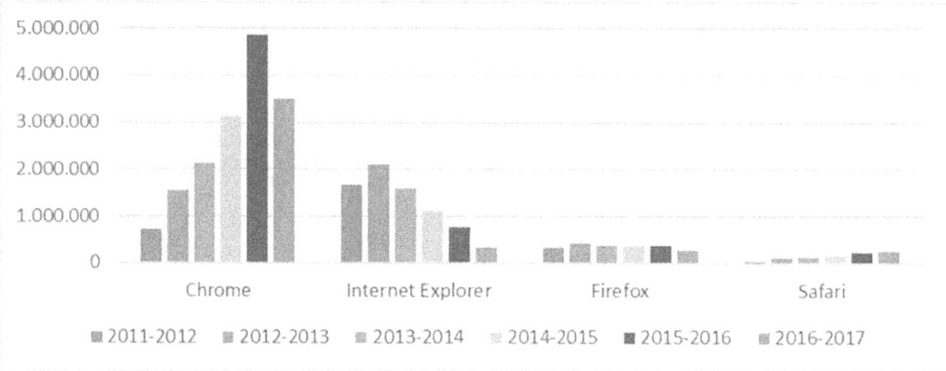

Table 3. Usage of different operating systems

Operating System	2011-2012	2012-2013	2013-2014	2014-2015	2015-2016	2016-2017	Change Ratio in 6 Years
Windows	2.712.352	4.052.209	4.081.375	4.297.522	5.415.006	3.592.449	1,32
Android	17.186	94.732	231.526	510.788	969.253	724.007	42,30
iOS	27.314	88.212	127.270	167.004	226.444	282.258	10,33
macOS	18.406	33.729	39.455	47.549	54.065	59.225	3,22
Linux	8.541	17.345	24.601	36.197	57.228	52.472	6,14

Table 5 indicates the devices which are used by visitors to access to EAEP on yearly basis. Last column shows that mobile usage has increased approximately 24 times, while desktop is 1,3 times and tablet is 4,8 times. There is a significant difference between desktop access and mobile & tablet access. Change ratio is calculated by dividing usage amount of 2016-2017 to 2011-2012.

Table 6 shows the changes in screen resolution by year. The highest six screen resolution for each year are taken into consideration for the comparison. 1366x768 screen resolution is top of the list in each year. 360x640 screen resolution wasn't used between 2011-2014. And also in Table 6, it is noteworthy that the resolution ratio of 1920x1080 which is a high definition ratio is increased about 3 times.

As seen in Table 7, 32-bit is the most preferred color option. However, 24-bit usage has increased and has become the most preferred color option between 2014 and 2017.

Table 8 indicates usage according to mobile device brands and models between 2011 and 2017. iPhone is top of the list in each year. iPad is the second mobile

Table 4. Usage of mobile device brand change between 2011-2012 and 2016-2017

Mobile Device Brand	2011-2012	Ratio (%)	2016-2017	Ratio (%)	Change Ratio in 6 Years
Samsung	7.271	13,09	367.015	34,43	50,47
Apple	34.732	62,54	298.290	27,98	8,58
LG	496	0,89	98.128	9,20	197,83
General Mobile	0	0	60.106	5,64	-
(not set)	4.747	8,55	48.012	4,50	10,11
Sony	30	0,05	44.903	4,21	1.496,76
HTC	2.238	4,03	21.257	1,99	9,49
Asus	45	0,08	19.875	1,86	441,66
Lenovo	0	0	16.810	1,58	-
Casper	0	0	14.166	1,33	-

Table 5. Change in usage for platform preferences of visitors

Device	2011-2012	2012-2013	2013-2014	2014-2015	2015-2016	2016-2017	Change Ratio in 6 Years
Desktop	2.740.420	4.111.593	4.145.766	4.358.575	5.492.194	3.685.559	1,34
Mobile	40.784	156.040	317.550	637.156	1.154.746	980.785	24,05
Tablet	10.035	30.632	53.135	80.131	88.846	48.284	4,81

Table 6. Usage of screen resolution preferences of visitors

Screen Resolution	2011-2012 (%)	2012-2013 (%)	2013-2014 (%)	2014-2015 (%)	2015-2016 (%)	2016-2017 (%)
1366x768	27,03	32,86	37,97	41,37	43,9	40,46
360x640	-	-	-	5,03	9,73	12,20
1920x1080	2,78	3,8	5,04	6,4	7,74	8,81
1280x800	13,92	10,36	7,77	5,94	4,31	3,34
1440x900	6,08	6,13	5,43	4,46	4,16	3,50
1024x768	16,19	11,96	7,75	5,34	3,82	2,16
1280x1024	7,51	6,47	5,42	4,7	3,53	3,17

Table 7. Usage for screen color preferences of visitors

Screen Color	2011-2012 (%)	2012-2013 (%)	2013-2014 (%)	2014-2015 (%)	2015-2016 (%)	2016-2017 (%)
32-bit	84,36	80,91	50,91	19,76	20,49	22,54
24-bit	13,98	17,76	48,31	79,85	79,33	77,38
16-bit	1,58	1,22	0,71	0,3	0,12	0,06
8-bit	0,02	0	0	0	0	0
4-bit	0,05	0,09	0,07	0,09	0,05	0,01

Table 8. Usage according to mobile device brands by models

2011-2012		2012-2013		2013-2014		2014-2015		2015-2016		2016-2017	
Brand	**Visits (%)**	**Brand**	**Visits (%)**	**Brand**	**Visits (%)**	**Brand**	**Visits (%)**	**Brand**	**Visits (%)**	**Brand**	**Visits (%)**
Apple iPhone	42,5	Apple iPhone	30,7	Apple iPhone	24,2	Apple iPhone	18,1	Apple iPhone	16,1	Apple iPhone	25,7
Apple iPad	15,1	Apple iPad	13	Apple iPad	10	not set	15,8	not set	8,22	not set	3,87
not set	10,6	not set	8,98	not set	9,93	Samsung Galaxy S IV	5,6	Samsung Galaxy S IV	3,12	Samsung Galaxy J7	3,28
Samsung Galaxy S II	4,88	Samsung Galaxy SIII	4,85	Samsung Galaxy S IV	6,68	DOOGEE DG500 Discovery	5,28	LG G3	2,83	LG G3	2,44
SonyEricsson Xperia Arc	4,03	Samsung Galaxy S II	4,5	Samsung Galaxy Note II	6,25	Apple iPad	5,15	Samsung Galaxy Note 3	2,54	Samsung Galaxy Note 4	2,18

device brand in the first three years, while its rank fall back to the fifth place in 2014-2015. Sony Ericsson and DOOGEE is only seen in 2011-2012 and in 2014-2015, respectively. LG has become more popular in the last two years.

SOLUTIONS AND RECOMMENDATIONS

The following suggestions can be considered in order to improve e-learning services in terms of content management and technical infrastructure. It is believed that these will also increase satisfaction of LMS users.

- It is clear that the bounce rate has almost tripled itself when 2011-2012 (5,74) and 2016-2017 (17,63) time periods are compared. Therefore, the LMS provider could try to figure out the reason(s) for this sharp increase.

It could be considered to design more interactive or attractive pages to solve this problem.

- From 2011-2012 to 2015-2016 total number of the visits have increased. Due to this change, the LMS providers could consider the future projections about technological investments such as server and bandwidth. In spite of this steady increase, it can be seen that there is a decrease of the total number of visits from 2015-2016 to 2016-2017. This change is remarkable so that provider should consider the reasons for such sharp changes. Average visit duration is remained nearly the same in each year.
- EAEP is being used by the users from different cities all around the Turkey. As the usage spreads around many different locations, technical issues or due to the increase in the number of the visitors, content management would be a harder issue for EAEP or for the other similar providers mainly. For this reason, technical support could be divided into regions and more qualified customer services could be provided. İstanbul is the city where the platform is used most intensively. So İstanbul might has more maintenance or service needs then the other cities.
- Although Chrome usage has decreased in the last year, Chrome has always been the most preferred browser through all years. It means that LMS providers could check the appearance of their pages especially on Chrome and they could make changes in design or structure of the page(s) if it is necessary. It is not known what kind of problems users face while using the other browsers. However, the reason why users do not prefer the others might be because of compatibility problems.
- Usage of the different operating systems is also examined. According to the results, Windows has always been the most preferred operating system. Although, the change ratio in the six years period of Windows preference is 1,32, the change ratio of Android is 42,30 which is worth noting. This may indicate that in the past six years Android based devices has spread to the market very quickly.
- Table of usage according to mobile device brands by models shows that while in 2011-2012 Apple was at the top of list, in 2016-2017 Samsung took its place with the change ratio of 34,43. So the LMS provider should focus on the compatibility of the system with Samsung devices.
- According to the findings it can be seen that visitors prefer mostly desktop access to use EAEP. However, it is predicted that mobile technologies will be on the top of the list in the near future according to the projections. In the light of these data, an LMS service provider should ensure that web pages are displayed properly on mobile and portable devices. A separate design for each platform will increase both the workforce and cost, causing slowdowns

in services such as updating and maintenance. This indicates that LMS providers should design their websites in responsive concept. It means that their pages can be reached on any platform without any required/technical changes by using responsive design methods. This change will increase user satisfaction and competitive power in the market.

- Moreover, another advantage that tablet and mobile devices has over desktop PC platforms is that they are not only portable, but also touch-enabled. If LMS service providers take advantage of this feature and produce more interactive content, bounce rate will be decreased. LMS users will be more interested in the course content and will spend more time on the system. The interactive course content developed through the touch feature prevents users from getting bored quickly, and also has a positive effect on learning situations.
- Observing the screen resolution and color preferences, users mostly prefer 1366x768 resolution and 24-bit color in 2016-2017 period. These preferences are mostly bounded with the used technologies.

FUTURE RESEARCH DIRECTIONS

Even though there has been a huge focus on the technology usage in the learning Wheeler (2012) states that e-learning studies *should "concentrate more on the learning than on the technology"*. This refers to the importance of the learner's needs while developing e-learning tools. The emphasis here is on the needs of learner instead of technological advances. This indicates that the development of such tools or platforms should focus on the learner's needs while adapting technological solutions. While providers should follow technological advances, they should take the learners need into account. In this context future analyses can focus more on the needs of the learners.

Users can be divided into groups so that differences between each group can be monitored according to their usage habits. Google Analytics data of the e-learning platform can be analyzed monthly, weekly, or even hourly intervals instead of annual so that more sensitive features of visitor behaviors can be examined. The results can be interpreted based on different time intervals can provide useful information for the providers in order to understand the usage behaviors of the learners.

Future researches can focus also on the effects of the different technologies. Implementation of the up-to-date technologies, new devices, new platforms etc. can be observed in terms of user acceptance. Furthermore, web mining techniques can be used to analyse web access data of LMS.

CONCLUSION

Today e-learning is one of the most important issues for both individuals and organizations. Also managing an LMS can be hard if some technical details are forgotten. It also has the advantage of being able to manage the LMS properly as well as teaching users the use of an LMS. It is now possible to track users online in order to improve LMS, rather than receiving feedback from users through classical methods such as questionnaires. LMS providers can develop their own web analytics tool or there are many web analytics tools developed for this job and LMS providers can prefer to use one of them.

In this study, comparison of usage behaviors which belongs to the users of EAEP of six different time periods (between 2011 and 2017) is performed. Findings of this study could be enlightening for both the e-learning platform providers and companies to improve systems and contents. This case study has once again showed that Google Analytics, one of the web analytics tools, can provide useful information for e-learning platform or LMS providers in order to understand behaviors and needs of the visitors as well as their own strengths and weaknesses. Using web site analytics tools would be the best way to prevent and improve the existing systems before the problem occurs. Design of the website and course contents can be improved. ICT infrastructure and compatibility issues should be revised in order to make the platform more user friendly and attractive.

ACKNOWLEDGMENT

Authors would like to thank Enocta for sharing their Google Analytics data. And also the authors thank Utku Sözeri and Betül Doğru for proofreading.

REFERENCES

Arkorful, V., & Abaidoo, N. (2015). The role of e-learning, advantages and disadvantages of its adoption in higher education. *International Journal of Instructional Technology and Distance Learning*, *12*(1), 29–42.

Bekavac, I., & Praničević, D. G. (2015). Web analytics tools and web metrics tools: An overview and comparative analysis. *Croatian Operational Research Review*, *6*(2), 373–386. doi:10.17535/crorr.2015.0029

BuiltWith. (2016). *Google Analytics Usage Statistics*. Retrieved 3 November 2016, from http://trends.builtwith.com/analytics/Google-Analytics

Chan, E. (2008). *5 Goals For Online Businesses and The Strategies To Reach Them*. Retrieved 10 August 2017, from http://businesscatalyst.com/bc-blog/5-goals-for-online-businesses-and-the-strategies-to-reach-them

Chang, V. (2016). Review and discussion: E-learning for academia and industry. *International Journal of Information Management*, *36*(3), 476–485. doi:10.1016/j.ijinfomgt.2015.12.007

Dubois, L. (2010). *11 Best Web Analytics Tools*. Retrieved 12 December 2016, from http://www.inc.com/guides/12/2010/11-best-web-analytics-tools.html

Dykes, B. (2014). *Web Analytics Kick Start Guide A Primer on the Fundamentals of Digital Analytics*. Peachpit. Retrieved from http://ptgmedia.pearsoncmg.com/imprint_downloads/adobepress/promotions/analytics_ksg/WebAnalyticsKickStartGuide.pdf

Edwards, S. (2013). *10 Common LMS Questions You Can Answer with Google Analytics*. Retrieved 10 August 2017, from https://www.opensesame.com/blog/10-common-lms-questions-you-can-answer-google-analytics

Ehlers, U.-D., & Hilera, J. R. (2012). Special Issue on quality in e-learning [Editorial]. *Journal of Computer Assisted Learning*, *28*(1), 1–3. doi:10.1111/j.1365-2729.2011.00448.x

Ellis, D. (2012). *Google Analytics as a tool in development of e-learning artefacts: A case study*. Presented at the ascilite 2012, New Zealand. Retrieved from http://www.ascilite.org/conferences/Wellington12/2012/images/custom/ellis,_damon_-_google_analytics.pdf

Fang, W. (2007). *Using Google Analytics for improving library website content and design: a case study*. Academic Press. 10.7282/T3MK6B6N

Google Analytics. (2013). *Fairmont Hotels harness the power of Google Analytics to optimize social media marketing*. Retrieved from https://static.googleusercontent.com/media/www.google.com/en/us/analytics/customers/pdfs/fairmont-raffles-hotels-international.pdf

Google Analytics. (2015a). *AlfaStrakhovanie Doubles Transaction Rates With Google Analytics and Enhanced Ecommerce*. Retrieved from https://static.googleusercontent.com/media/www.google.com/en/us/analytics/customers/pdfs/alfastrakhovanie.pdf

Google Analytics. (2015b). *Cancer.org Donations Rise 5.4% With Help From Google Analytics*. Retrieved from https://static.googleusercontent.com/media/www.google.com/en/us/analytics/customers/pdfs/americancancersociety.pdf

Google Analytics. (2015c). *Mumzworld Reaches 300% ROAS With Enhanced Ecommerce and Data Import From Google Analytics*. Retrieved from https://services.google.com/fh/files/misc/mumzworld-enhanced-ecommerce.pdf

Google Analytics Solutions. (2017). *Success Stories*. Retrieved from https://www.google.com/analytics/success-stories/#?modal_active=none

Hasan, L., Morris, A., & Probets, S. (2009). Using Google Analytics to Evaluate the Usability of E-Commerce Sites. In M. Kurosu (Ed.), *Human Centered Design* (pp. 697–706). Springer Berlin Heidelberg; doi:10.1007/978-3-642-02806-9_81

Henderson, A. J. (2003). *The e-learning question and answer book: a survival guide for trainers and business managers*. New York: American Management Association.

Hewitt, J. (2015). E-Learning. In R. Gunstone (Ed.), *Encyclopedia of Science Education* (pp. 362–363). Dordrecht, The Netherlands: Springer Netherlands; doi:10.1007/978-94-007-2150-0_36

Holloway, G. (2014). *The Business Benefits of Google Analytics*. Retrieved 10 August 2017, from https://www.koozai.com/blog/analytics/google-analytics-business-benefits/

Hošková-Mayerová, Š., & Rosická, Z. (2015). E-Learning Pros and Cons: Active Learning Culture? *Procedia: Social and Behavioral Sciences*, *191*, 958–962. doi:10.1016/j.sbspro.2015.04.702

Marketingland. (2015). *How Many Websites Use Google Analytics? Estimates Suggest 30-50 Million*. Retrieved 3 November 2016, from http://marketingland.com/as-google-analytics-turns-10-we-ask-how-many-websites-use-it-151892

Mc Guckin, C., & Crowley, N. (2012). Using Google Analytics to Evaluate the Impact of the CyberTraining Project. *Cyberpsychology, Behavior, and Social Networking*, *15*(11), 625–629. doi:10.1089/cyber.2011.0460 PMID:23075266

Medved, J. P. (2015). *LMS Industry User Research Report*. Retrieved 12 December 2016, from http://www.capterra.com/learning-management-system-software/user-research

Mihalca, R., Uta, A., Andreescu, A., & Intorsureanu, I. (2008). Knowledge management in e-learning systems. *Revista Informatica Economică*, *46*(2), 60–65.

Özen, Z., Bakioğlu, F. Ö. K., & Beden, Ş. (2014). The Examination of User Habits through the Google Analytic Data of Academic Education Platforms. *International Journal of E-Adoption*, *6*(2), 31–45. doi:10.4018/ijea.2014070103

Pakkala, H., Presser, K., & Christensen, T. (2012). Using Google Analytics to measure visitor statistics: The case of food composition websites. *International Journal of Information Management*, *32*(6), 504–512. doi:10.1016/j.ijinfomgt.2012.04.008

Pappas, C. (2015a). *The Top 8 Open Source Learning Management Systems*. Retrieved 12 December 2016, from https://elearningindustry.com/top-open-source-learning-management-systems

Pappas, C. (2015b). *The Top LMS Statistics and Facts for 2015 You Need to Know*. Retrieved 12 December 2016, from https://elearningindustry.com/top-lms-statistics-and-facts-for-2015

Pappas, C. (2016). *8 Types of eLearning Analytics to Obtain from Google and Your LMS*. Retrieved 3 October 2016, from https://elearningindustry.com/types-elearning-analytics-obtain-google-lms

Plaza, B. (2009). Monitoring web traffic source effectiveness with Google Analytics: An experiment with time series. *Aslib Proceedings*, *61*(5), 474–482. doi:10.1108/00012530910989625

Plaza, B. (2011). Google Analytics for measuring website performance. *Tourism Management*, *32*(3), 477–481. doi:10.1016/j.tourman.2010.03.015

Rodriguez, C. O. (2012). MOOCs and the AI-Stanford like Courses: Two Successful and Distinct Course Formats for Massive Open Online Courses. *European Journal of Open, Distance and E-Learning, 15*(2). Retrieved from http://www.eurodl.org/index.php?p=archives&year=2013&halfyear=2&article=516

Rouse, M. (2005). *What is Web analytics? - Definition from WhatIs.com*. Retrieved 10 August 2017, from http://searchcrm.techtarget.com/definition/Web-analytics

StatCounter. (2016). *Mobile and tablet internet usage exceeds desktop for first time worldwide | StatCounter Global Stats*. Retrieved 14 December 2016, from http://gs.statcounter.com/press/mobile-and-tablet-internet-usage-exceeds-desktop-for-first-time-worldwide

Statista. (2016a). *Global smartphone shipments market forecast 2010-2020 | Statistic*. Retrieved 14 December 2016, from https://www.statista.com/statistics/263441/global-smartphone-shipments-forecast/

Statista. (2016b). *Laptop, PC, tablet sales 2010-2020 | Statistic*. Retrieved 14 December 2016, from https://www.statista.com/statistics/272595/global-shipments-forecast-for-tablets-laptops-and-desktop-pcs/

Tîrziu, A.-M., & Vrabie, C. (2015). Education 2.0: E-Learning Methods. *Procedia: Social and Behavioral Sciences*, *186*, 376–380. doi:10.1016/j.sbspro.2015.04.213

Tutorials Point. (2015). *Web Analytics*. Retrieved from https://www.tutorialspoint.com/web_analytics/web_analytics_tutorial.pdf

Waisberg, D., & Kaushik, A. (2009). Web Analytics 2.0: Empowering customer centricity. *The Original Search Engine Marketing Journal*, *2*(1), 5–11.

Watson, W. R., & Watson, S. L. (2007). What are learning management systems, what are they not, and what should they become. *TechTrends*, *51*(2), 29.

Web Analytics Association. (2008). *Web Analytics Definitions*. Retrieved from https://www.digitalanalyticsassociation.org/Files/PDF_standards/WebAnalyticsDefinitions.pdf

Wheeler, S. (2012). e-Learning and Digital Learning. In Encyclopedia of the Sciences of Learning (pp. 1109–1111). Boston, MA: Springer US.

Wohlers, S. D., & Jamieson, J. (2014). "What in me is Dark, Illumine": developing a semantic URL learning analytics solution for Moodle. In B. Hegarty, J. McDonald, & S. K. Loke (Eds.), Rhetoric and reality: Critical perspectives on educational technology – Conference proceedings (pp. 110–118). Academic Press. Retrieved from http://ascilite.org/conferences/dunedin2014/files/fullpapers/105-Wohlers.pdf

xyleme.com. (2014). *Taking a Google Analytics Approach to Measuring Learning*. Retrieved 3 October 2016, from http://resources.xyleme.com/h/i/63857530-taking-a-google-analytics-approach-to-measuring-learning

Yamamoto, G. T., Özen, Z., & Beden, Ş. (2012). *Examination of LMS Logs in Academic Education Platforms: A Case Study*. Presented at the 4th International Future-Learning Conference on Innovations in Learning for the Future 2012: e-Learning, İstanbul, Turkey.

Yılmaz, Y. (2012a). Knowledge management in e-learning practices. *TOJET: The Turkish Online Journal of Educational Technology, 11*(2).

Yılmaz, Y. (2012b). Transition to knowledge society in Turkey: Current state and future perspectives. *Turkish Studies*, *13*(3), 509–522. doi:10.1080/14683849.2012.717439

KEY TERMS AND DEFINITIONS

Bounce Rate: Rate of the single-page views of the web site, which is interpreted as a negative thing since it indicates that the users leave the web site quickly.

E-Learning: Integration of the web-based ICT technologies with the learning processes. It is possible to implement e-learning processes in both academic and commercial institutions.

Google Analytics: Web analytics tools to analyse the usage habits of the users.

Learning Management System: A learning management system (LMS) is a system that provides its users an online learning environment.

Unique Visitors: Displaying a web page by different visitors.

Visit: Displaying a web page by visitors.

Web Analytics: Analysis of the web sites in order to understand the traffic, user behaviors, and to optimize. Different web analysis tools enable researchers to make such analyses.

Chapter 10
Value of Adaptation of Methodologies Between Different Knowledge Areas:
As Applied in the Context of Project-Based Learning

Eduardo Manchado-Pérez
University of Zaragoza, Spain

Ignacio López-Forniés
University of Zaragoza, Spain

Luis Berges-Muro
University of Zaragoza, Spain

ABSTRACT

Project-based learning (PBL) is a powerful tool for teaching that helps students to get the best in terms of ratio effort/learning outcomes, especially in studies with a very practical basis, such as university degree studies in engineering. A way of getting even more out of this is by means of the adaptation of methodologies from different knowledge areas, because this allows the launch of innovative ways of working with certain guarantees of success from the very first moment, and at the same time to integrate skills from different fields within a shared context. Furthermore, it helps to put into practice some transversal competences, which are very useful for future professionals. The chapter also includes some case studies on the successful adaptation of different methodologies coming from different fields such as graphic design, biology, and social sciences in the context of a university engineering degree in industrial design and product development.

DOI: 10.4018/978-1-5225-5279-6.ch010

INTRODUCTION

Since the launch of the European Higher Education Area (EHEA), Project Based Learning (PBL) methodologies have experienced a noticeable development in the context of university studies, increasingly replacing the traditional master class.

A more extended use of learning methodologies based on the principles of "learning by doing", or "learning through experience", together with an increase in scientific papers and publications dedicated to communicating the obtained results, (since Anzai & Simon, 1979; or Barron *et al,* 1998) is helping to expand the realization of different and innovative learning experiences. This not only contributes to the showcasing of these techniques, but at the same time expands their potential. Moreover, the results of the students in terms of the ratio effort/learning outcomes are steadily improving (Manchado *et al,* 2015). Consequently, the number of teachers who are enthusiastic about teaching innovation based on PBL is growing, and new experiences are proposed and developed every day.

Nevertheless, sometimes the degree of student success implied in the first-time application of a new methodology is affected by the need for adjustments, required to make some techniques fully develop their potential. Meanwhile, sometimes, only after several editions do these prove to be truly useful. For this reason, it seems convenient to show experiences where the development time has been reduced, and the speed of implementation increased, to sooner obtain the expected, good results.

These PBL experiences are particularly well exemplified when a series of teaching activities is organized via a modular coordination of selected subjects from different knowledge areas yet within the same degree. This modular approach can help the student to achieve a number of global learning objectives and competencies, through the acquisition of partial objectives for each subject in an overall context, allowing improved academic outcomes and reducing abandonment of subjects (Manchado & López, 2012).

By working with modules of subjects, the aim is to ensure learning outcomes, complementing other subjects without overlapping content, to make sense of all subjects within the same degree (regardless of their particular area of expertise), and to optimize the use of implied resources and efforts of the student. The achievement of these objectives is through a series of activities planned by the team of teachers involved in the module, in which the best practices are introduced. Some of these practices include continuous and group assessment, normally basing the learning experience on the development of a common project involving all areas of knowledge, (which at the same time prepares the student for the reality of professional work).

This strategy has been broadly developed in the context of the Engineering Degree in Industrial Design and Product Development, providing not only good results for students, but also other important achievements. One of these achievements has been the opening of new and active communication pathways between teachers from very different fields of expertise, which finally improve the overall framework and help to create goodwill among all members of the university community involved in the degree. Thus, one of the most noticeable benefits is a high level of interchange of knowledge between teachers of different areas, and the discussion of the potential interest of certain methodologies and theory principles when applied not only to a precise, expected learning environment, but also to different professional fields or disciplines (Manchado & López, 2012).

In this respect, it makes sense in this context to think more deeply about the general interest in - and potential of - adapting professional methodologies used as learning resources in some knowledge area, to some other areas where (regardless of their degree of theoretical affinity), some kind of analogies could be found. In fact, the identification of analogies between apparently unrelated elements is a powerful tool which has been used broadly across different fields for decades to generate creative and highly feasible solutions (Gordon, 1963).

More specifically, the objectives of this chapter are to show how by adapting methodologies which are commonly utilized in a different field of knowledge, and applying them to a specific subject, an opportunity is created for launching learning processes that are highly innovative and alternative to this subject. These processes are, however, based on the knowledge, experience, scientific reports and guidance of colleagues from the disciplines of origin. This helps to reduce development and adjustment terms and allows more certain guarantees of success from the very first moment of work with students.

Background

Learning by doing proves to be a powerful tool for reaching teaching objectives, especially in very practical fields such as engineering and design. Project Based Learning is natural to design schools and implies some difficulties but also some big opportunities. It can help students to obtain different skills in a very effective way, by the integration of some knowledge traditionally linked to different separate areas into an interdisciplinary environment. The main learning result is that students will not only acquire some specific skills or professional methods, but will also understand them and be able to adapt them to specific real situations, integrating knowledge from different areas of expertise in a context where keeping the continuous overall view of the projects (involving technology, ergonomics, or human behavior matters) is especially required. The theory bases are more valuable when the students understand their applicability to real cases.

Despite the previous know-how that can be considered, the way in which the practices are guided in these subjects and projects is something that can be reviewed and must be continuously revised and improved, as new literature appears every day. Not only because some new learning methods can be introduced, but also because professional environments and the working procedures in real life change and evolve very quickly. It is therefore a duty to reflect on this fact especially in the university context. Besides, universities must play their research role as a service to society as a whole, being responsible for producing and testing innovative working methods to be later used by industry and manufacturing companies.

This chapter describes how different tools for product design were obtained by the adaptation of knowledge and techniques coming from different areas such as graphic design, biology, and social sciences. The original techniques were chosen because their value was broadly tested on their respective fields, and because these proved some relevant analogies to product design.

The resulting methodologies are easy to understand and apply, and proved useful in driving the creativity of product design students towards the generation of innovative, functional, user-oriented new concepts of products. The testing of these methods by their application to several projects proved useful also to industry, and this is another remarkable achievement.

As previously stated, the main input of this chapter to the general development of PBL experiences is to consider the value that can be found in the adaptation of techniques between different fields of expertise. Thus, the generic formulation of the approach would mainly comprise:

- The selection of an existing method. This needs to be widely verified on a prior area of knowledge and different from that of its destination.
- The collection and identification of analogies, own characteristics and differences between both fields.
- The generation of new techniques by means of adapting original ones to their new context.

Thanks to the existing experience, bibliography and know-how, the development, adjustment and optimization of the new tools can be completed in a shorter term, obtaining good results sooner. This general proposal can be interesting to several cases of PBL in any knowledge field. Nevertheless, the cases shown in this chapter, from a product design area, are the result of specific experiences that might not be directly applied to any other field; the aim is merely to illustrate some successful cases that could be inspiring to other teachers and researchers in the development of their own experiences.

ADAPTION OF METHODOLOGIES FROM DIFFERENT AREAS TO PRODUCT DESIGN ENGINEERING

The main purpose of the university Engineering Degrees in Industrial and Product Design is to generate professionals who are able to understand and improve the process of designing a product, in its technical, functional, communicative, social and cultural aspects. They should be able to attend to the needs of manufacturing companies, markets, users and society as a whole. Thus, the main skills to be practiced during their learning period include scientific and humanistic knowledge, combined with the growth of their creative aptitudes and other strengths such as team working or project management. All of these are required for a correct performance as professionals whose main duty is to propose objects with a potential for acceptation in the market.

The success of a product within the market depends on the balance between a series of closely related factors. Considered as a whole, these factors can be broadly defined as follows:

- The efficiency of the product as an artifact; this depends on its practical functionality and the appropriate definition of its properties, materials, processes, etc.
- The efficiency of the product in terms of its communicative capacity, to successfully transmit to a potential user its functional characteristics and use, but also its values.
- The efficiency of the product in terms of its communicative capacity, to successfully be able to get users to identify themselves with the product, even as a reflection of their own personalities or ethical and moral values, more than with other products on the market.
- The efficiency of the product to be recognized by the user as a characteristic element within a social and cultural context.

Adaption of Techniques From Graphic Design to Product Design

The development of a product clearly differs from the conception of an artistic object. The latter is a result of free expressiveness without practical functionality and for that reason will not be considered in the same way as a product, where expressiveness is closely related to function (Stangos, 1987). Meanwhile, a technological artifact lacking in the relation between its functional and communicative capacity, whose form could not be appropriately interpreted by its user, would not be considered a well-designed product because the user would not know how to use it to solve a specific need. And it would have no chance in the market since the user would not recognize it, would not consider it and obviously would not choose it (Norman, 2005).

For the above reasons, to develop a product with guarantees of success it is necessary to have the means to control the way in which a technological, functional device is able to communicate abstract concepts of different natures such as brand, overall quality, technology, cultural matters or social status. The communicative capacity of products, thus considered, is of great relevance and should be emphasized in product specification sheets prior to technical development.

Even if the close relationship between form and function is broadly accepted, aesthetics and communication development skills have been commonly studied too separately from technical knowledge at schools of design. Thus, the majority of product specifications traditionally use definitions oriented to strictly functional or practical aspects (Munari, 1983), or to project management (Viladás, 2008).

However, more recently some new methodologies have appeared that give greater relevance to the emotional relationship between product and user; among them, variations of Kansei Engineering, which applies mainly to aesthetic design of products (Helander & Tham, 2003), (Khalid, 2006). KE helps efficiently to define some aspects of the communicative capacity of the product, but:

- It is very limited as used mainly to the design of the aesthetics of the product.
- It implies the use of difficult mathematical methods even if its conclusions are qualitative, and is therefore not attractive to many designers.
- It is based on results obtained from the opinions of users only on existing products.

For these reasons it seemed convenient to develop a new methodology for product design, suitable for application in a teaching context. In trying to obtain certain guarantees of success from the very beginning, it was decided not to develop a brand new methodology, but instead to adapt the technique of working with Systems of Layouts, as used in the Graphic Design professional and teaching fields.

Graphic design uses techniques that focus on the development of the communicative capacity of the graphic product, whether it be a book, a newspaper, a poster or magazine. This quality is effortlessly expressed in definitions with technical characteristics that relate the practical functionality (legibility, printing requirements, logistics, environmental factors, shipment requirements, etc.) to the control of the communication act by means of so-called "layouts" (Swan, 1990). These may be defined as "the previous exposition of specifications for the composition of pages and texts that condition their definitive characteristics" (Ambrose, 2005).

This definition of layout simultaneously develops the communicative and functional aspects of the graphic design, directly relating these to materials, processes and means, and affecting factors such as costs, price, logistics, etc., in a process analogue to that required in product design.

A concept generation technique in graphic design is to use a matrix in order to relate messages and communicate with the available layouts, applying common creative techniques such as brainstorming, among others to provide partial conceptual solutions; different combinations configure possible global and controlled overall solutions to the communicative problem, implying aesthetic, technical and functional aspects, as in Table 1.

Moles (1972) was one of the first specialists in the study of the formal aspect of products using the theory of communication, writing about the interpretation of products as communicative signs, framed within a culture, that possesses certain features and values, and that transmits information which is intelligible to a particular society, being valued mainly for its form.

When users are asked to assess a product, they tend to use abstract values that describe the character of the product, and often talk more about the sensations they experience when using the product than about its functional characteristics (Shi & Xie, 2010). Products are more than tools with a practical aim. They influence, and are influenced by, the perception which users have of themselves, of others, and of their surroundings.

Thus, during the development of a product, it is fundamental to identify: first, what characteristics are important for the user and how these can be expressed; and secondly, how the product is able to "explain" these characteristics to the possible buyer in such a way that he or she can perceive these characteristics from precise abstract formal codes.

Table 1. Concept generation (grey, dark grey) from groups of potential partial solutions (Adapted from DYNA; Año 80, Nº 181. pp 16-24)

Project: Design of labelling for wine brand						
	Applicable layouts in order of relevance					
Messages	**Illustrations**	**Typographies**	**Colors**	**Format**	**Printing**	**Finishing**
Origin Denomination	Map		Violet			
Quality	Landscape with vineyards	Bodonis	Golden	Elongated, vertical	Serigraph on laid paper	Stampings, metal
Intense flavour	Splash	Decorative	High saturation, dark colors	Square, small size	Offset on thick paper, mate, textured	
Young		Helvetic		Non ortogonal	Offset on *couche* paper	Plastified
Innovative			Cold colours	Non ortogonal		Plastified
Unique		Handscript	Cold colours	Non ortogonal		Limited edition, hand-crafted

Possible concepts
Concept 1
Concept 2

The definition of all aspects of the product (technical features, selection of materials, geometry, etc.) should facilitate to the maximum the perception of those sensations by the user. So the definition of the product can be understood as the definition of an act of communication in which all elements take part, transmitting to the user the messages needed in order to make a purchasing decision, as in graphic design.

These messages not only include those that can influence the purchase, but also those that reassure the buyer that the decision to buy was the correct one and that help him/her to get the most out of the product. It is common to hear users saying that they did not use certain functions because *they did not know how to* or even because *they did not know that the product included them* (Norman). These situations generate user dissatisfaction with the product. Thus, the product must transmit information about abstract meanings such as its own identity (what it is, what it is for), of its manufacturer, its quality, or its historical value, but also practical information about the way in which it should be used or handled.

Physical structure of the product must enable it to function satisfactorily and to transmit messages that support or reinforce the development of its function, together with a series of abstract meanings. The development of the product must take into account that these factors are closely related and equally relevant.

The language the product can use to communicate is not only its physical form; its performance, its sounds, and all its combined elements contribute to its understanding on the part of the user. By a process of recognition, the user can make use of the functionality of the product and place a value on it. I.e., the user who drives an automobile perceives information by means of multiple signals coming from multiple sources, whether visual, tactile, olfactory, or sonorous stimuli.

Consequently, a product can be considered not only as a functional structure, but also as a collection of signs or expressive signals in whose interpretation the environment or the socio-cultural moment in time plays a fundamental role, and without whose, its instrumental capacity as a tool would be impossible to exploit. It is thus possible to focus on the definition of the different aspects that make up the product, including the technical aspects, from an appraisal of its communicative potential as its most relevant feature, as in graphic design.

In order to define the elements of a product as communicative elements, these must be considered as potential emitters whose configuration produces expressions. The receiver of the message is the user. For the user to perceive the signal, the codes and channels must be appropriate for the message to be intelligible. It is essential to know what kind of message each element must transmit or contribute to, considering the surroundings in which this communication takes place and the possible presence of noise.

From these principles, a method is proposed to define the product as a series of parts of a communicative process which is organized, coordinated, structured, and oriented to achieving user satisfaction.

A definition is sought for each of the product's characteristics and components, not only so that its production is possible and that it functions correctly, but also so that appropriate expressive values are provided for each case. The estimation of all the components as potential communicative elements can facilitate the development of the product, thinking not solely in technical terms, but also incorporating into the development of its structure some criteria oriented towards obtaining the maximum return from its perception by the user.

In order to satisfy these objectives, a general work procedure is proposed that will be adapted to each particular project (Manchado & Berges, 2013).

- **Phase One:** Definition of user expectations. By means of different techniques, the expectations of the user are identified and prioritized. In communicative terms, the aim is to identify the message expected by the user, or the message that it is of interest to transmit to the user.
- **Phase Two:** Organization and classification of the parts of the message. Possibly the content of the message will be complex, and will consist of more abstract characteristics which pertain to the character of the product and other more practical characteristics related, for example, to learning how to use it.
- **Phase Three:** Identification of the different aspects of the product subject to intervention, and the possible degree of intervention. Each one of the aspects which it is possible to control is denominated a product layout. The definition process of a product layout concentrates on determining the elements that compose the layout, and their contribution to the communicative structure as a whole.

The key to innovation thus lies in interpreting the structure of the product, with the greatest possible creativity. Not only must the messages be transmitted appropriately, but also other new messages must be found and/or new forms to express them.

This task of product configuration, equivalent to the work of compositional layout in graphic design, must be sufficiently limited to ensure the attainment of the predefined objective but at the same time sufficiently open in order to leave space for the development of creative solutions. The user must also enjoy the maximum functionality of the product with minimum effort, thanks to its communicative capacity. An accurate "compositional layout" can help the user to use a product of high technological complexity very easily.

- **Phase Four:** Distribution, using a matrix, of the messages among the Layouts of product (Table 2).

This simple matrix comprises the following: The first column lists the messages intended for transmission, and the first row contains the elements that make up each of the layouts.

Next, each cell is completed with the possible configuration of each of these elements with the aim of transmitting each of the messages. Sometimes the relation is obvious; other times cultural icon analysis can be applied. Creative techniques such as brainstorming are also to be used.

Each cell includes information referring only to the way in which an element can be configured to explain one of the messages, so that the relation between both can be clearly seen. The aim is to divide a complex problem into small problems whose resolutions will be easier, in order to later integrate the set of partial solutions. Working with matrices or tables as a method for the decomposition of problems is a system commonly used in industrial design (QFD, DAFO). The combination of the partial solutions, following brand, market or production strategies, will define different conceptual options. The greater the number of cells completed, the greater the number and potential of the possible options, even if specialists in different fields might be necessary in order to complete the cells appropriately (Li, 2011).

Table 2. Matrix Messages/Expressive resources. (Adapted from DYNA; Año 80, Nº 181. pp 16-24)

	LAYOUT NAME Available resources:		
MESSAGES	**Resource 1**	**Resource 2**	**...**
Message 1	Possible configuration		
Message 2			
...			

- **Phase Five:** Assessment of the possible concepts of product for subsequent development. Suitability of the possible concepts is strategically assessed, considering their potential functionality and communicative capacity. Differing from Kansei Engineering, the proposed technique aims not for a reduction in the number of conceptual options, but for the generation of a maximum number of suitable possibilities for their subsequent valuation.
- **Phase Six:** Selection of the most suitable options and initiation of the subsequent product development phases.

The more suitable options serve as a departure point for writing detailed design specification sheets that will govern the subsequent development, technical definition, and production phases.

Among other projects, this technique has been used on the thesis work of a team of students completing the University Engineering Degree in Industrial Design and Product Development, in order to produce and design a series of product concepts for ecological motorcycles (Abella, Jericó & Martínez-Hansen, 2012). This theoretical case (developed without any relation to the company) was proposed because it related aesthetic, constructive, technological and functional matters, with very strong emotional, cultural and brand values, within a growing and innovative market. Development is taken to a conceptual level, which is valid because the objective of the method is to define general characteristics from which to obtain detailed design specifications.

The identification and classification of messages which could satisfy user expectations, was by means of an online survey amongst 240 members of different motorcycling clubs, who were asked about their reasons for choosing the product they owned, and its features they appreciated most. Results were shown using word clouds and were statistically nested by the number of citations, unifying similar expressions by means of clusters. The value "ecology" was added as a strategic requirement for the new product, never existing before in the brand nor suggested to the users.

Messages to be transmitted were related to the configurations of available resources using a matrix of layouts. The team discussed the strategic market potential of the possible combinations, and proposed some highly innovative rupturing products, which maintained brand key visuals but included the ecological message in the most relevant layouts in order to produce a strong marketing impact.

The concepts developed had been shown to 125 potential users by means of an online survey; users assessed on a scale of 0 to 7 the presence in the concept of the features collected on the message board, camouflaged among others so as not to be perceived; almost 85% perceived more clearly (between 5 and 7) the desired aspects, which suggests the potential for success in such concepts, as well as highlighting their suitability for possible review, and showing that the method is useful.

Regarding the results obtained in terms of the value of product concepts generated using the new technique, the participant students evaluated the experience positively; they considered remarkable the integration of interdisciplinary knowledges, as well as the achievements in terms of coordination, leadership and role assignment within the development team.

Some other good considerations included those related to information management, and about participants' own self-confidence, when using a tested and well known method of working. In their own words, these helped to manage some very subjective matters, such as those related to the aesthetic and communicational capacities of the product. Something that is common on the graphic design field and that has been implemented in the field of product design by means of the adaptation of methodologies from there on.

Adaption of Techniques From Biology to Product Design

As was said before, the success of a product in the market depends on the balance between a series of related factors. One of these factors is to achieve a good balance between its functional capacity, the resources involved in terms of maximum efficiency and costs, and the correct resolution of the needs of the user.

The definition of practical functions, and more specifically aspects such as the operation of the product, as well as considerations about its technology, structure, materials or production, may be based on the study of nature as a model to find potential innovation and sources of creative solutions, as well as alternatives to current knowledge.

On the other hand, it is possible to improve the communicative capacity of the product by introducing features or characteristics of living beings that condition the perception of the product by the user. The development of the product's formal language from the study of nature can be made by means of very basic and essential elements such as shape, color, texture, geometry, etc., or by means of more subtle ones such as the perception of smells and aromas, sounds, touch and visual quality of surfaces, etc. These elements can be used as a whole, sending a complete and well-referenced message from the living being or as a part by selecting a portion of the message and applying the references in the form of details.

A bioinspired design project is, in essence, a design project in which a number of constraints have been introduced related to existing information and knowledge in the field of biology. This information allows to introduce improvements and innovations into different aspects of the product, such as the functional benefits or the aesthetic-formal characteristics. Bio-inspired design is a complementary method that improves the traditional design process, since it adds to rather than replacing it. It can be used as a creative resource thanks to its demonstrated potential in combination with other design methodologies (Viñolas i Marlet, 2005).

These conditions and biological information can appear in the different phases of the design project, from the initial and most creative conceptualization, in which the work is done in a more intuitive and abstract way, to the final stages of product development and visualization, where the characteristics of nature have a greater definition and product concepts appear as replicates of the natural environment.

There are already several methodologies of bioinspired design, some of which set a structure for a unique process, while others present different processes depending on the approach of each project. The latter is an example of adapting the methodology to the requirements of each case. However, it is possible to define in general two bioinspired design processes (Roshko 2010) clearly differentiated:

- **A "Direct" Process:** Starting from a design problem research is carried out on biological knowledge to find a specific solution.
- **A Second "Indirect" Process:** Information specific to the field of biology is used as inspiration applicable to a specific design project.

In both cases the information that is transferred from the field of biology to that of the technique can have a conceptual, functional, structural, material, or performance, application.

Within these two large groups, several examples of application can be found; thus, in the direct method, BioTriz, whose methodology developed by Vincent (2007), combines the knowledge of biology with the TRIZ innovation technique. BioTriz is considered as a systematic use tool for solving technical problems, basing the solutions on the relationships with similar solutions found in nature, in the so-called natural patent base, as described by its creators.

Regarding the indirect method two cases can highlighted: that of Lodato, called BIODESIGN, which is essentially an observation of how the bionic application establishes the relationship between nature and the design project; and the one from Benyus (2012) at the Institute of Biomimicry, which promotes biomimetic learning by emulating natural forms, processes and ecosystems to create healthier and more sustainable designs and technologies.

However, in the bibliographic review it is observed that, in most cases, authors describe methodologies with both approaches combined and that it is the opportunity for innovation or the need for resolution of design problems which determines which method is chosen. This has been the case in the methodology developed at the University of Freiburg by Milwich (2006), where the basic biological research in biomechanics and functional morphology is made available to the technology for

TOP-DOWN processing. In other cases, an alternative strategy is followed in order to search for possible solutions in the biological model which will solve specific technical problems - BOTTOM-UP.

In a similar way, Helms (2009) poses two processes, problem-driven and solution-based processes of BID Bio Inspired Design, whose application has been raised in the learning of young designers as an alternative method of design. There are even methods that hybridize both approaches, such as the case of Di Bartolo (2000), whose experience in and contributions to the study of the methods and applications cases experienced in the Center for Research of Natural Structures constitute a valuable first step to understanding the various ways of accessing bionics in relation to the project.

According to Emmeche (1991),

...biosemiotics is a growing field that studies the production, action and interpretation of signs, such as sounds, objects, smells, movements, but also signs on molecular scales in an attempt to integrate the findings of biology and semiotics to form a new vision of life and meaning as immanent traits of the natural world.

This definition focuses on the physical part of the biology of recognition, memory, categorization, mimicry, learning and communication for application in semiotic tools such as interpretation, semiosis, sign types and meaning.

From the perspective of design it is interesting to adopt this vision of the physical part in order to investigate biological principles and to be able to transform and apply them in the design project as elements of perception and product communication. Although three basic elements of design such as shape, color and texture have been defined as the main candidates for observation and application in design projects, there are others of great interest when product communication aims to configure a precise and particular message. So, in this sense, can be cited acoustic signals, vibrations, movements, chemical (odor molecules), light, visual (beyond shape, color and texture), geometric patterns, etc.

The proposed method pursues the search for a series of functional improvements which can be applied to the product. These in turn are based on the observation of biological principles. As in the case of layout systems from graphic design, a combinatorial system of the chosen functions is used to enhance the creative process and to obtain a wider range of conceptual proposals. To achieve this improvement and functional innovation, research in the field of biology is structured, (again using the creative method of analogy) to make the translation from the natural to the technological field, so that a new field of study allows for the creation of multiple new solutions.

It is difficult to obtain valuable information from the chosen biological references, and this is essential since the quality of the information obtained determines the result of its application. A solution could be the interdisciplinary work between biologists and designers (López-Forniés & Berges, 2012). However, the development of a transversal research capacity, together with the study of information in a new area of knowledge, is more effective.

On the other hand, it has to be considered that it is possible to make a functional improvement based on a more effective representation of messages addressed to the user through the development of the formal and communicative capabilities of the product. This implies that the methodology must integrate both approaches: the work with analysis of functions and with the characterization through signs coming from the natural scope.

In order to achieve this adaptation of methods to the design process for the functional and communicative improvement of the product, the following phases are proposed:

- **Phase One:** Establishing the framework, the design space. Firstly, the analysis of the problem is carried out, determining the opportunities for functional improvement which will form the basis of the work. The functions are studied; define the design requirements and the expectations regarding the final solution, thus establishing a criterion for evaluating the design.

Given that, the combination of functions for the generation of concepts is represented by a morphological table (ZWICKY 1969), where a matrix is established in which the functions to be studied are positioned in rows and the different living beings, possible solution candidates, are positioned in columns. This initially empty morphological table defines the framework, so that for each function can be found in nature different individuals capable of developing the same function in alternative ways. For example, the function "structure folding" may be solved in various ways in nature, by plants and animals, so that the analysis of the living beings allows the table to be completed to subsequently make the combination.

- **Phase Two:** State of the art. A state of the art study is carried out in both the technological and natural scope. In the technological area it is intended to have a clear idea of the existing solutions for each of the functions, so that the potential for innovation is determined. If the intended function is already solved in multiple ways it will be difficult to generate innovation, whereas if the function has not yet been developed or only in alternative ways, there is potential for improvement. In the biological field a search for information from the possible candidates is done. Initially, through basic information, the objective is to understand the mechanism or biological principle that fulfills the desired function. Subsequently, in the case of identifying a candidate or model capable of solving the function, a more complete and exhaustive search will be carried out.

Once the functions and living beings are selected, the matrix is filled out, as can be seen in Figure 1, ready as it is to establish the different combinations.

- **Phase Three:** Creative and combinatorial process. Morphological analysis is a technique belonging to random or combinatorial creative methods (Marín 1991), similar to the "discovery matrix"; it consists of a double entry table

Figure 1. Matrix of combinations, morphological table of functions / living beings

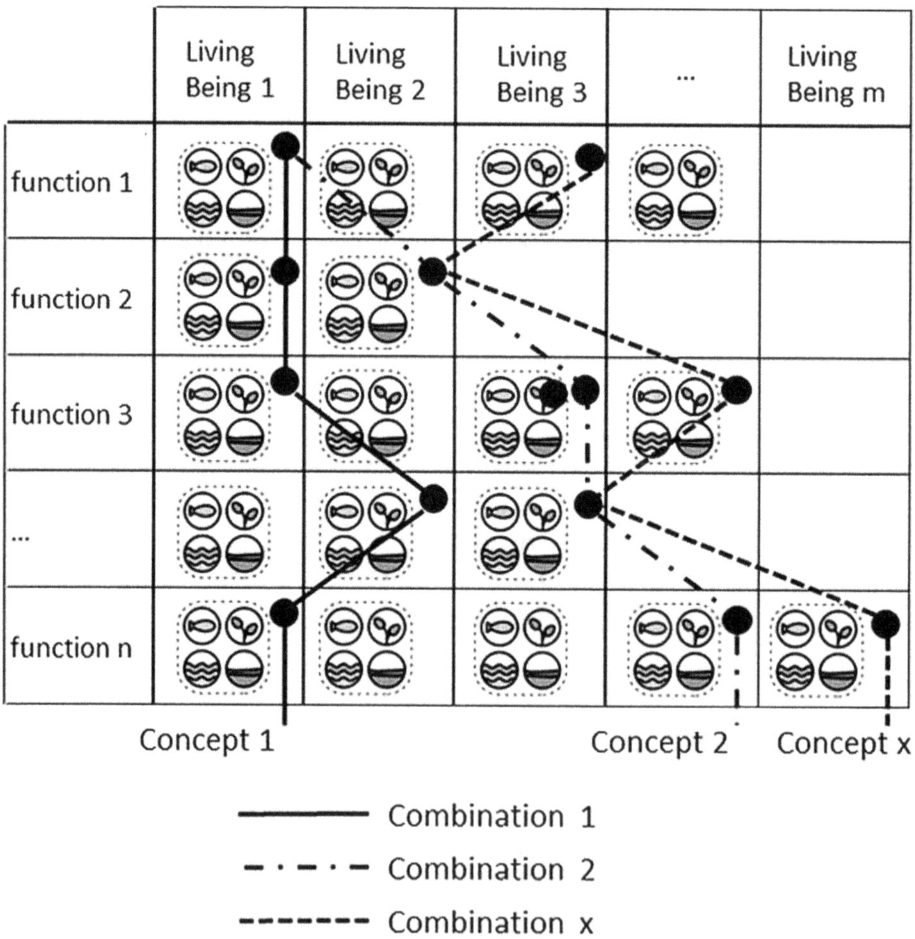

in which various elements that have been selected by a particular feature and the potential solutions to be applied are crossed. Within the table connection lines can drown between different solutions that integrate each of the characteristics, forming conceptual combinations, as can be seen in figure 3. These organized meetings will allow to reveal the innovation opportunities, that is to say, the little gaps in which the original solutions are more likely to arise. The matrix is organized to give rise to the combination of elements susceptible to be associated and the work with every element is systematic, since the defined crosses are explored one by one. With the most suitable combinations, a concept is chosen that will be developed in more detail; it may be necessary to expand the information of each individual to achieve a more valuable translation from the natural to the technological.

- **Phase Four:** Biological research. In order to obtain more detailed and quality information a documentary research exercise is carried out, with which to acquire enough knowledge to later apply to the design of an object. An analysis and description of the information is carried out in order to define the features that characterize the principle or foundation of nature and how they would apply to the product. From this point these features or principles are related by means of mechanical, physical or chemical characteristics, which are described and outlined. Using this information and the knowledge acquired, the design of the product is proposed at the conceptual level.
- **Phase Five:** Conceptual design, product definition and visualization. In this last phase of conceptualization the different partial solutions are integrated with all the information of the individuals, and a product representation is generated that shows the final solution with the functional characteristics to be incorporated. At this moment, semantic aspects are also integrated to strengthen the communication of the product; as previously indicated, the formal language itself observed in nature can be applied to the product.

In an example of methodological application, the following process can be observed, initially analyzing problems related to the world of recycling, domestic waste and the impact on the environment. Then a series of functions which could improve the management of domestic waste were defined, such as the filtration of liquids, or compaction or storage of waste. The taxonomy map of functions, from the biomimicry institute (Ask Nature database, 2017), is used to identify functions related to: filtration, indication, detection, compression, and conversion of chemical energy, protection against biotics and non-biotics, biological control of populations, pests or diseases, erosion and sediment control, modification and conversion of energy, feeling signals, distribution, ejection and hermeticity.

In this way, in the 1st phase, the functions of compressing, containing and filtering liquids are chosen, with a focus on the concept of domestic waste. For this concept, a series of ideas intuitively appears which could provide some solution; at this moment it is necessary to define which individuals or living beings can give solutions to these functions, starting the 2nd phase; the matrix is finally completed in order to perform the morphological analysis, as can be seen in Table 3.

In the 3rd phase a series of combinations are established and the one that involves the bromelia, the honey pot ant and the chilopoda is chosen. With these individuals, a more detailed search phase is started, looking for specific knowledge of the characteristics that make them suitable for the selected functions. For each of them the biological principle is analyzed and it is established how the translation could be applied to a technological principle, as observed in Table 4. (In this example the information of the biological scope has been extracted again from The Ask Nature database).

Finally, in the 5th phase, the conceptual definition is made, and it specifically details its application to the development of a container for plastic waste from domestic waste, intended for an environmentally responsible user. It is observed that a large number of users use shopping bags to store waste, so the design is directed towards the creation of a reusable shopping bag that adapts to what it contains. The container collects possible residues of liquids that have remained in the containers thanks to the shape of the bag and the funnel at the bottom. In Figure 2 the bag and the funnel are shown, which are joined and detachable, so that the liquid falls into a reservoir located in the lower area of the container.

Due to its form and use of materials, the container can be compressed, along with the waste inside, and the residues will be reduced to a smaller volume. The inner bag has holes so that no vacuum is created in the compression.

Some formal features are added to complete the message, such as lobes on the filter to stop coarse particles, reinforcing bands in the bag reminiscent of the thickest part of the ant's abdomen, or the zig-zag outline of the body of waste marking the sections of an invertebrate.

Table 3. Matrix of combinations, morphological table of functions / living beings

	Living Being 1	Living Being 2	Living Being 3
To capture, absorb, filter	Epifita	Sabella pavorina	Bromelia
To contain	Camel	Honey pot ant	Boa constrictor
To compress volume	Firefly	Pythonidae	Chilopoda

Table 4. Biological and engineering principles

Living Being	Biological Principle	Engineering Principle
BROMELLAS (Ask Nature 2017)	The leaves of some bromeliads have the capacity to absorb water and nutrients in a storage tank thanks to the shape of their leaves and hydrophobic qualities. Water collection takes place thanks to the convex shape of the leaves.	The concept is a silicon funnel with slits inside. It would be located in the interior and bottom of the container in order to collect liquids or debris that fall when the waste is disposed of. The slits help to direct the water acting as guides. The water collection is placed in a container at the bottom of the product.
HONEY ANTS (Ask Nature 2017)	The abdomen consists of rigid parts joined by a very flexible membrane. As the abdomen fills with nectar the membrane expands to the size of a grape.	The membrane has the capacity to expand depending on the volume of residues in its interior analogous to the membrane of the ant. The rings provide resistance to the bag.
CENTIPEDE / chilopoda	Very flexible structure that can be easily compressed to be introduced in smaller places.	Foldable structure that can be compressed to crush the waste with the arrangement of hoops.

Adaption of Techniques From Sociology to Product Design

The human being lives in communities that are characterized by certain differential characteristics. Moles (1972) indicated that the differential characteristics of a

Figure 2. Components of the domestic bin and use sequence

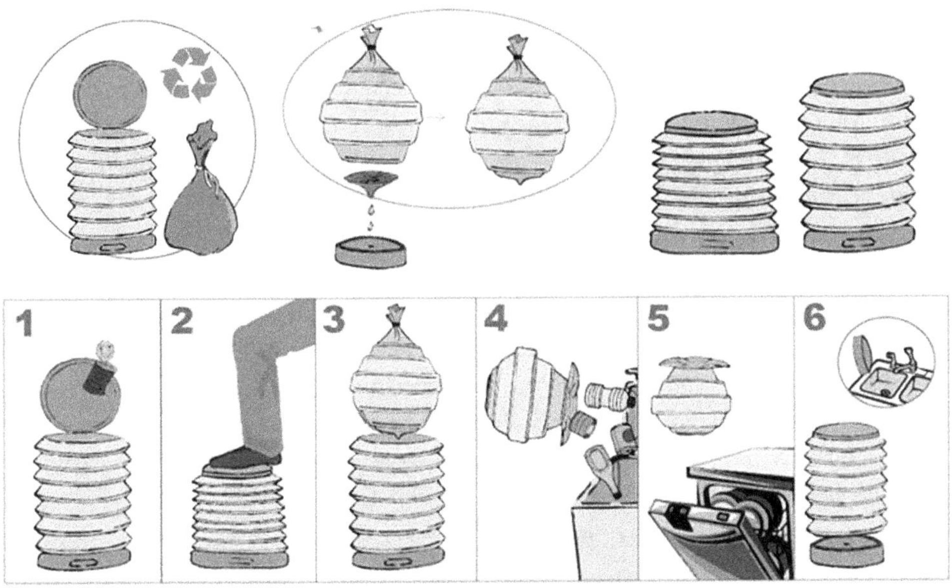

given human community constitute the so called cultural context, in which human communities are developed. This context results from the artificial modification of the natural environment, and is identified by means of a series of languages or sets of signs, a series of rituals and a certain number of objects; these objects are, in turn, the result of the manipulation of the natural elements in each cultural environment and, at the same time, the tool that allows for a continued adaptation of the environment.

Manzini (1991) asserts that it is precisely the artificial factor that is more characteristic of human communities, pointing out the fundamental value of objects as an agent of the social context, and the importance of maintaining an adequate balance between artificial intervention over nature, in order to ensure the survival of human communities while at the same time preserving the sustainability of natural ecosystems.

Later, Manzini (2015) himself refers more broadly to the contribution that industrial design activity can make as an agent within a social context, saying that its main feature is the systematic and methodologically ordered ability to analyze the needs of human populations so as to propose conceptual solutions to these needs, and provide the technical resources to make them feasible. This fact implies that it is essential to know in the first place what these human communities are like, the social environment in which they develop, and what their needs are.

Defining what the social environment consists of is complicated, since it entails multiple interrelated elements. Barnett and Casper (2001) indicate that

Human social environments encompass the immediate physical surroundings, social relationships, and cultural milieus within which defined groups of people function and interact. (…) Components of the social environment include built infrastructure; industrial and occupational structure; labor markets; social and economic processes; wealth; social, human, and health services; power relations; government; race relations; social inequality; cultural practices; the arts; religious institutions and practices; and beliefs about place and community. The social environment subsumes many aspects of the physical environment, given that contemporary landscapes, water resources, and other natural resources have been at least partially configured by human social processes. (…) Embedded within contemporary social environments are historical social and power relations that have become institutionalized over time. Social environments can be experienced at multiple scales, often simultaneously, including households, kin networks, neighborhoods, towns and cities, and regions. (…). Social environments are dynamic and change over time as the result of both internal and external forces. There are relationships of dependency among the social environments of different local areas, because these areas are connected through larger regional, national, and international social and economic processes and power relations.

Consequently, the in-depth study of such a complex medium must be carried out from rigorous and contrasted methodologies. Giddens & Shutton (2013) describe sociology as the scientific study of human life, social groups, and societies as a whole. They add that the field of sociology is extremely broad and ranges from the analysis of encounters between individuals on the street to the investigation of international relations. This proposal is extremely valuable for the field of industrial design, if incorporating only one more element: from the perspective of sociology, the aim of the study is to merely obtain knowledge about the characteristics of the community observed, but from an industrial design perspective, this knowledge is the starting point necessary to identify opportunities for later intervention and solving needs. The difference is, therefore, in providing a somewhat more proactive attitude, but methods of work and research can be adopted almost literally.

Giddens & Shutton also define ethnography as the systematic study, based on field work, of people and cultures through participant observation and various types of interview. It consists of observing the cultural practices of social groups and being able to participate in them, so that it is possible to contrast what people say and do.

This empathic approach is at the heart of the Design Thinking strategy formulated among others by Tim Brown (2010) as well as the User Centered Design strategies; the following paragraph is taken from the website www.ideo.org (2017):

Human-centered design consists of three phases. In the Inspiration Phase you'll learn directly from the people you're designing for as you immerse yourself in their lives and come to deeply understand their needs. In the Ideation Phase you'll make sense of what you learned, identify opportunities for design, and prototype possible solutions. And in the Implementation Phase you'll bring your solution to life, and eventually, to market. And you'll know that your solution will be a success because you've kept the very people you're looking to serve at the heart of the process(...).

Vinyets (2009) indicates that the current market demands a new relationship between producers and consumers, and a more direct connection with people: more empathic, creative, open and collaborative. He adds that organizations must develop new approaches to reach their clients and users, improve their understanding of their motivations and needs, and be able to design products and services that are relevant and that give them real value. It is necessary to promote an innovation based on people, and he proposes the incorporation of tools from the field of anthropology to achieve this goal.

Therefore, the incorporation of some sociological (and more specifically ethnographic and anthropological) tools for the study of product users is gaining importance as a complement to the analyses (such as those based on market trend statistics), which were traditionally considered sufficiently reliable. Furthermore, the active incorporation of users in experiences such as living labs gains relevance as opposed to a traditional approach that considered the user as a mere receiver of the process, incapable of contributing any value during its development.

From this perspective a critical assessment can be made of the contribution that some tools from sociology can make to the field of industrial design:

- **Factual Research:** This consists of the collection of quantitative data (i.e. how many people, for how long, at what time, in what places, what proportion of the population ...) It also involves laborious investigations (unless there is bibliography or access to previous reliable and relevant studies) of an eminently statistical nature, based on the cycle of scientific analysis: Formulation of theory / Design of experiment / Data analysis. Finally, it is also a starting point for comparative analyses (i.e. is it true in more places or moments?), as well as historical (i.e. since when is it so?, how long will it remain so?) and for the formulation of theoretical hypotheses (i.e. probably this is so for this reason).
- **Surveys:** These allow the collection of data from very large groups, although the answers may be very superficial on the part of the participants and not necessarily reflect the reality of the subject studied (since the participant responds from the conscious and in fact his or her behavior can be different from the one he or she self-perceives). On the other hand, open-ended questionnaires are difficult to analyze on a large scale.
- **Field Work (Ethnography):** This seeks to obtain rich and profound information obtained on small groups, which can sometimes be difficult to extrapolate.
- **Focus Groups:** Study of ethnographic format focused on a group of subjects, supported by an empathic in depth interview format, already widely used in the context of marketing.
- **Documentary Research and Obtaining of Graphic Documentation:** Filming, photography or graphic representation, or audio recording, of observable phenomena. The reduction to a unique graphic (and / or sound) representation of a complex event involves focusing on relevant elements, and if so, helps leading the subsequent analysis. It can be very valuable, although it can be easily decontextualized.

- **Comparative Methods:** Identification of analogies and differences between two populations of subjects and analysis of the same, supported by different techniques of data collection.
- **Life Stories:** Comprehensive documentation of specific cases, considered representative of a given population. The PERSONAS method is an adaptation of this tool.

For the effective application of these methods to product design it is necessary to consider that sociology understands that a single source of data hardly constitutes sufficient material for a valid analysis. What is effective therefore is to combine several methods in a triangulation process, to be defined at the beginning of the study, depending on the context to be studied, the existing difficulties, and the available time and resources.

This triangulation can be:

- Of data (collected in different areas)
- Of researchers (comprising different participating teams)
- Theoretical (starting from different approaches and contrasting results) or
- Methodological (different techniques are used).

The traditional validation of data from the sociological approach can also be reinterpreted from the perspective of social reflexivity: the *age of information* raises a high degree of expertise in a multitude of individuals and over a range of subjects, which can contribute to the verification of data, and give greater value to methods such as focus groups (comprised therefore of a panel of experts).

These tools have been successfully applied in experiences such as the so called "*Hybrid Projects*" (Manchado et al, 2015), where interdisciplinary teams composed of students of Engineering in Industrial Design and Product Development and by students of electronic engineering were able to successfully identify specific user needs, and to propose viable product concepts, in the context of a real collaboration with industrial companies. On this occasion, students from different fields (some of whom had no prior experience in user analysis for identifying product opportunities) were able to develop in a coordinated and effective manner different studies of great value for the project, documenting the entire process in a short time.

CONCLUSION

On a general basis, it is concluded that the adaptation of methodologies already used and tested on analogous fields to different knowledge areas is a valuable and useful tool for PBL development. This adaptation can be facilitated by the existence of bibliography, references and previous experiences and know-how on the field of origin. The results can be obtained on shorter terms and with some certain guarantees of success.

The opportunities for drawing on the expertise of colleagues from different fields enriches the interdisciplinary PBL experience, and reinforces the communication pathways and collaboration inside the teaching team, helping to understand the value of their particular contributions and the role of each of its members.

The adaptation of techniques from different areas helps the students to practice some skills such as team working, project and time management, leadership, collaboration and participation in multidisciplinary environments.

In the cases presented, all of them facilitates the integration of knowledge from different areas, focusing to the satisfaction of a global objective by means of the generation of multiple partial but coherent solutions. Their application is easy and does not require the use of complex methods or expensive resources, so it can be attractive to learning environments and to industry experimentation. Moreover, it can help to introduce technical engineering students to more humanistic matters such as emotional and aesthetic design.

REFERENCES

Abella, J., Jericó, S., & Martínez-Hansen, M. (2012). *Validación de una metodología de trabajo para diseño industrial y desarrollo de producto mediante el diseño de 3 versiones de un producto: moto ecológica. Trabajo Fin de Grado. EINA, Graduado en Ingeniería en Diseño Industrial y Desarrollo de Producto. TAZ-TFG-2012-682*. Universidad de Zaragoza.

Ambrose, H. (2005). *Layout (2nd ed.)*. Barcelona, Spain: Parramón.

Anzai, Y., & Simon, H. A. (1979). The theory of learning by doing. *Psychological Review*, *86*(2), 124–140. doi:10.1037/0033-295X.86.2.124 PMID:493441

Barnett, E., & Casper, M. (2001). A Definition of Social Environment. *American Journal of Public Health*, *91*(3), 465. doi:10.2105/AJPH.91.3.465a PMID:11249033

Barron, B. J., Schwartz, D. L., Vye, N. J., Moore, A., Petrosino, A., Zech, L., & Bransford, J. D. (1998). Doing with understanding: Lessons from research on problem-and project-based learning. *Journal of the Learning Sciences*, *7*(3-4), 271–311. doi:10.1080/10508406.1998.9672056

Benyus, J. M. (2002). *Biomimicry: Innovation Inspired by Nature.* New York: Perennial.

Brown, T., & Wyatt, J. (2010). *Design thinking for social innovation.* IDEO. *Development Outreach, 12*(1), 29–31. doi:10.1596/1020-797X_12_1_29

Di Bartolo, C. (2000). Naturaleza Como Modelo, Naturaleza Como Sistema. *Experimenta: Ediciones de Diseño,* (31), 39-45.

Emmeche, C., & Hoffmeyer, J. N. (1991). From language to nature: The semiotic metaphor in biology. *Semiotica, 84*(1/2), 1–42. doi:10.1515/semi.1991.84.1-2.1

Giddens, A., & Sutton, P. W. (2013). *Sociology* (7th ed.). Cambridge, UK: Polity Press.

Gordon, W. (1963). *Sinéctica. El desarrollo de la capacidad creadora.* México DF, Mexico: Editorial Herrero.

Helander, M. G., & Tham, M. P. (2003). Hedonomics – affective human factors design. *Ergonomics, 46*(13-14), 1269–1272. doi:10.1080/00140130310001610810 PMID:14612318

Helms, M., Vattam, S. S., & Goel, A. K. (2009). Biologically Inspired Design: Process and Products. *Design Studies, 30*(5), 606–622. doi:10.1016/j.destud.2009.04.003

Khalid, H. M. (2006). Embracing diversity in user needs for affective design. *Applied Ergonomics, 37*(4), 409–418. doi:10.1016/j.apergo.2006.04.005 PMID:16764817

Li, S. (2011). A matrix-based clustering approach for the decomposition of design problems. *Research in Engineering Design, 22*(4), 263–278. doi:10.1007/s00163-011-0111-z

Lodato, F. (2000). Biónica: La Naturaleza como herramienta de innovación. *Experimenta: Ediciones de Diseño,* (31), 46-51.

López-Forniés, I., & Berges Muro, L. (2012). A Top-down Biomimetic Design Process For Product Concept Generation. *International Journal of Design & Nature and Ecodynamics, 7*(1), 27–48. doi:10.2495/DNE-V7-N1-27-48

Manchado, E., & Berges, L. (2013). Sistemas de retículas: Un método para diseñar nuevos conceptos de producto hacia el usuario. *DYNA. Año, 80*(181), 16–24.

Manchado, E., & López, I. (2012). Coordinación por módulos de asignaturas en el Grado de Ingeniería de Diseño Industrial y Desarrollo de Producto de la UZ. *REDU, 10*(3), 195–207.

Manchado, E.; Romero, C.; Blanco, T.; Casas, R.; López, J.M. (2015): Una experiencia de aprendizaje colaborativo, basada en la adaptación de metodologías de Design Thinking. *La sociedad del aprendizaje. Actas del iii congreso internacional sobre aprendizaje, innovación y competitividad. Cinaic 2015, 1*, 434-439.

Manzini, E. (1991). *Artefacts: vers une nouvelle écologie de l'environnement artificial*. Paris, France: Editions du Centre Pompidou.

Manzini, E., & Coad, R. (2015). *Design, when everybody designs: An introduction to design for social innovation*. MIT Press.

Marín, R., & De la Torre, S. (1991). *Manual de la creatividad*. Barcelona: Vicens Vives.

Milwich, M., Speck, T., Speck, O., Stegmaier, T., & Planck, H. (2006). Biomimetics and Technical Textiles: Solving Engineering Problems with the Help of Nature's Wisdom. *American Journal of Botany, 93*(10), 1455–1465. doi:10.3732/ajb.93.10.1455 PMID:21642093

Moles, A. (1972). *Théorie des objets*. Paris, France: Ed. Universitaires.

Munari, B. (1983). *Cómo nacen los objetos? Apuntes para una metodología proyectual*. Barcelona, Spain: Gustavo Gili.

Norman, D. A. (2005). *El diseño emocional; por qué nos gustan (o no) los objetos cotidianos*. Barcelona, Spain: Paidós Ibérica.

Roshko, T. (2010). The Pedagogy of Bio-Design: Methodology Development. *WIT Transactions on Ecology and the Environment., 138*, 545–558. doi:10.2495/DN100491

Shi, L., & Xie, Z. (2010). Designing product to improve affective communication. *Intelligent Systems and Applications (ISA), 2010 2nd International Workshop on*, 1-4.

Stangos, N. (1987). Conceptos de arte moderno. Madrid, Spain: Ed.Alianza, Col. Alianza y Forma nº 53.

Swan, A. (1990). *Cómo diseñar retículas*. México DF, Mexico: Gustavo Gili.

Viladás, X. (2008). *Diseño rentable: diez temas a debate*. Barcelona, Spain: Index Book.

Vincent, J. (2007). *Biomimetics: Strategies for Product Design Inspired by Nature - A Mission to the Netherlands and Germany. Report of A DTI Global Watch Mission*. Pera.

Viñolas i Marlet, J. (2005). *Diseño ecológico: Hacia un diseño y una producción en armonía con la naturaleza*. Barcelona, Spain: Blume.

Vinyets, J. (2009). Showing the Value of Ethnography in Business. *Ethnographic Praxis in Industry Conference Proceedings, (1),* 162-169.

Zwicky, F. (1969). *Discovery, invention, research through the morphological analysis*. New York: McMillan.

KEY TERMS AND DEFINITIONS

Bio-inspired Design: It is a design method that allows finding solutions in nature that can be applied to design in a creative way; inspiration can be introduced into the different phases throughout the design process.

Biosemiotic: It is an emerging field of study that can help researchers to find new ways of communication applicable to many areas of science; in design, it allows the improvement of communication and product interaction mimicking the features of the natural world.

Coordination by Modules of Subjects: It is a manner of relating the subjects within a learning process, where these are coordinated not only vertically (with subjects of the same knowledge area, on a chronologic basis), but also horizontally (with subjects of different knowledge areas that are offered to students at the same time).

Engineering Principle: It is the justification and demonstration of the technical principle obtained from the transformation of the biological principle; it is a reasoned and useful technical explanation that takes as base of understanding a foundation of the nature.

Hybrid Projects: It is a learning experience based on the principles of learning by doing, where students and teachers from different origins (i.e., different university degrees) do team work on a project with shared objectives.

Morphological Analysis: It is a technique that belongs to the group of combinatorial methods in creativity. Combination is a key element in creativity, and methods using combination set the potential of high numbers of possible solutions.

Social Reflexivity: It is a phenomenon caused by the great increase of access to information that modern technologies implies; some users become super informed, expert users, able to do very critical reflections on many topics.

System of Layouts: It is an organized, hierarchical, collection of elements suitable to be configured in order to provide an object of design with a specific identity or features.

User-Centered Design: It is a design strategy that places the study of the user's needs at the core of the product development process, very often incorporating real user experiences through different specific techniques.

Chapter 11
Students Skills in a Blended Learning Environment:
A Qualitative Research

Ana Mª Pinto-Llorente
University of Salamanca, Spain

Mª Cruz Sánchez-Gómez
University of Salamanca, Spain

Francisco José García-Peñalvo
University of Salamanca, Spain

ABSTRACT

The objective of this chapter is to determine participants' perceptions about the characteristics that students must have to be successful in a blended learning environment. The authors adopted a qualitative approach in which 91 learners participated. Data collection was carried out through guided interviews over three months. The results of the research emphasized the need to have a good level of English and new digital skills as essential requirements to study an English degree in a blended learning modality. Participants were strongly aware of the necessity to master synchronous and asynchronous technological tools to communicate online with their virtual classmates, and to do assessment and self-assessment e-activities. Learners also considered that it was essential to be mature enough to be able to reflect and evaluate their own learning. Results showed that students must be organised and tenacious, and must have the ability to innovate and collaborate inside the created e-learning community.

DOI: 10.4018/978-1-5225-5279-6.ch011

INTRODUCTION

The development of technology has allowed the convergence of two learning environments: traditional face-to-face learning and distributed learning systems (Coaten, 2003; García-Peñalvo, 2008; Güzer & Caner, 2014; López & Matesanz, 2009). The combination of these two teaching-learning models has led to the development of the blended learning modality that according to Graham (2006) is defined as:

(...) part of the ongoing convergence of two archetypal learning environments. On one hand, we have the traditional face-to-face learning environment that has been around for centuries. On the other hand, we have distributed learning environments that have begun to grow and expand in exponential ways as new technologies have expanded the possibilities for distributed communication and interaction. (p.5)

Other authors, such as Garrison and Kanuka (2004), emphasized the same idea and stated that "blended learning is the thoughtful integration of classroom face-to-face learning experiences with online learning experiences" (p.96). Yen & Lee (2011) also added that "blended learning, thoughtfully combining the best elements of online and face-to-face education, is likely to emerge as the predominant teaching model of the future" (p.138). The direct consequence of the convergence of these two learning environments is the development of a different kind of education that exploits the advantages of both types of instructions (García & García, 2005; Graham, 2006; Osguthorpe & Graham, 2003). On one hand, there are face-to-face interactions between teachers and students due to the different face-to-face activities in which they meet in a certain place and in a strict timetable in order to carry out the different lessons previously defined. And, on the other hand, the different synchronous and asynchronous tools (chat, forum, wiki, online glossaries or online questionnaires) facilitate the distribution of contents, learning experiences and interaction between all the participants of the teaching-learning process without sharing the same time and physical place. As Thorne (2003) points out:

(...) blended learning represents a real opportunity to create learning experiences that can provide the right learning at the right time and in the right place for each and every individual, not just at work, but in schools, universities and even at home. It can be truly universal, crossing global boundaries and bringing groups of learners together through different cultures and time zones. In this context blended learning could become one of the most significant developments of the 21st century. (p.18)

Blended learning instruction has become the kind of education that responds to the educational demands of the current knowledge society in which people need to carry out lifelong learning, since this instruction eliminates inflexibility and spatial and temporal barriers, producing a change in the learning approach towards a more open education in which students are more involved and more active in the construction of knowledge (Arteaga & Duarte, 2010; Pinto-Llorente, Sánchez-Gómez & García-Peñalvo, 2017). Moreover, there are studies that conclude similar learning outcomes in content knowledge acquisition between people receiving blended instruction and others receiving face-to-face instruction, specifically in the English language teacher scope (Kocoglu, Ozek & Kesli, 2011). Lifelong learning also implies the necessity to develop some skills that allow citizens to keep job skills up-to-date and enable them to adapt to different situations and changes in the current society (González, Rodríguez, Olmos, Borham & García, 2013).

The European Union in the Recommendation of the European Parliament and of the Council of 18 December 2006 points out the key competences for lifelong learning. They defined the competences as *a* "combination of knowledge, skills and attitudes appropriate to the context. Key competences are those which all individuals need for personal fulfilment and development, active citizenship, social inclusion and employment", and set out a Reference Framework of eight competences: Communication in the mother tongue; Communication in foreign languages; Mathematical competence and basic competences in science and technology; Digital competence; Learning to learn; Social and civic competences; Sense of initiative and entrepreneurship; and Cultural awareness and expression.

In this study, researchers highlighted the key competence that refers to communication in foreign languages. Spanish Primary Education Schools are trying to give answer to this necessity and for this reason most of them offer a bilingual education. Under this kind of programme pupils do not only study the subject of English as a second language but also other ones as Natural Science, Social Science, Physical Education, Art & Crafts or Music which are taught in English. The direct consequence of the implementation of that kind of bilingual project is the necessity of having teachers with a high level of English as a foreign language. However, most of them do not have the adequate level to be able to teach in that second language and need to prepare and improve their foreign language skills in order to be able to teach the subjects mentioned before, using English as the language of communication. As a result, teachers must continue their studies and need to find a kind of instruction that allow them to study and work at the same time. In other words, a type of instruction, different from the traditional one, that adapts to their personal and professional duties (Sánchez-Gómez, Pinto-Llorente & García-Peñalvo, 2010).

On account of this necessity, there is a great educational demand for the Bachelor's Degree in Primary Education: English taught in the blended learning modality.

Nowadays, there are a great amount of researches about blended learning programmes that confirm the incredible possibilities and advantages that this kind of education provides to study English or any other language (Ferriman, 2013; Pinto-Llorente, Sánchez-Gómez & García-Peñalvo 2016, 2017). However, it is essential to carry out research that not only concentrates on the academic potentiality of blended learning modality, but also on the main human factors that are basic to be successful on this kind of education. This is the main goal of this research. It is a more humanistic investigation that professes to help develop and implement more adequate blended learning models for learners. To carry out the research a hypermedia modular model was designed via the e-learning platform Moodle. The authors decided to choose this platform and its project, inspired by the social constructivist pedagogy, because it facilitates remote or blended learning educational models through the use of different technological tools such as online questionnaires, online glossaries, wikis and forums that allow the development of e-activities and the communication between the participants of the teaching-learning process (Carroll, 2009; Collis, 1998; Pinto-Llorente, Sánchez-Gómez & García-Peñalvo, 2016). In this study, researchers implemented a blended learning model in the subject of Foreign Language and its Didactics I (English) of the Degree in Primary Education: English. They gathered information on the participants' perceptions about the characteristics that students must have to be successful in a blended learning environment based on their personal experience in the model implemented.

In the following parts of the paper the authors provide an overview of the study. First of all, the authors introduce the research by providing the qualitative method used, a descriptive study with a qualitative approach, following the process of the Grounded Theory for the categorization of dimensions. This study uses a qualitative instrument to collect data including a semi-structured interview. The population and the sample was composed of the students of the Faculty of Education at the Pontifical University of Salamanca enrolled in the subject Foreign Language and its Didactics I (English) of the Degree in Primary Education: English taught in the blended learning modality. Secondly, researchers present the results of the data analysis. The authors followed the scheme proposed by Miles & Huberman (1994): a) data reduction, b) data display, and c) drawing and verifying conclusions. The findings of the study were set out in different subsections according to learners' characteristics: 1. Digital Skills; 2. Ability to Communicate Online; 3. Being Organised; 4. Being Tenacious; 5. Being Innovative; 6. Capacity to Work Collaboratively; 7. Knowledge of Synchronous and Asynchronous Tools; 8. Maturity; and 9. Level of English. The authors also present the main gender and age differences found on the overall evaluation of participants' opinions as well as the results of a cluster analysis, which described the coding similarities that existed between participants' discourses. The formation of the clusters was strongly influenced by age and gender. Finally, the authors present the main conclusions obtained.

METHOD

Based on the main goal of the research, to determine participants' perceptions about the characteristics that students must have to be successful in a blended learning environment, researchers decided to use a descriptive study with a qualitative approach, following the process of the Grounded Theory for the categorization of dimensions. Strauss & Corbin (1994) defines this method as "a general methodology for developing theory that is grounded in data systematically gathered and analyzed. Theory develops and evolves during the research process due to the interplay between data collection and analysis phases." (p.273). The theory emerges and develops inductively from the research data, not deductively from theoretical frameworks. Strauss & Corbin (1998) sum up its characteristics in:

A. The ability to step back and critically analyze situations.

B. The ability to recognize the tendency toward bias.

C. The ability to think abstractly.

D. The ability to be flexible and open to helpful criticism.

E. Sensitivity to the words and actions of respondents.

F. A sense of absorption and devotion to the work process. (p.7)

The process of analysis developed was dynamic and creative, and two fundamental strategies were distinguished: theoretical sampling and a method of constant comparison. On the one hand, the theoretical sampling refers to the process of data collection, its analysis and categorization on the part of the researcher. Glaser & Strauss (1967) define theoretical sampling as:

(...) the process of data collection for generating theory whereby the analyst jointly collects, codes, and analyses his data and decides what data to collect next and where to find them, in order to develop his theory as it emerges. This process of data collection is controlled by the emerging theory, whether substantive or formal. (p.45)

Data collection and analysis are carried out until theoretical saturation occurs, "when no additional data are found that advance, modify, qualify, extend or add to the theory develop" (Glaser & Strauss, 1967, p.61). On the other hand, the constant comparison method generates a theory from a constant comparative analysis of data. Thus, researchers encode and reflect on the type of data they are collecting from the beginning (Trinidad, Carretero & Soriano, 2006; Glaser & Strauss, 2009). Four phases can be distinguished: 1) comparing incidents and data that are applicable to each category; 2) integrating these categories and their properties; 3) bounding the theory; and 4) setting out the theory (Glaser & Strauss, 2009).

The qualitative design was divided into two phases as seen in Figure 1 and Figure 2.

Instrument

A semi-structured individual interview was used to operationalize the variables and collect data. Creswell (2005) considers that the qualitative interview should allow participants to express their opinions freely, without being influenced by researchers. The categories should not be established a priori, but interviewees should generate them. As Rogers & Bouey (2005) state the main characteristics of a qualitative interview are:

1. The beginning and end of the interview are not predetermined and clearly defined; even the interviews can be designed in several phases. It is flexible.

Figure 1. Phase I Exploratory analysis (Source: Adapted from Sánchez, Delgado & Santos, 2012)

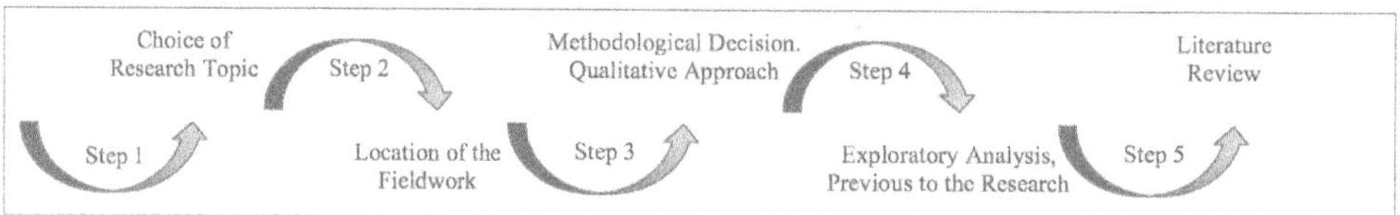

Figure 2. Phase II Adapted from Miles & Huberman (1994); Rodríguez, Gil & García (1999); Rodríguez & Gómez (2010); Sánchez & Revuelta (2004)

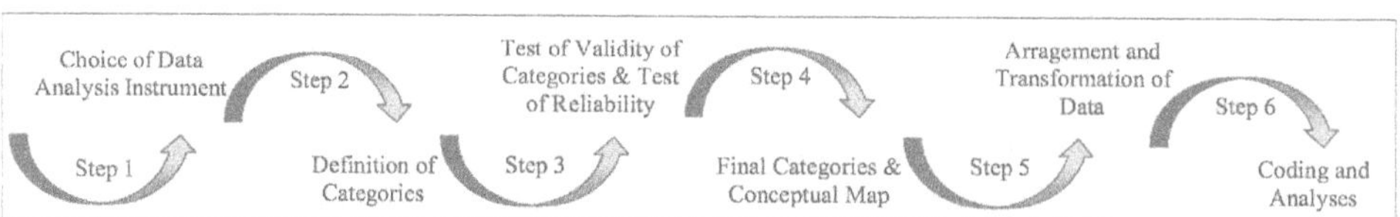

2. The questions and the order in which they are done are adapted to the participants in the study.
3. The qualitative interview is, to a certain extent, anecdotal.
4. The interviewee shares the pace and direction of the interview with the interviewer.
5. The social context is considered fundamental for the interpretation of meanings.
6. The interviewer adapts his communication to the interviewee's language.
7. The qualitative interview is friendlier than other kinds of interviews.

Ninety-one interviews were conducted for three months, from May to July, at the Faculty of Education at the Pontifical University of Salamanca, and each one lasted between 35 and 51 minutes. They were videotaped with the consent of the participants in order to transcribe them later more accurate analysis. Researchers recorded a total of 60 hours and did interviews until theoretical saturation occurred. The internal validity was achieved because the criteria established by Coleman & Unrau (2010), and Hernández, Fernández & Baptista (2010) were met. Referring to the reliability of the coding, 16 experts in education, educational research methodology, English, and technology collaborated in the study, coding the interviews. Once the experts had coded the interviews, researchers compared the different encoding, to identify the agreements that existed between them, with an agreement of 81%. This level of reliability suggests that the encodings were valid. The high agreement between encoders determined that each category was part of the final conceptual map.

Population and Sample

The population of the study was composed of the students of the Faculty of Education at the Pontifical University of Salamanca enrolled in the subject Foreign Language and its Didactics I (English) of the Degree in Primary Education: English provided in the blended learning modality. According to the data provided by the Office of the University Secretary, there were 451 students in the course. 91 learners, 45 women and 46 men aged between 20 and 58, composed the qualitative sample, which was non-probabilistic, cumulative and sequential as the sample needed to get enough information, reach the theoretical saturation and meet the objective of the research. Glaser & Strauss (1967) define theoretical sampling as:

(…) the process of data collection for generating theory whereby the analyst jointly collects, codes, and analyses his data and decides what data to collect next and where to find them, in order to develop his theory as it emerges. This process of data collection is controlled by the emerging theory, whether substantive or formal. (p.45)

Participants had already finished at least a university Degree, and the majority (70.70%) had completed a Degree in Primary Education and had a job related to education (86.73%).

ANALYSES AND RESULTS

Once the fieldwork was finished, researchers ordered the data and followed the scheme propose by Miles & Huberman (1994) to process qualitative data analysis. Three basic tasks can be distinguished: a) data reduction, b) data display, and c) drawing and verifying conclusions. It is a nonlinear design but convergent and recursive, that finishes when saturation of information is reached (See Figure 3).

In the first phase, the data reduction phase, researchers carried out the separation of units according to thematic criteria, considering the units in relation the theme tackled. They used the line as textual unit in order not to distort the meaning of the text. They also identified and classified the elements through categorizing and coding of data units in order to recognize their thematic components, and to classify them in a certain content category. In this study, the categories were defined deductively, with a priori categories, and inductively, introducing modification when data were examined, so categories were renamed or eliminated. Researchers used NVIVO11 to carry out that categorizing and coding. The choice of the program was due to its foundations, which allowed them to carry out the content analysis according to some of the principles of the grounded theory (Strauss & Corbin, 1994). The last task of that first phase was synthesis and clustering, leading to a physical grouping of the

Figure 3. Basic tasks of qualitative data analysis. Adapted from Miles & Huberman (1994, p.12)

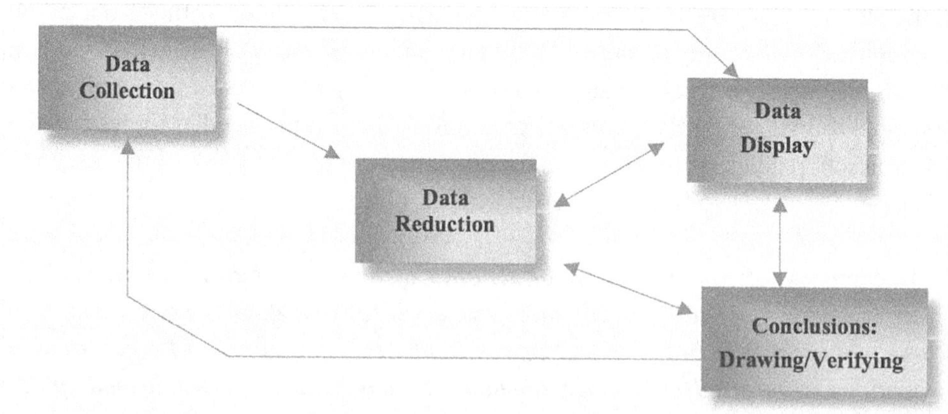

units that belong to the same category, and synthesizing in a meta-category all the information that was contained in different categories with points in common. Once researchers finished the data reduction phase, they carried out the second basic task of analysis, the data display phase. They arranged qualitative data in different charts and crosstabs, using Microsoft Office Word 2013 to illustrate the results. In the last phase, the phase of drawing and verifying conclusions, researchers presented the results, their interpretation and the extraction of the conclusions from the study.

Researchers carried out the semantic analysis in order to know the words that occurred more frequently in the semi-structured interviews. Figure 4 shows the results obtained in that analysis. It can be appreciated in the word cloud that the words that appeared more often were: technology (601 references), innovative (434 references), blended learning (428 references), English (422 references), organised (421 references), collaborative (395 reference), communication (389 references) and active (379 reference).

Following the procedure described before, researchers determined a series of categories and subcategories organized hierarchically (See Figure 5) according to participants' perspectives. Nine themes or subcategories emerged from the data analysis with reference to the objective of the research. The findings of the study emphasized the following students' characteristics to be successful in a blended learning environment: 1. Digital Skills; 2. Ability to Communicate Online; 3.

Figure 4. Word Cloud

Being Organised; 4. Being Tenacious; 5. Being Innovative; 6. Capacity to Work Collaboratively; 7. Knowledge of Synchronous and Asynchronous Tools; 8. Maturity; and 9. Level of English.

Table 1 shows the results of each subcategory of the qualitative research.

The verbatim quotes from the study participants are labelled in terms of their gender and age.

Subcategory 1: Digital Skills

The majority of students considered that one of the essential requirements for those who wanted to study English in a blended learning modality was to have adequate digital skills, since they had to use an e-learning platform and its technological tools. They were strongly aware of the necessity of being updated in technology. The data analysis also revealed that technology was a relevant aspect of that kind of education, which had a significant impact on the development of the activities. The model created a favorable environment for learning and updating participants'

Figure 5. Conceptual Map

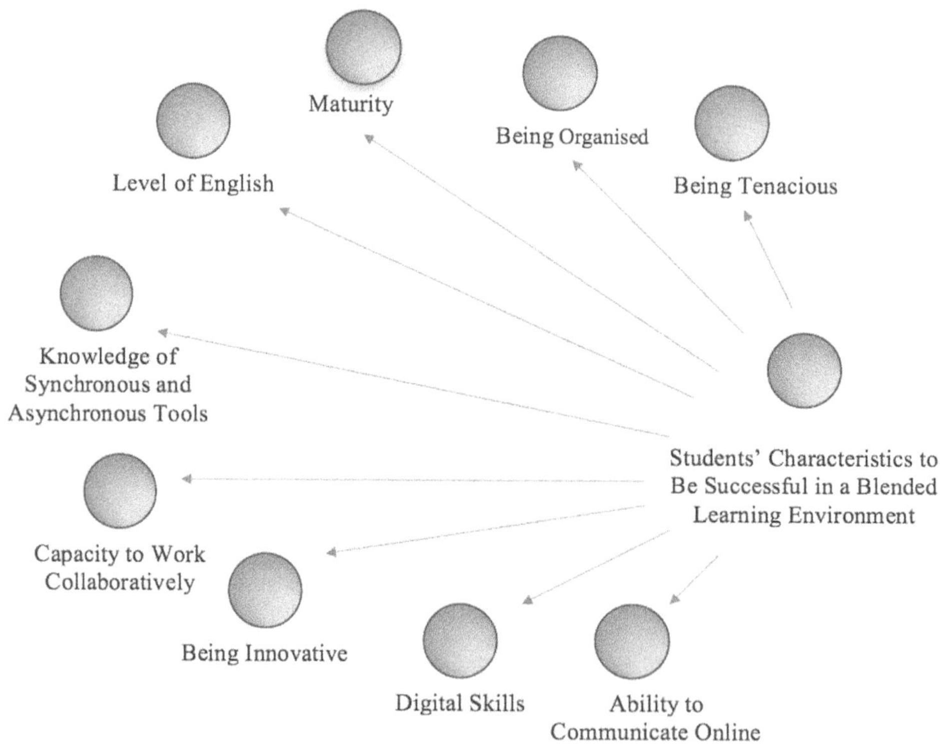

Table 1. From Global Theme to Organizing Theme to Basic Theme

Students' Characteristics to Be Successful in a Blended Learning Environment	A	B	C	D	E	F
1. Digital Skills	2026	86	95%	18438	11%	10%
2. Ability to Communicate Online	1610	69	76%	14709	11%	8.3%
3. Being Organised	511	46	51%	10094	5.1%	2.6%
4. Being Tenacious	704	63	69%	13691	5.1%	3.6%
5. Being Innovative	120	21	23%	4407	2.7%	0.62%
6. Capacity to Work Collaboratively	1668	70	77%	14960	11%	8.6%
7. Knowledge of Synchronous and Asynchronous Tools	2429	87	96%	18548	13%	12%
8. Maturity	168	21	23%	4433	3.8%	0.86%
9. Level of English	79	11	12%	2386	3.3%	0.41%

A = total number of text units (lines) coded of each category
B = number of interviews in which students refer to each category
C = percentage of the number of interviews in which students refer to each category in relation to the total number of interviews
D = number of text units (lines) of B
E = percentage of A in relation to the total text units (lines) of the interviews in which students refer to each category
F = percentage of A in relation to the total text units (lines) of all individual interviews

knowledge of the technological resources. The use of different technological tools permitted the digital literacy of the different generations that made up the sample. Although the use of these technologies meant an extra effort on the part of students at the beginning of the course, the constant use of them allowed learners to get up to date with technologies, and to improve their confidence in their digital competence.

Well, I think that students who decide to study a degree in blended learning education must have good digital skills and must be able to know how to use different technological tools provided during the course in the platform. When I started my studies in the blended learning modality, I had little experience in the use of technological tools and I had to make a great effort to learn how to use them. Moreover, I think that the use of technological tools has been a determining factor in some of my final grades since the fact of not using them in an adequate way did not let me do the activities correctly. If people do not have an adequate level in technology, it is a barrier that they must overcome if they want to be successful in this kind of education (31-year-old woman, participant 24).

I consider that people who study a degree in blended learning modality must be quite innovative and learn how to use technology in their training. It would be a mistake to choose a degree taught in that modality and not to know how to handle an e-learning platform, and its tools. During this blended learning course there have been just a few face-to-face lessons so e-learning platform has become an essential element for us, since we could not have interacted with our classmates and teachers, and we could not have done the different e-activities of the subject without it and its technological tools (38-year-old man, participant 86).

People who are thinking about studying a degree in the blended learning modality must not be afraid of using technology. They must be aware that the e-learning platform is going to be their virtual classroom and the technological tools provided in that virtual environment must be used to communicate with their classmates and teachers. There are not many face-to-face lessons, so the platform turns into an essential element for students (34-year-old man, participant 45).

Subcategory 2: Ability to Communicate Online, and Subcategory 7: Knowledge of Synchronous and Asynchronous Tools

The ability to communicative online and the use of different synchronous and asynchronous tools were two subcategories with a high percentage of the discourse coded (8.3% and 12%). In our model, learning was embedded in social experience through the e-activities develop by means of tools such as online questionnaires, online glossary, chat, wikis and forums.

The technological tools used to carry out different e-activities promoted collaborative and individual learning as well as interaction and the creation of a learning community made up of all the participants in the teaching-learning process. These activities managed to eliminate the isolation that could occur as a result of the existing geographical distance between the participants, since they did not share the same physical space, but shared the same virtual space.

The participants considered that the ignorance of the use of those synchronous and asynchronous tools to communicate and do assessment and self-assessment activities had some negative aspects. On one hand, those students that were not used to using technological tools to communicate could find it difficult to use them since they did not have an adequate ability to communication with other people without any face-to-face interaction. The direct consequence of that lack of ability to communicate accurately and effectively was isolation. On the other hand, students considered that those people who did not know how to use synchronous and asynchronous tools properly could see them as barriers for their learning and to perform some of the activities in which their grades were worse than in a traditional education without the use of technological tools.

One of the things that people who want to study in this modality must master are the different communication tools because they need to use them to communicate with their virtual classmates and also to do some of the activities. If they do not have ability to communicate through them, they will feel isolated (26-year-old woman, participant 13).

Students must master the synchronous and asynchronous tools that the e-learning platform provides to work as a team, avoiding an individualistic learning. We have had, for example, the forums to work together and develop wikis of English culture or share methodological ideas for teaching English. These tools have allowed us to communicate online with our classmates with whom we have shared the platform and work together, so it is necessary to know how to use them (28-year-old woman, participant 17).

I also emphasize that it is necessary to know how the use the asynchronous tools to communicate online and work as a team. We, for example, have designed together different wikis and have used different forums to share our ideas. This has improved or promoted collaborative learning and teamwork so it is essential to master technological tools to be able to participate (28-year-old man, participant 77).

I will highlight the necessity of having the necessary skills to be able to interact in a virtual learning environment. Students must master Moodle and its synchronous and asynchronous tools to communicate when they have to do and participate in different activities or when they have doubts and need help from their teachers or classmates (33-year-old man, participant 53).

During the course we have used different synchronous and asynchronous tools to do activities and communicate. It is essential that students acquire experience on how to use them since they are necessary in blended learning instruction (37-year-old woman, participant 62).

Subcategory 3: Being Organized, and Subcategory 4: Being Tenacious

The students also emphasised the necessity of being organized and tenacious (2.6% and 3.6%). They thought that it was important to be determined in their studies from the beginning of the course. The participants considered that having more independence in the blended learning education could lead to a separation from the university environment and their studies, leading to failure if the students did not follow through with their responsibilities from the beginning of the course.

In this modality we have been totally free to choose what we wanted to do every day. The problem that could be derived from that freedom is the failure of the subject. If learners are not persevering enough and connect to the platform frequently, they are not going to be successful in this instruction. This modality implies a strong commitment and effort on students' side to try to make compatible their personal and professional lives, and the necessity of carrying on studying (26-year-old man, participant 36).

This kind of education implies that learners must make more effort than in traditional instruction since there are just few face-to-face lessons so they must study syllabus and do some of the tasks by themselves or as part of a group. This means that they must plan their learning very well and must connect to the e-learning platform very often. If they are not persevering, they will feel lost and will probably fail the course (29-year-old woman, participant 75).

This education implies a greater commitment and perseverance. Learners must devote all the free time they have to prepare contents, of the virtual activities and participate in the learning community created in the e-learning platform. The fact of not having to attend face-to-face lessons everyday, it implies to have more freedom to decide what to study and when to do it, but also implies to make a bigger effort on learners' side. If they do not commit themselves to make that effort from the beginning of the course, this modality and the e-learning platform can become barriers to pass the subjects and, of course, to study a university degree (32-year-old man, participant 54).

This kind of education requires a strong personal commitment on the part of students. They must be determined and dedicate enough time to learn the contents of the course and to do the activities. The flexibility of this education does not mean that the students do not have to work and study every day. They will have to work much harder in the blended learning education than in face-to-face one (33-year-old woman, participant 57).

Learners who decide to study a degree in blended learning modality must be aware of the kind of instruction that they have choosen. This kind of education is very flexible and most of the time students are free to decide what to study, when to do it and so on. So if they want to be successful, they must be persevering and must devote time every day to organizing their learning, the tasks they have to do, etc. (34-year-old woman, participant 24).

Subcategory 5: Being Innovative

The empirical analysis also provided evidence on the positive link between blended learning education and its use of technological tools, and the innovative behaviour of students. This was not just because of the knowledge and learning generated in the face-to-face context, but also for the knowledge generated online in a learning community. Moreover, that link was stronger for those participants with higher digital skills. In addition to that ability, researchers also found that characteristics, such as age, ability to use synchronous and asynchronous tools, ability to communicative and work online or a positive attitude to technology, were also relevant factors of the likelihood of innovation.

Results showed that students' innovative capacity was strongly related with his or her technological abilities. More concretely, researchers found that digital skills increased the probability of being innovative not directly, but because it raised the capacity of communicate and work in different ways through synchronous and asynchronous tools.

There were several reasons to expect that the grade of being innovative in that kind of modality was related to teachers' work, the design of the virtual learning environment, or the integration of new tools and method in their instruction. Indeed, according to different studies the way in which activities and materials were presented would be endogenous to the decision to innovate, which was one of the most important educational issues of the 21st century education.

People who want to study English or another second language course in the blended learning modality must be innovative and be willing to change the traditional way of learning since this kind of education and the use of different technological tools imply a completely different way to communicate and work with teachers and classmates (26-year-old woman, participant 13).

This kind of education also implies the use of new technology and we must try to use new ways to do and learn things in a blended learning environment. It is clear that everything is not new, but we need to be innovative and develop different ways to use the existing resources to work, study and learn (36-year-old man, participant 70).

Subcategory 6: Capacity to Work Collaboratively

Research findings also indicated that the students believed that it was essential to have the capacity to work in groups and to work collaboratively. They thought that collaborative learning was one of the advantages of the blended learning education, since it contributed to the creation of a learning community. Learners added that they needed to have the ability to communicate online with all the participants of the teaching-learning process. That characteristic forced the students to interact through the different asynchronous tools (wiki and forum). The students believed that collaborative work reinforced the need for strong interpersonal skills such as respecting opinions, remaining flexible and appreciating teammates' contributions. According to the concept of the zone of proximal development (ZPD) developed by the Russian psychologist Lev Vygotsky (1978), students' learning is greater when they study in a community, in a collaborative way that provides them a supportive framework that allows them to do different activities that cannot be accomplished individually (Brown, Collins & Duguid, 1989; Duffy & Cunningham, 1996; Roschelle & Teasley, 1995).

Collaborative learning is essential in this type of instruction, so students must know how to work as a team. For example we have done a wiki of culture of the United Kingdom and the United States and have been able to communicate with our classmates through different forums. We have used this tool to establish a more fluent interaction and decide the points we wanted to include and our roles in the process (28-year-old woman, participant 46).

The blended learning instruction is linked to the use of an e-learning platform, which facilitates collaborative learning between the participants of the course. Learners must know that they must work together with their peers and must be willing to do it, to communicate between them, to share information and to assume their responsibilities, respecting and thanking the work of others (28-year-old man, participant 77).

They must know how to work as a team and collaborate with their classmates. During the course there have been several group activities developed through technological tools. It is necessary that the students respect the ideas of others and become involved in the teamwork that they have to do (32-year-old woman, participant 55).

Finally, I believe that people who are thinking about studying a second language in that modality have to have the necessary skills to work as a team and carry out collaborative tasks in which they must respect each other and support the work done by their peers. They must involve in all the activities since a part of the grades of their classmates depend on them (32-year-old man, participant 80).

The students must know that they will have to work together. They will have to share knowledge, to take charge of their responsibilities in the group activities, to respect the opinions of others, and to thank for their work (34-year-old man, participant 51).

Students who attend a blended learning course must work collaboratively with the rest of their virtual classmates. Therefore, it is important that they cooperate in the development of the tasks, take different roles, share their knowledge and ideas, respect each other and connect frequently to the e-learning platform in order to carry out all the tasks of the course (36-year-old man, participant 88).

Subcategory 8: Maturity

Hence, they believed that it was better for those students to have more face-to-face lessons and get their independence from teachers little by little to become independent, mature and self-directed learners. Patterson, Crooks & Lunyk-Child (2002) define self-directed learning as a "process in which individuals take the initiative, with or without the help of others, in diagnosing their learning needs, formulating learning goals, identifying human and material resources for learning, choosing and implementing appropriate learning strategies, and evaluating learning outcomes." (p.1) Therefore, this implies the development of individual or intrapersonal skills to be able to plan their learning and the goals they want to achieve.

Participants also added that students must be mature enough, that is to say, they must have a high level of academic maturity, to reflect and evaluate their own learning and the teaching-learning process. Moreover, the fact of not being mature enough and the flexibility of that kind of education may lead to students to not work enough and to not persevere in their learning. Hence, they do not do the work required to get the goals of the course and fail.

I believe that learners must have the maturity to be engaged in their own learning in a blended learning course. They must demonstrate the maturity to devote the necessary time to do the virtual tasks and study (34-year-old man, participant 90).

In this kind of modality, we are placed at the centre of the teaching-learning process, and the teachers become guides or counsellors, and provide us with opportunities to learn independently. For all these reasons, I think that people who study in this kind of education must have the enough maturity and be responsible of their learning (36-year-old man, participant 88).

I do not know very well how to explain it, but sometimes you do not feel that you are studying a university degree since you do not have to go to class every day. Blended learning instruction is a type of modality very different from the face-to-face one. There can be a disconnection on the part of learners of what they are studying if they do not devote time to review the learning platform, download the new materials or do the virtual tasks every day. For all this, I think it is a modality that implies maturity. It is a good way to continue training for all those people who have different responsibilities, either familiar or personal, and cannot attend face-to-face lessons every day. I would not recommend young students who begin university, or at least during the first year, to study in this modality since their immaturity or lack of experience could lead to failure (38-year-old man, participant 86).

I think that the students must have certain maturity. I do not think that this education is appropriate for the students that start university, since its flexibility and freedom can be misunderstood, and perhaps they do not do the work required (40-year-old woman, participant 61).

Subcategory 9: Level of English

The last characteristic highlighted by the students was related to studying a second language in the blended learning modality. The data showed that learners who wanted to study a university English degree in that modality must have a reasonable competence in English in order to be autonomous in their learning. They pointed out that students with a low level of English could not be sufficiently independent because they still needed the constant support of teachers.

I would like to recommend people who want to study a university degree in English as a second language in the blended learning modality to have a good level in that language. They should have at least an intermediate level of English because they will have to work and collaborate in a learning community using that second language. Moreover, sometimes they must work individually, doing some virtual activities or learning the contents, so they must have the adequate level that allows them to do it (24-year-old woman, participant 9).

The fact of having a low level in English can also be a barrier because the students must work independently. If they do not have the adequate level, this independence cannot exist because they need that the teachers help them continuously (28-year-old woman, participant 49).

If students have a low level of English, this can become a barrier to study in blended learning instruction. Sometimes they need to learn and do the virtual activities independently. However, If they do not have a really good level of the second language, they cannot be independent and they will need teachers' help all the time. This is not the most appropriate modality for them (28-year-old woman, participant 47).

I do not think that people who have a low level in English can study this language in this modality because it requires much individual work on the part of the students. They have to use this second language to do the activities and to communicate with their classmates and teachers and this can be a barrier for all of the students who do not have the necessary level (33-year-old man, participant 43).

I think that it is very important to have a good level of English. Studying a university degree in this modality, it often means to be independent. It is true that there are face-to-face and virtual lessons to help students, but they must have an adequate level in English to be able to understand them and participate. Moreover, they must be able to learn and improve their level by themselves. If learners do not have a good level of English, it will be quite difficult to pass the course (38-year-old man, participant 67).

Gender Differences

The research on students' perceptions about the characteristics that learners must have to be successful in a blended learning environment found gender differences in aspects such as adequate digital skills, capacity of being innovative, capacity to work collaboratively and knowledge of synchronous and asynchronous tools (See Figure 6). Although, there were no significant differences in men and women's opinions about those characteristics, the research results showed that men put more emphasis in the necessity of having an adequate knowledge of technology and the synchronous and asynchronous tools to be successful in a blended learning education. The use of these tools is essential in order to learn and practice English, and to communicate with all the protagonists of the teaching-learning process. On the other hand, men put more emphasis in the necessity to be innovative and adapt to the new ways of learning that afford the blended learning modality and that are essential in the 21st century education. Finally, they put the accent on the capacity to work collaboratively to improve group's efficiency, the respect for virtual classmates' different opinions and their work, and the development of different e-activities among all the participants of the course.

Figure 6. Gender Differences

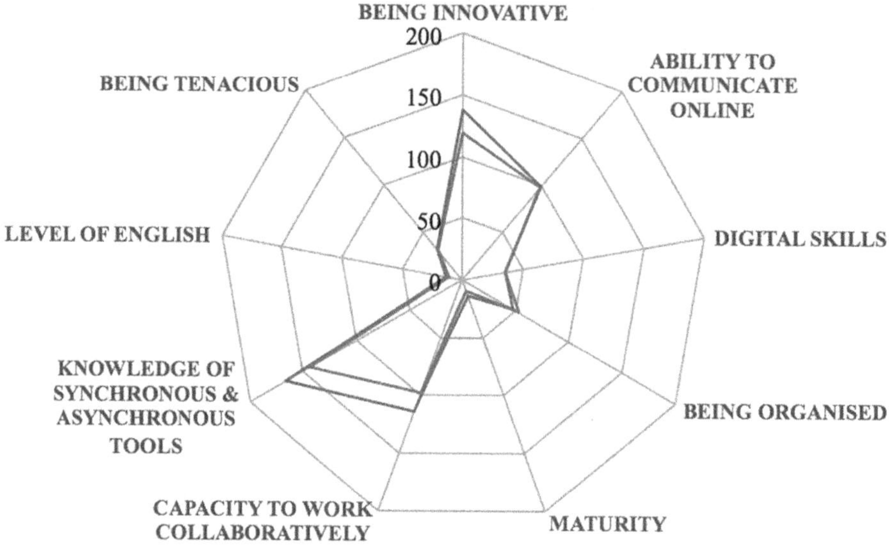

Age Differences

Based on the overall evaluation of participants' opinions there were some differences between the different age groups (See Figure 7). First of all, the oldest students (aged between 30 and 34, and more than 35) put more emphasis in those characteristics that referred to the ability to communicate online and the knowledge of synchronous and asynchronous tools to learn, communicate and develop the e-activities. They considered that they had little or no ICT experience and believed that the lack of knowledge and skills in the different synchronous and asynchronous technological tools could become barriers to learning. They pointed out that they had had difficulties in developing some of the e-activities due to the lack of knowledge of the tools that they had to use to do them. However, the youngest participants, although they also emphasized the necessity of having those characteristics to study an English course in the blended learning modality, did not consider technological tools as barriers but as essential elements to be successful in that kind of education. Secondly, the oldest students (aged between 30 and 34, and more than 35) also emphasized the necessity of being organised and being tenacious. They believed that the blended learning modality was very flexible and if students did not plan their time and study from the beginning of the course, they could not be successful. They also linked that opinion with the necessity of a certain academic maturity since they believed that students that started university or were too young were very dependent in their learning and

Figure 7. Age Differences

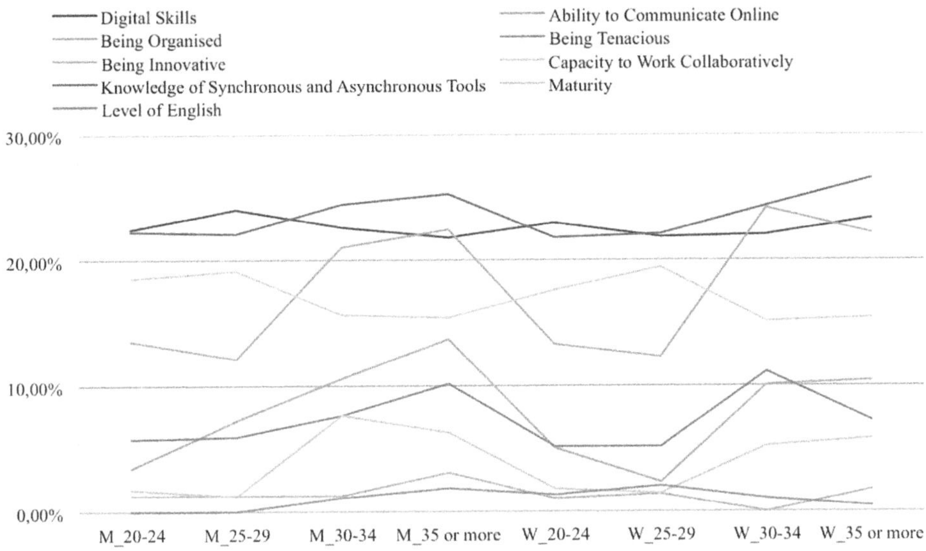

were not mature enough to be able to study in that kind of education. They thought that the youngest students had a good and adequate knowledge of technology, but they thought that it was not enough for that modality. Finally, there was a significant difference between the youngest students (aged between 20 and 24, and between 25 and 29) and the oldest ones (aged between 30 and 34, and more than 35) in the subcategory that referred to the capacity to work collaboratively. Although all of the participants believed that it was essential, the youngest students thought that it was easy for them to carry it out thanks to the use of technological tools.

Clusters by Coding Similarity

This section shows the results ofa cluster analysis. It describes the coding similarities that existed between participants' discourses. Researchers decided the column as the criterion to be used for measuring it since the objective of that cluster analysis was to form similar groups. Figure 8 shows the patterns in the research by grouping nodes that were coded similarly, providing a graphical representation to see those similarities. The formation of the clusters was strongly influenced by age and gender.

The first cluster was made up of 11 people, all of them women aged between 22 and 24. The coding similarity happened in the subcategory that referred to knowledge of synchronous and asynchronous tools. They considered that it was essential that students mastered those tools in order to communicate online and do assessment and

Figure 8. Global Cluster by Coding Similarity

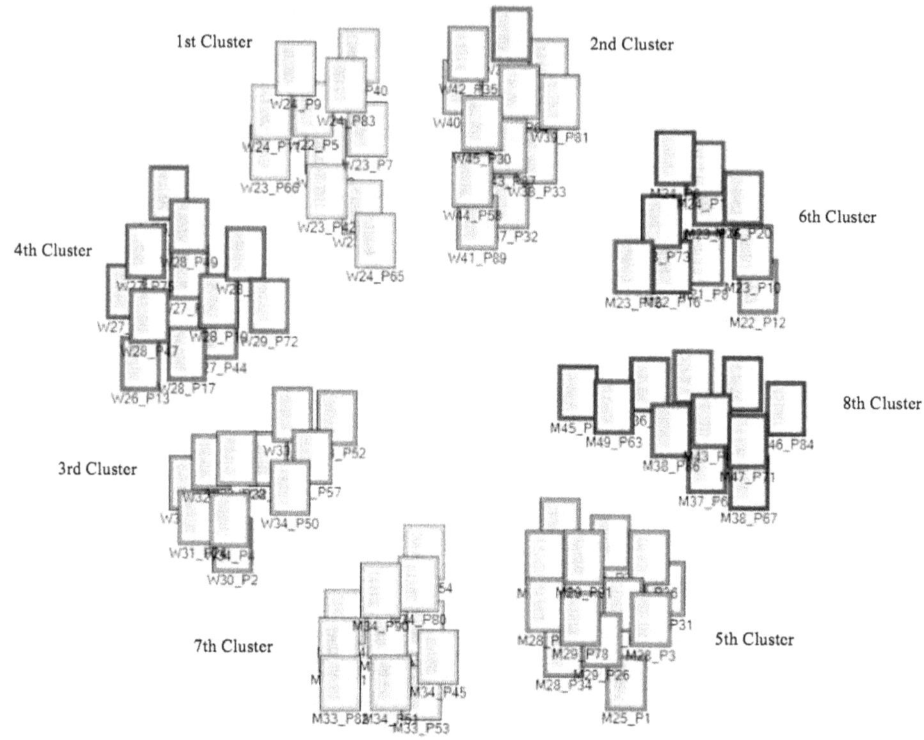

self-assessment e-activities. They pointed out that they were actively engaged in using these tools, and were aware of their advantages in the blended learning education.

The second cluster was made up of 12 people, women aged between 36 and 45. The closeness in their speech was in the subcategories that referred to adequate digital skills and the knowledge of synchronous and asynchronous tools. Although they emphasised the positive points of ICT and, particularly of those tools, they believed that they could turn into barriers to study in the blended learning modality.

The third cluster was made up of 11 people, women aged between 30 and 34. There coded similarities in four of the subcategories: digital skills, ability to communicate online, capacity to work collaboratively, and knowledge of synchronous and asynchronous tools. Generally speaking, they assessed very positively on the use of ICT to study in the blended learning modality, to communicate with their classmates and teachers, and to work in groups. However, they thought that they needed to improve their skill in order to use them correctly.

The fourth cluster was made up of 12 people, women aged between 26 and 29. There were similarities in the subcategory that referred to adequate computer skills. They pointed out that an effective blended learning model was only achieved by the use of technologies. Moreover, they considered that the activities designed in which technology was used were very innovative and efficient to improve their oral and written skills, and to learn the contents of the subject.

The fifth cluster was made up of 13 people, men aged between 25 and 29. There was closeness in their speech in the subcategories that referred to digital skills, being organized, and being tenacious. They strongly agreed with the statement that technology was essential in blended learning modality and they emphasised their advantages. They valued their digital skills very positively. However, they considered that having good digital skills was not enough for that kind of education. They pointed out that students needed to work harder than in the traditional education and for that reason they must organise their time and work from the beginning, and devote certain hours every day to studying and doing the activities of the subject.

The sixth cluster was made up of 10 people, men aged between 21 and 24. There was closeness in their speech in the subcategories that referred to adequate digital skills and knowledge of synchronous and asynchronous tools. In contrast with women's opinions expressed in the second cluster, men of this group pointed out that technology and the synchronous and asynchronous tools created better opportunities to interact and learn.

The seventh cluster was made up of 11 people, men aged between 31 and 34. There were coding similarities in three subcategories: digital skills, ability to communicate online, and capacity to work collaboratively. They emphasised the necessity of having good digital skills to study in that kind of education. They thought that the fact of not having a good level was a barrier when they had to work and communicate online. Finally, they agreed with the necessity of having the abilities to work in groups, but they emphasised the barriers that they found when they had to do it through technological tools.

The eighth cluster was made up of 11 people, men aged between 36 and 49. There was closeness in their speech in the subcategory that referred to the knowledge of synchronous and asynchronous tools. In contrast with women's opinions in the first cluster, the men of this group pointed out that the synchronous and asynchronous tools used had positive aspects, but sometimes they created barriers to communicate. They believed that their grades in some of the e-activities were worse because they had to do them through these technological tools.

CONCLUSION AND DISCUSSION

To meet the aim of the study, researchers analyzed participants' perceptions about the characteristics that students must have to be successful in a blended learning environment. The results of the research supported the idea that students needed to have reasonable digital skills and master certain technological tools that allowed them to participate online in the multiple e-activities designed and implemented in the hypermedia modular model (Edel-Navarro, 2004; Luna, 2002; Reed & Sharp, 2003; Tang & Chaw, 2016). The results also highlighted that students must learn how to use different synchronous and asynchronous tools (Simonson, Smaldino, Albright, & Zvacek, 2003) to be able to communicate with all the participants of the teaching-learning process since the direct consequence of not using them was isolation (Croft, Dalton & Grant, 2015; Pinto-Llorente, Sánchez-Gómez & García-Peñalvo, 2015a). Moreover, students had to participate in different online e-activities through forums, wikis, online questionnaires, and online glossaries, and the use of these tools played a decisive role in their final grades.

Innovativeness cannot be taught as an independent subject at university. This quality needs to be worked and developed through the whole educational process. Nowadays, teachers and universities have the necessary tools to design courses, which help students to develop this characteristic that, as qualitative results showed, was emphasized as essential to be successful in the blended learning modality (Pinto-Llorente, Sánchez-Gómez, & García-Peñalvo, (2015a). A hypermedia modular model that promotes the development of innovativeness is favourable to constructivist learning in a constructivist environment that involves students in the development of knowledge by having an active role in the teaching-learning process in which they learn interacting with others, and being placed at the center of that process, taking an active role (Pinto, Sánchez & García, 2015b).

Findings emphasized how learners must have the ability to work both individually and in-group. In the implemented model there were e-tasks in which participants must share ideas and reflect about the best solution to do each e-activity. They had to share their points of view about the contents of the e-activities, the different perspectives in which they could address the contents, and decided the roles and functions of each member of the group (Pinto-Llorente, Sánchez-Gómez & García-Peñalvo, 2016). These were e-activities that required negotiation on the part of the students, as well as the exposition of different opinions and perspectives, and, of course, the respect of them. Collaborative learning was related, therefore, to the experience of professional roles: teamwork, leadership, negotiation, and cooperation (Honebein, 1996). Students addressed a problem from different perspectives and also developed their ability of reflection, and valued the benefits and richness of considering the different perspectives suggested (Walsh & Kahn, 2010).

Students must have the ability to manage their learning process (Díaz & Hernández, 2002; Vat, 2010) and focus on those aspects that they found more difficult both individually and in a collaborative way. That led to the creation of a learning community in which the students respected opinions and works of their virtual classmates, and shared their knowledge with others (Lee, 2010; Twu, 2010).

Researchers concluded that the success of team work lay in the organization, in inter and intra interactions and in planning of the team as well as in decision-making and developing consensus to create knowledge (Salinas, 2000), but it also required an adequate plan to support students and assess the learning process (Gros, García & Lara, 2009). Students must be organized and tenacious in their learning. They must plan their time from the beginning of the course so that their personal and professional duties were compatible with their studies. The model implemented in the blended learning modality was characterized by its flexibility and freedom. However, these characteristics were negative aspects for those students who did not make the required effort. Moreover, the study demonstrated that the blended learning education required a high level of academic maturity on the part of the students, so it was not the appropriate modality for young students starting a university program (Hawkes & Cambre, 2000; Kravtsova & Galiguzova, 2015).

Finally, the data showed that participants who wanted to study an English course in the blended learning modality had to have a reasonable competence in English in order to be autonomous in their learning (Gan, Menkhoff & Smith, 2015; Tang & Chaw, 2016). Researchers concluded that students who were not competent in English could not study independently because they still needed the constant support of teachers. Having a basic level of a second language increased students' dependence on teachers and the necessity of having a greater number of face-to-face lessons that helped them make progress in that second language. Learner autonomy and independence increased as they improved their competence in English. Researchers emphasized that the majority of students in our study had an appropriate level of English that allowed them to learn without having constant support from their teachers.

The results helped to establish a link between the blended learning model implemented and the learner-centered pedagogy, since students were placed at the center of the teaching-learning process and took a principal role in its development (Cortiella & Horowitz, 2014; Evangelidou, 2017; Yeh & Swinehart, 2017). In this pedagogy learners take an active role in their education and create knowledge. All this is opposed to students' passive role in the tutor-centered pedagogy in which teachers had the knowledge, and were experts and responsible for transmitting that knowledge to students who had to learn it, memorize it and repeat it (Allan, 2007; Oliver, 2000).

REFERENCES

Allan, B. (2007). *Blended Learning tools for teaching and training*. London: Facet Publishing.

Arteaga, R., & Duarte, A. (2010). Motivational factors that influence the acceptance of Moodle using TAM. *Computers in Human Behavior, 26*(6), 1632–1640. doi:10.1016/j.chb.2010.06.011

Brown, J. S., Collins, A., & Duguid, P. (1989). Situated Cognition and the Culture of Learning. *Educational Researcher, 18*(1), 32–42. doi:10.3102/0013189X018001032

Carroll, J. M. (2009). *Learning in Communities. Interdisciplinary Perspectives on Human Centered Information Technology*. London: Springer.

Coaten, N. (2003). Blended e-learning. *Educaweb, 69*. Retrieved July 20, 2017, from http://www.educaweb.com/esp/servicios/monografico/formacionvirtual/1181076.asp

Coleman, H., & Unrau, Y. A. (2010). Qualitative Data Analysis. In R. M. Grinnell & Y. A. Unrau (Eds.), *Social Work: Research and Evaluation. Foundations of Evidence-Based Practice* (pp. 447–464). New York: Oxford University Press.

Collis, B. (1998). WWW-Based Environments for Collaborative Group Work. *Education and Information Technologies, 3*(3/4), 231–245. doi:10.1023/A:1009697415029

Cortiella, C., & Horowitz, S. H. (2014). *The State of learning disabilities: facts, trends and emerging issues*. New York: National Center for Learning Disabilities.

Creswell, J. (2005). *Educational Research: Planning, Conducting, and Evaluating Quantitative and Qualitative Research*. Upper Saddle River, NJ: Pearson Education.

Croft, N., Dalton, A., & Grant, M. (2015). Overcoming Isolation in Distance Learning: Building a Learning Community through Time and Space. *Journal for Education in the Built Environment, 5*(1), 27–64. doi:10.11120/jebe.2010.05010027

Díaz, F., & Hernández, G. (2002). *Estrategias docentes para un aprendizaje significativo*. McGraw-Hill.

Duffy, T. M., & Cunningham, D. J. (1996). Constructivism: Implications for the Design and Delivery of Instruction. In D. H. Jonassen (Ed.), *Handbook of Research for Educational Communications and Technology* (pp. 170–198). New York: Macmillan Library Reference.

Edel-Navarro, R. (2004). Educación a distancia y eficiencia terminal exitosa: El caso de la sede Tejupilco en la Universidad Virtual del Tecnológico de Monterrey. *Revista de Educación a Distancia, III*(12), 1–23.

European Union. (2006). *Recommendation of the European Parliament and of the Council on Key Competences for Lifelong Learning*. Official Journal of the European Union L394 of 30.12.2006, pp.0010-0018.

Evangelidou, F. (2017). Kavala's 6th Graders towards Mastering Mathematics: A Blended Learning Proposal at Disabled Students' Side. In A. Palalas & P. Pawluk (Eds.), *Blended Learning Theory and Practice Proceedings of the 2nd World Conference on Blended Learning (IABL2017)* (pp. 79-84). Toronto: International Association for Blended Learning and George Brown College.

Ferriman, N. (2013). The impact of blended e-learning on undergraduate academic essay writing in English (L2). *Computers & Education*, *60*(1), 243–253. doi:10.1016/j.compedu.2012.07.008

Gan, B., Menkhoff, T., & Smith, R. (2015). Enhancing students' learning process through interactive digital media: New opportunities for collaborative learning. *Computers in Human Behavior*, *51*, 652–663. doi:10.1016/j.chb.2014.12.048

García, F. J., & García, J. (2005). Educational Hypermedia Resources Facilitator. *Computers & Education*, *44*(3), 301–325. doi:10.1016/j.compedu.2004.02.004

García-Peñalvo, F. J. (2008). *Advances in E-Learning: Experiences and Methodologies*. Hershey, PA: Information Science Reference. doi:10.4018/978-1-59904-756-0

Garrison, D. R., & Kanuka, H. (2004). Blended learning: Uncovering its transformative potential in higher education. *The Internet and Higher Education*, *7*(2), 95–105. doi:10.1016/j.iheduc.2004.02.001

Glaser, B. G., & Strauss, A. L. (1967). *The Discovery of Grounded Theory: Strategies for Qualitative Research*. New York: Aldine Publishing Company.

Glaser, B. G., & Strauss, A. L. (2009). *The Discovery of Grounded Theory: Strategies for Qualitative Research*. New York: Aldine Publishing Company.

González, A. B., Rodríguez, M., Olmos, S., Borham, M., & García, F. (2013). Experimental evaluation of the impact of b-learning methodologies on engineering students in Spain. *Computers in Human Behavior*, *29*(2), 370–377. doi:10.1016/j.chb.2012.02.003

Graham, C. R. (2006). Blended Learning Systems. Definition, Current Trends, and Future Directions. In C. R. Graham & C. J. Bonk (Eds.), *The Handbook of Blended Learning. Global Perspectives, Local Design* (pp. 3–21). San Francisco: Eds. Pfeiffer.

Gros, B., García, I., & Lara, P. (2009). El desarrollo de herramientas de apoyo para el trabajo colaborativo en entornos virtuales de aprendizaje. *RIED*, *12*(2), 115–138.

Güzer, B., & Caner, H. (2014). The Past, Present and Future of Blended Learning: An in Depth Analysis of Literature. *Procedia: Social and Behavioral Sciences*, *116*, 4596–4603. doi:10.1016/j.sbspro.2014.01.992

Hawkes, M., & Cambre, M. (2000). The cost factor. *Technological Horizons in Education*, *28*(1), 26–32.

Hernández, R., Fernández, C., & Baptista, P. (2010). *Metodología de la investigación*. McGraw Hill.

Honebein, P. C. (1996). Seven Goals for the Design of Constuctivist Learning Environment. In B. G. Wilson (Ed.), *Constructivist learning environments: Case studies in instructional design* (pp. 11–24). Englewood Cliffs, NJ: Educational Technology Publications.

Kocoglu, Z., Ozek, Y., & Kesli, Y. (2011). Blended learning: Investigating its potential in an English language teacher training program. *Australasian Journal of Educational Technology*, *27*(7), 1124–1134. doi:10.14742/ajet.908

Kravtsova, O. A., & Galiguzova, A. V. (2015). Blended Learning: In Search of the Right Blend. In D. A. Kryachkov, E. B. Yastrebova, & O. A. Kravtsova (Eds.), *The Magic of Innovation. New Techniques and Technologies in Teaching Foreign Languages* (pp. 67–83). Newcastle, UK: Cambridge Scholars Publishing.

Lee, L. (2010). Exploring wiki-mediated collaborative writing: A case-study in an elementary Spanish course. *CALICO Journal*, *27*(2), 260–276. doi:10.11139/cj.27.2.260-276

López, C., & Matesanz, M. (2009). *Las plataformas de aprendizaje. Del mito a la realidad*. Madrid: Editorial Biblioteca Nueva.

Luna, E. (2002). Requerimientos básicos de los planes de estudio para la educación a distancia. In *Proceedings of the II Congreso Internacional de Educación Abierta y a Distancia: CICESE, ANUIES*. Universidad Autónoma de Baja California.

Miles, M. B., & Huberman, A. (1994). *Qualitative Data Analysis: An Expanded Sourcebook*. Newbury Park, CA: SAGE.

Oliver, R. (2000). When Teaching Meets Learning: Design, Principles and Strategies for Web-based Learning Environments that Support Knowledge Construction. In R. Sims, M., O'Reilly & S. Sawkins (Eds.), *Learning to Choose: Choosing to Learn. Proceedings of the 17th Annual ASCILITE Conference* (pp. 17-28). Lismore, Australia: Southern Cross University Press.

Osguthorpe, R. T., & Graham, C. R. (2003). Blended Learning Environments: Definitions and Directions. *Quarterly Review of Distance Education*, *4*(3), 227–233.

Patterson, C., Crooks, D., & Lunyk-Child, O. (2002). A New Perspective on Competencies for Self-Directed Learning. *The Journal of Nursing Education*, *41*(1), 25–31. PMID:11843104

Pinto, A. M., Sánchez, M. C., & García, F. J. (2015b). Developing a VLE to Enable the Innovative Learning of English Pronunciation. In *Proceedings of the 3rd International Conference on Technological Ecosystems for Enhancing Multiculturality* (pp. 83-89). New York: ACM. doi:10.1145/2808580.2808594

Pinto-Llorente, A. M., Sánchez-Gómez, M. C., & García-Peñalvo, F. J. (2015a). To Be or Not to Be Successful? That Does Not Only Depend on Technology, But Also on Human Factors. *Journal of Cases on Information Technology*, *17*(1), 51–69. doi:10.4018/JCIT.2015010104

Pinto-Llorente, A. M., Sánchez-Gómez, M. C., & García-Peñalvo, F. J. (2016). Assessing the Effectiveness of Interactive and Collaborative Resources to Improve Reading and Writing in English. *International Journal of Human Capital and Information Technology Professionals*, *7*(1), 66–85. doi:10.4018/IJHCITP.2016010105

Pinto-Llorente, A. M., Sánchez-Gómez, M. C., & García-Peñalvo, F. J. (2017). A Mixed Methods Research of Pre-service Teachers' Perceptions about the Benefits of Wiki-Based Tasks and Discussion Boards. In A. P. Costa, L. P. Reis, F. N. De Souza, & A. Moreire (Eds.), *Computer Supported Qualitative Research, Advances in Intelligent Systems and Computing* (pp. 260–276). Springer International Publishing.

Reed, E., & Sharp, D. (2003). Minimum Technical Competencies for Distance Learning Students. *Journal of Research on Technology in Education*, *34*(3), 319–325.

Rodríguez, G., Gil, J., & García, E. (1999). *Metodología de la investigación cualitativa*. Málaga: Aljibe.

Rodríguez, G., & Gómez, M. A. (2010). Análisis de contenido y textual de datos cualitativos. In S. Nieto (Ed.), *Principios, métodos y técnicas esenciales para la investigación educativa* (pp. 447–469). Madrid: Dykinson.

Rogers, G., & Bouey, E. (2005). Participant Observation. In R. M. Grinnell & Y. A. Unrau (Eds.), *Social Work: Research and Evaluation. Quantitative and Qualitative Approaches* (pp. 231–244). New York: Oxford University Press.

Roschelle, J., & Teasley, S. D. (1995). The Construction of Shared Knowledge in Collaborative Problem Solving. In C. O'Malley (Ed.), *Computer-supported collaborative learning* (pp. 69–97). Berlin: Springer Verlag. doi:10.1007/978-3-642-85098-1_5

Salinas, J. (2000). El aprendizaje colaborativo con los nuevos canales de comunicación. In J. Cabero, J. Salinas, A. Duarte, A. & J. Domingo (Eds), Nuevas Tecnologías aplicadas a la Educación (pp. 199-228). Madrid: Síntesis.

Sánchez, M. C., Delgado, M. C., & Santos, M. C. (2012). *El proceso de la investigación cualitativa. Manual de procedimiento: ejemplificación con una tesis doctoral.* Valladolid: Fundación Intras.

Sánchez, M. C., & Revuelta, I. F. (2004). El proceso de transcripción en el marco de la metodología cualitativa actual. *Enseñanza*, *23*, 367–386.

Sánchez-Gómez, M. C., Pinto-Llorente, A. M., & García-Peñalvo, F. J. (2010). Blended learning University students' perception of digital competence. In *Proceedings of the Conference on European Applied Business Research (EABR) and European College Teaching & Learning (ETLC)* (pp. 1-11). Dublin, Ireland: The Clute Institute for Academic Research.

Simonson, M., Smaldino, S., Albright, M., & Zvacek, S. (2003). *Teaching and learning at a distance foundations of distance education.* Person.

Strauss, A., & Corbin, J. (1994). Grounded Theory Methodology. An Overview. In N. K. Denzin & Y. S. Lincoln (Eds.), *Handbook of Qualitative Research* (pp. 273–285). Thousand Oaks, CA: SAGE.

Strauss, A., & Corbin, J. (1998). *Basics of Qualitative Research: Techniques and Procedures for Developing Grounded Theory.* Thousand Oaks, CA: Sage.

Tang. C. M., & Chaw, L. Y. (2016). Digital Literacy: A Prerequisite for Effective Learning in a Blended Learning Environment? *The Electronic Journal of e-Learning, 14*(1), 54-65.

Thorne, K. (2003). *Blended Learning: How to Integrate Online and Traditional Learning.* London: Kogan.

Trinidad, A., Carretero, V., & Soriano, R. M. (2006). *Teoría fundamentada "Grounded Theory" La construcción de la teoría a través del análisis interpretacional.* Madrid: CIS.

Twu, H. (2010). A predictive study of Wiki interaction: Can attitude toward Wiki predict Wiki interaction in High-Context Cultures groups? *Journal of Educational Technology Development and Exchange*, *3*(1), 57–68. doi:10.18785/jetde.0301.05

Vat, K. H. (2010). Virtual Organizaing Professional Learning Communities through a Servant-leader Model of Appreciative Coaching. In Y. Inoue (Ed.), Cases on Online and Blended Learning Technologies in Higher Education: Concepts and Practices (pp.183-206). Hershey, PA: Information Science Reference, IGI Global.

Vygotsky, L. S. (1978). *Mind in Society: The Development of Higher Psychological Processes*. Cambridge, MA: Harvard University Press.

Walsh, L., & Kahn, P. (2010). *Collaborative Working in Higher Education. The Social Academy*. New York: Routledge.

Yeh, E., & Swinehart, N. (2017). A Learner-Centered Approach to Technology Integration: Online Geographical Tools in the ESL Classroom. In J. Keengwe & G. Onchwari (Eds.), Handbook of Research on Learner-Centered Pedagogy in Teacher Education and Professional Development (pp. 1-22). Hershey, PA: Information Science Reference, IGI Global.

Yen, J. C., & Lee, C. Y. (2011). Exploring problem solving patterns and their impact on learning achievement in a blended learning environment. *Computers & Education*, *56*(1), 138–145. doi:10.1016/j.compedu.2010.08.012

KEY TERMS AND DEFINITIONS

Autonomous Learning: It refers to a situation in which learners are responsible for their learning. They take charge of their own learning and are actively involved, taking individual decisions according to their necessities or preferences focused on the goals they need to achieve.

Blended Learning: A kind of education modality that combines both face-to-face and online lessons.

Collaborative Learning: It refers to a situation in which a group of students learn together and have an active role in creating and sharing knowledge.

Distributed Learning Systems: A general term used to describe a multi-media instructional methods including variety of web-based tools, streaming video conferencing, face-to-face classroom time, television or video, or other combinations of electronic and traditional educational methods.

Face-to-Face Learning: A traditional educational method where students and teachers are located in the same place at the same time.

Higher Education: It is the third level of education. It is beyond secondary education and is normally provided by universities or colleges.

Learning Community: It refers to a group of students who work collaboratively, share knowledge, and have common learning objectives.

Learner-Centred Pedagogy: It refers to a pedagogy that places students at the centre of the teaching-learning process. Students are more active and participative, and the process turns knowledge into a negotiation between teachers and students. Furthermore, this pedagogy starts from students' interests and objectives, and it carries out a learning process based on real situations and materials, using the current technologies to favour this process.

Qualitative Research: It is a methodological approach based on an interpretative perspective focused on the understanding of human behaviour. The techniques used to collect qualitative data are observation, interviews (semi-structured interviews, unstructured interviews, structured interviews), group discussions, evaluation of personal experiences, interaction with different communities, or recordings of life stories.

Related References

To continue our tradition of advancing academic research, we have compiled a list of recommended IGI Global readings. These references will provide additional information and guidance to further enrich your knowledge and assist you with your own research and future publications.

Adeyemo, O. (2013). The nationwide health information network: A biometric approach to prevent medical identity theft. In *User-driven healthcare: Concepts, methodologies, tools, and applications* (pp. 1636–1649). Hershey, PA: IGI Global. doi:10.4018/978-1-4666-2770-3.ch081

Adler, M., & Henman, P. (2009). Justice beyond the courts: The implications of computerisation for procedural justice in social security. In A. Martínez & P. Abat (Eds.), *E-justice: Using information communication technologies in the court system* (pp. 65–86). Hershey, PA: IGI Global. doi:10.4018/978-1-59904-998-4.ch005

Aflalo, E., & Gabay, E. (2013). An information system for coping with student dropout. In L. Tomei (Ed.), *Learning tools and teaching approaches through ICT advancements* (pp. 176–187). Hershey, PA: IGI Global. doi:10.4018/978-1-4666-2017-9.ch016

Ahmed, M. A., Janssen, M., & van den Hoven, J. (2012). Value sensitive transfer (VST) of systems among countries: Towards a framework. *International Journal of Electronic Government Research*, *8*(1), 26–42. doi:10.4018/jegr.2012010102

Aikins, S. K. (2008). Issues and trends in internet-based citizen participation. In G. Garson & M. Khosrow-Pour (Eds.), *Handbook of research on public information technology* (pp. 31–40). Hershey, PA: IGI Global. doi:10.4018/978-1-59904-857-4.ch004

Aikins, S. K. (2009). A comparative study of municipal adoption of internet-based citizen participation. In C. Reddick (Ed.), *Handbook of research on strategies for local e-government adoption and implementation: Comparative studies* (pp. 206–230). Hershey, PA: IGI Global. doi:10.4018/978-1-60566-282-4.ch011

Aikins, S. K. (2012). Improving e-government project management: Best practices and critical success factors. In *Digital democracy: Concepts, methodologies, tools, and applications* (pp. 1314–1332). Hershey, PA: IGI Global. doi:10.4018/978-1-4666-1740-7.ch065

Akabawi, M. S. (2011). Ghabbour group ERP deployment: Learning from past technology failures. In E. Business Research and Case Center (Ed.), Cases on business and management in the MENA region: New trends and opportunities (pp. 177-203). Hershey, PA: IGI Global. doi:10.4018/978-1-60960-583-4.ch012

Akabawi, M. S. (2013). Ghabbour group ERP deployment: Learning from past technology failures. In *Industrial engineering: Concepts, methodologies, tools, and applications* (pp. 933–958). Hershey, PA: IGI Global. doi:10.4018/978-1-4666-1945-6.ch051

Akbulut, A. Y., & Motwani, J. (2008). Integration and information sharing in e-government. In G. Putnik & M. Cruz-Cunha (Eds.), *Encyclopedia of networked and virtual organizations* (pp. 729–734). Hershey, PA: IGI Global. doi:10.4018/978-1-59904-885-7.ch096

Akers, E. J. (2008). Technology diffusion in public administration. In G. Garson & M. Khosrow-Pour (Eds.), *Handbook of research on public information technology* (pp. 339–348). Hershey, PA: IGI Global. doi:10.4018/978-1-59904-857-4.ch033

Al-Shafi, S. (2008). Free wireless internet park services: An investigation of technology adoption in Qatar from a citizens' perspective. *Journal of Cases on Information Technology, 10*(3), 21–34. doi:10.4018/jcit.2008070103

Al-Shafi, S., & Weerakkody, V. (2009). Implementing free wi-fi in public parks: An empirical study in Qatar. *International Journal of Electronic Government Research, 5*(3), 21–35. doi:10.4018/jegr.2009070102

Aladwani, A. M. (2002). Organizational actions, computer attitudes and end-user satisfaction in public organizations: An empirical study. In C. Snodgrass & E. Szewczak (Eds.), *Human factors in information systems* (pp. 153–168). Hershey, PA: IGI Global. doi:10.4018/978-1-931777-10-0.ch012

Aladwani, A. M. (2002). Organizational actions, computer attitudes, and end-user satisfaction in public organizations: An empirical study. *Journal of Organizational and End User Computing, 14*(1), 42–49. doi:10.4018/joeuc.2002010104

Allen, B., Juillet, L., Paquet, G., & Roy, J. (2005). E-government and private-public partnerships: Relational challenges and strategic directions. In M. Khosrow-Pour (Ed.), *Practicing e-government: A global perspective* (pp. 364–382). Hershey, PA: IGI Global. doi:10.4018/978-1-59140-637-2.ch016

Alshawaf, A., & Knalil, O. E. (2008). IS success factors and IS organizational impact: Does ownership type matter in Kuwait? *International Journal of Enterprise Information Systems, 4*(2), 13–33. doi:10.4018/jeis.2008040102

Ambali, A. R. (2009). Digital divide and its implication on Malaysian e-government: Policy initiatives. In H. Rahman (Ed.), *Social and political implications of data mining: Knowledge management in e-government* (pp. 267–287). Hershey, PA: IGI Global. doi:10.4018/978-1-60566-230-5.ch016

Amoretti, F. (2007). Digital international governance. In A. Anttiroiko & M. Malkia (Eds.), *Encyclopedia of digital government* (pp. 365–370). Hershey, PA: IGI Global. doi:10.4018/978-1-59140-789-8.ch056

Amoretti, F. (2008). Digital international governance. In A. Anttiroiko (Ed.), *Electronic government: Concepts, methodologies, tools, and applications* (pp. 688–696). Hershey, PA: IGI Global. doi:10.4018/978-1-59904-947-2.ch058

Amoretti, F. (2008). E-government at supranational level in the European Union. In A. Anttiroiko (Ed.), *Electronic government: Concepts, methodologies, tools, and applications* (pp. 1047–1055). Hershey, PA: IGI Global. doi:10.4018/978-1-59904-947-2.ch079

Amoretti, F. (2008). E-government regimes. In A. Anttiroiko (Ed.), *Electronic government: Concepts, methodologies, tools, and applications* (pp. 3846–3856). Hershey, PA: IGI Global. doi:10.4018/978-1-59904-947-2.ch280

Amoretti, F. (2009). Electronic constitution: A Braudelian perspective. In F. Amoretti (Ed.), *Electronic constitution: Social, cultural, and political implications* (pp. 1–19). Hershey, PA: IGI Global. doi:10.4018/978-1-60566-254-1.ch001

Amoretti, F., & Musella, F. (2009). Institutional isomorphism and new technologies. In M. Khosrow-Pour (Ed.), *Encyclopedia of information science and technology* (2nd ed.; pp. 2066–2071). Hershey, PA: IGI Global. doi:10.4018/978-1-60566-026-4.ch325

Andersen, K. V., & Henriksen, H. Z. (2007). E-government research: Capabilities, interaction, orientation, and values. In D. Norris (Ed.), *Current issues and trends in e-government research* (pp. 269–288). Hershey, PA: IGI Global. doi:10.4018/978-1-59904-283-1.ch013

Anderson, K. V., & Henriksen, H. Z. (2005). The first leg of e-government research: Domains and application areas 1998-2003. *International Journal of Electronic Government Research, 1*(4), 26–44. doi:10.4018/jegr.2005100102

Anttiroiko, A. (2009). Democratic e-governance. In M. Khosrow-Pour (Ed.), *Encyclopedia of information science and technology* (2nd ed.; pp. 990–995). Hershey, PA: IGI Global. doi:10.4018/978-1-60566-026-4.ch158

Baker, P. M., Bell, A., & Moon, N. W. (2009). Accessibility issues in municipal wireless networks. In C. Reddick (Ed.), *Handbook of research on strategies for local e-government adoption and implementation: Comparative studies* (pp. 569–588). Hershey, PA: IGI Global. doi:10.4018/978-1-60566-282-4.ch030

Becker, S. A., Keimer, R., & Muth, T. (2010). A case on university and community collaboration: The sci-tech entrepreneurial training services (ETS) program. In S. Becker & R. Niebuhr (Eds.), *Cases on technology innovation: Entrepreneurial successes and pitfalls* (pp. 68–90). Hershey, PA: IGI Global. doi:10.4018/978-1-61520-609-4.ch003

Becker, S. A., Keimer, R., & Muth, T. (2012). A case on university and community collaboration: The sci-tech entrepreneurial training services (ETS) program. In Regional development: Concepts, methodologies, tools, and applications (pp. 947-969). Hershey, PA: IGI Global. doi:10.4018/978-1-4666-0882-5.ch507

Bernardi, R. (2012). Information technology and resistance to public sector reforms: A case study in Kenya. In T. Papadopoulos & P. Kanellis (Eds.), *Public sector reform using information technologies: Transforming policy into practice* (pp. 59–78). Hershey, PA: IGI Global. doi:10.4018/978-1-60960-839-2.ch004

Bernardi, R. (2013). Information technology and resistance to public sector reforms: A case study in Kenya. In *User-driven healthcare: Concepts, methodologies, tools, and applications* (pp. 14–33). Hershey, PA: IGI Global. doi:10.4018/978-1-4666-2770-3.ch002

Bolívar, M. P., Pérez, M. D., & Hernández, A. M. (2012). Municipal e-government services in emerging economies: The Latin-American and Caribbean experiences. In Y. Chen & P. Chu (Eds.), *Electronic governance and cross-boundary collaboration: Innovations and advancing tools* (pp. 198–226). Hershey, PA: IGI Global. doi:10.4018/978-1-60960-753-1.ch011

Borycki, E. M., & Kushniruk, A. W. (2010). Use of clinical simulations to evaluate the impact of health information systems and ubiquitous computing devices upon health professional work. In S. Mohammed & J. Fiaidhi (Eds.), *Ubiquitous health and medical informatics: The ubiquity 2.0 trend and beyond* (pp. 552–573). Hershey, PA: IGI Global. doi:10.4018/978-1-61520-777-0.ch026

Borycki, E. M., & Kushniruk, A. W. (2011). Use of clinical simulations to evaluate the impact of health information systems and ubiquitous computing devices upon health professional work. In *Clinical technologies: Concepts, methodologies, tools and applications* (pp. 532–553). Hershey, PA: IGI Global. doi:10.4018/978-1-60960-561-2.ch220

Buchan, J. (2011). Developing a dynamic and responsive online learning environment: A case study of a large Australian university. In B. Czerkawski (Ed.), *Free and open source software for e-learning: Issues, successes and challenges* (pp. 92–109). Hershey, PA: IGI Global. doi:10.4018/978-1-61520-917-0.ch006

Buenger, A. W. (2008). Digital convergence and cybersecurity policy. In G. Garson & M. Khosrow-Pour (Eds.), *Handbook of research on public information technology* (pp. 395–405). Hershey, PA: IGI Global. doi:10.4018/978-1-59904-857-4.ch038

Burn, J. M., & Loch, K. D. (2002). The societal impact of world wide web - Key challenges for the 21st century. In A. Salehnia (Ed.), *Ethical issues of information systems* (pp. 88–106). Hershey, PA: IGI Global. doi:10.4018/978-1-931777-15-5.ch007

Burn, J. M., & Loch, K. D. (2003). The societal impact of the world wide web-Key challenges for the 21st century. In M. Khosrow-Pour (Ed.), *Advanced topics in information resources management* (Vol. 2, pp. 32–51). Hershey, PA: IGI Global. doi:10.4018/978-1-59140-062-2.ch002

Bwalya, K. J., Du Plessis, T., & Rensleigh, C. (2012). The "quicksilver initiatives" as a framework for e-government strategy design in developing economies. In K. Bwalya & S. Zulu (Eds.), *Handbook of research on e-government in emerging economies: Adoption, e-participation, and legal frameworks* (pp. 605–623). Hershey, PA: IGI Global. doi:10.4018/978-1-4666-0324-0.ch031

Cabotaje, C. E., & Alampay, E. A. (2013). Social media and citizen engagement: Two cases from the Philippines. In S. Saeed & C. Reddick (Eds.), *Human-centered system design for electronic governance* (pp. 225–238). Hershey, PA: IGI Global. doi:10.4018/978-1-4666-3640-8.ch013

Camillo, A., Di Pietro, L., Di Virgilio, F., & Franco, M. (2013). Work-groups conflict at PetroTech-Italy, S.R.L.: The influence of culture on conflict dynamics. In B. Christiansen, E. Turkina, & N. Williams (Eds.), *Cultural and technological influences on global business* (pp. 272–289). Hershey, PA: IGI Global. doi:10.4018/978-1-4666-3966-9.ch015

Capra, E., Francalanci, C., & Marinoni, C. (2008). Soft success factors for m-government. In A. Anttiroiko (Ed.), *Electronic government: Concepts, methodologies, tools, and applications* (pp. 1213–1233). Hershey, PA: IGI Global. doi:10.4018/978-1-59904-947-2.ch089

Cartelli, A. (2009). The implementation of practices with ICT as a new teaching-learning paradigm. In A. Cartelli & M. Palma (Eds.), *Encyclopedia of information communication technology* (pp. 413–417). Hershey, PA: IGI Global. doi:10.4018/978-1-59904-845-1.ch055

Charalabidis, Y., Lampathaki, F., & Askounis, D. (2010). Investigating the landscape in national interoperability frameworks. *International Journal of E-Services and Mobile Applications*, *2*(4), 28–41. doi:10.4018/jesma.2010100103

Charalabidis, Y., Lampathaki, F., & Askounis, D. (2012). Investigating the landscape in national interoperability frameworks. In A. Scupola (Ed.), *Innovative mobile platform developments for electronic services design and delivery* (pp. 218–231). Hershey, PA: IGI Global. doi:10.4018/978-1-4666-1568-7.ch013

Chen, I. (2005). Distance education associations. In C. Howard, J. Boettcher, L. Justice, K. Schenk, P. Rogers, & G. Berg (Eds.), *Encyclopedia of distance learning* (pp. 599–612). Hershey, PA: IGI Global. doi:10.4018/978-1-59140-555-9.ch087

Chen, I. (2008). Distance education associations. In L. Tomei (Ed.), *Online and distance learning: Concepts, methodologies, tools, and applications* (pp. 562–579). Hershey, PA: IGI Global. doi:10.4018/978-1-59904-935-9.ch048

Chen, Y. (2008). Managing IT outsourcing for digital government. In A. Anttiroiko (Ed.), *Electronic government: Concepts, methodologies, tools, and applications* (pp. 3107–3114). Hershey, PA: IGI Global. doi:10.4018/978-1-59904-947-2.ch229

Chen, Y., & Dimitrova, D. V. (2006). Electronic government and online engagement: Citizen interaction with government via web portals. *International Journal of Electronic Government Research*, *2*(1), 54–76. doi:10.4018/jegr.2006010104

Chen, Y., & Knepper, R. (2005). Digital government development strategies: Lessons for policy makers from a comparative perspective. In W. Huang, K. Siau, & K. Wei (Eds.), *Electronic government strategies and implementation* (pp. 394–420). Hershey, PA: IGI Global. doi:10.4018/978-1-59140-348-7.ch017

Chen, Y., & Knepper, R. (2008). Digital government development strategies: Lessons for policy makers from a comparative perspective. In H. Rahman (Ed.), *Developing successful ICT strategies: Competitive advantages in a global knowledge-driven society* (pp. 334–356). Hershey, PA: IGI Global. doi:10.4018/978-1-59904-654-9.ch017

Cherian, E. J., & Ryan, T. W. (2014). Incongruent needs: Why differences in the iron-triangle of priorities make health information technology adoption and use difficult. In C. El Morr (Ed.), *Research perspectives on the role of informatics in health policy and management* (pp. 209–221). Hershey, PA: IGI Global. doi:10.4018/978-1-4666-4321-5.ch012

Cho, H. J., & Hwang, S. (2010). Government 2.0 in Korea: Focusing on e-participation services. In C. Reddick (Ed.), *Politics, democracy and e-government: Participation and service delivery* (pp. 94–114). Hershey, PA: IGI Global. doi:10.4018/978-1-61520-933-0.ch006

Chorus, C., & Timmermans, H. (2010). Ubiquitous travel environments and travel control strategies: Prospects and challenges. In M. Wachowicz (Ed.), *Movement-aware applications for sustainable mobility: Technologies and approaches* (pp. 30–51). Hershey, PA: IGI Global. doi:10.4018/978-1-61520-769-5.ch003

Chuanshen, R. (2007). E-government construction and China's administrative litigation act. In A. Anttiroiko & M. Malkia (Eds.), *Encyclopedia of digital government* (pp. 507–510). Hershey, PA: IGI Global. doi:10.4018/978-1-59140-789-8.ch077

Ciaghi, A., & Villafiorita, A. (2012). Law modeling and BPR for public administration improvement. In K. Bwalya & S. Zulu (Eds.), *Handbook of research on e-government in emerging economies: Adoption, e-participation, and legal frameworks* (pp. 391–410). Hershey, PA: IGI Global. doi:10.4018/978-1-4666-0324-0.ch019

Ciaramitaro, B. L., & Skrocki, M. (2012). mHealth: Mobile healthcare. In B. Ciaramitaro (Ed.), Mobile technology consumption: Opportunities and challenges (pp. 99-109). Hershey, PA: IGI Global. doi:10.4018/978-1-61350-150-4.ch007

Comite, U. (2012). Innovative processes and managerial effectiveness of e-procurement in healthcare. In A. Manoharan & M. Holzer (Eds.), *Active citizen participation in e-government: A global perspective* (pp. 206–229). Hershey, PA: IGI Global. doi:10.4018/978-1-4666-0116-1.ch011

Cordella, A. (2013). E-government success: How to account for ICT, administrative rationalization, and institutional change. In J. Gil-Garcia (Ed.), *E-government success factors and measures: Theories, concepts, and methodologies* (pp. 40–51). Hershey, PA: IGI Global. doi:10.4018/978-1-4666-4058-0.ch003

Cropf, R. A. (2009). ICT and e-democracy. In M. Khosrow-Pour (Ed.), *Encyclopedia of information science and technology* (2nd ed.; pp. 1789–1793). Hershey, PA: IGI Global. doi:10.4018/978-1-60566-026-4.ch281

Cropf, R. A. (2009). The virtual public sphere. In M. Pagani (Ed.), *Encyclopedia of multimedia technology and networking* (2nd ed.; pp. 1525–1530). Hershey, PA: IGI Global. doi:10.4018/978-1-60566-014-1.ch206

D'Abundo, M. L. (2013). Electronic health record implementation in the United States healthcare industry: Making the process of change manageable. In V. Wang (Ed.), *Handbook of research on technologies for improving the 21st century workforce: Tools for lifelong learning* (pp. 272–286). Hershey, PA: IGI Global. doi:10.4018/978-1-4666-2181-7.ch018

Damurski, L. (2012). E-participation in urban planning: Online tools for citizen engagement in Poland and in Germany. *International Journal of E-Planning Research, 1*(3), 40–67. doi:10.4018/ijepr.2012070103

de Almeida, M. O. (2007). E-government strategy in Brazil: Increasing transparency and efficiency through e-government procurement. In M. Gascó-Hernandez (Ed.), *Latin America online: Cases, successes and pitfalls* (pp. 34–82). Hershey, PA: IGI Global. doi:10.4018/978-1-59140-974-8.ch002

de Juana Espinosa, S. (2008). Empirical study of the municipalitites' motivations for adopting online presence. In A. Anttiroiko (Ed.), *Electronic government: Concepts, methodologies, tools, and applications* (pp. 3593–3608). Hershey, PA: IGI Global. doi:10.4018/978-1-59904-947-2.ch262

de Souza Dias, D. (2002). Motivation for using information technology. In C. Snodgrass & E. Szewczak (Eds.), *Human factors in information systems* (pp. 55–60). Hershey, PA: IGI Global. doi:10.4018/978-1-931777-10-0.ch005

Demediuk, P. (2006). Government procurement ICT's impact on the sustainability of SMEs and regional communities. In S. Marshall, W. Taylor, & X. Yu (Eds.), *Encyclopedia of developing regional communities with information and communication technology* (pp. 321–324). Hershey, PA: IGI Global. doi:10.4018/978-1-59140-575-7.ch056

Devonshire, E., Forsyth, H., Reid, S., & Simpson, J. M. (2013). The challenges and opportunities of online postgraduate coursework programs in a traditional university context. In B. Tynan, J. Willems, & R. James (Eds.), *Outlooks and opportunities in blended and distance learning* (pp. 353–368). Hershey, PA: IGI Global. doi:10.4018/978-1-4666-4205-8.ch026

Di Cerbo, F., Scotto, M., Sillitti, A., Succi, G., & Vernazza, T. (2007). Toward a GNU/Linux distribution for corporate environments. In S. Sowe, I. Stamelos, & I. Samoladas (Eds.), *Emerging free and open source software practices* (pp. 215–236). Hershey, PA: IGI Global. doi:10.4018/978-1-59904-210-7.ch010

Diesner, J., & Carley, K. M. (2005). Revealing social structure from texts: Meta-matrix text analysis as a novel method for network text analysis. In V. Narayanan & D. Armstrong (Eds.), *Causal mapping for research in information technology* (pp. 81–108). Hershey, PA: IGI Global. doi:10.4018/978-1-59140-396-8.ch004

Dologite, D. G., Mockler, R. J., Bai, Q., & Viszhanyo, P. F. (2006). IS change agents in practice in a US-Chinese joint venture. In M. Hunter & F. Tan (Eds.), *Advanced topics in global information management* (Vol. 5, pp. 331–352). Hershey, PA: IGI Global. doi:10.4018/978-1-59140-923-6.ch015

Drnevich, P., Brush, T. H., & Luckock, G. T. (2011). Process and structural implications for IT-enabled outsourcing. *International Journal of Strategic Information Technology and Applications*, *2*(4), 30–43. doi:10.4018/jsita.2011100103

Dwivedi, A. N. (2009). *Handbook of research on information technology management and clinical data administration in healthcare* (Vols. 1–2). Hershey, PA: IGI Global. doi:10.4018/978-1-60566-356-2

Elbeltagi, I., McBride, N., & Hardaker, G. (2006). Evaluating the factors affecting DSS usage by senior managers in local authorities in Egypt. In M. Hunter & F. Tan (Eds.), *Advanced topics in global information management* (Vol. 5, pp. 283–307). Hershey, PA: IGI Global. doi:10.4018/978-1-59140-923-6.ch013

Eom, S., & Fountain, J. E. (2013). Enhancing information services through public-private partnerships: Information technology knowledge transfer underlying structures to develop shared services in the U.S. and Korea. In J. Gil-Garcia (Ed.), *E-government success around the world: Cases, empirical studies, and practical recommendations* (pp. 15–40). Hershey, PA: IGI Global. doi:10.4018/978-1-4666-4173-0.ch002

Esteves, T., Leuenberger, D., & Van Leuven, N. (2012). Reaching citizen 2.0: How government uses social media to send public messages during times of calm and times of crisis. In K. Kloby & M. D'Agostino (Eds.), *Citizen 2.0: Public and governmental interaction through web 2.0 technologies* (pp. 250–268). Hershey, PA: IGI Global. doi:10.4018/978-1-4666-0318-9.ch013

Estevez, E., Fillottrani, P., Janowski, T., & Ojo, A. (2012). Government information sharing: A framework for policy formulation. In Y. Chen & P. Chu (Eds.), *Electronic governance and cross-boundary collaboration: Innovations and advancing tools* (pp. 23–55). Hershey, PA: IGI Global. doi:10.4018/978-1-60960-753-1.ch002

Ezz, I. E. (2008). E-governement emerging trends: Organizational challenges. In A. Anttiroiko (Ed.), *Electronic government: Concepts, methodologies, tools, and applications* (pp. 3721–3737). Hershey, PA: IGI Global. doi:10.4018/978-1-59904-947-2.ch269

Fabri, M. (2009). The Italian style of e-justice in a comparative perspective. In A. Martínez & P. Abat (Eds.), *E-justice: Using information communication technologies in the court system* (pp. 1–19). Hershey, PA: IGI Global. doi:10.4018/978-1-59904-998-4.ch001

Fagbe, T., & Adekola, O. D. (2010). Workplace safety and personnel well-being: The impact of information technology. *International Journal of Green Computing*, *1*(1), 28–33. doi:10.4018/jgc.2010010103

Fagbe, T., & Adekola, O. D. (2011). Workplace safety and personnel well-being: The impact of information technology. In *Global business: Concepts, methodologies, tools and applications* (pp. 1438–1444). Hershey, PA: IGI Global. doi:10.4018/978-1-60960-587-2.ch509

Farmer, L. (2008). Affective collaborative instruction with librarians. In S. Kelsey & K. St.Amant (Eds.), *Handbook of research on computer mediated communication* (pp. 15–24). Hershey, PA: IGI Global. doi:10.4018/978-1-59904-863-5.ch002

Favier, L., & Mekhantar, J. (2007). Use of OSS by local e-administration: The French situation. In K. St.Amant & B. Still (Eds.), *Handbook of research on open source software: Technological, economic, and social perspectives* (pp. 428–444). Hershey, PA: IGI Global. doi:10.4018/978-1-59140-999-1.ch033

Fernando, S. (2009). Issues of e-learning in third world countries. In M. Khosrow-Pour (Ed.), *Encyclopedia of information science and technology* (2nd ed.; pp. 2273–2277). Hershey, PA: IGI Global. doi:10.4018/978-1-60566-026-4.ch360

Filho, J. R., & dos Santos, J. R. Jr. (2009). Local e-government in Brazil: Poor interaction and local politics as usual. In C. Reddick (Ed.), *Handbook of research on strategies for local e-government adoption and implementation: Comparative studies* (pp. 863–878). Hershey, PA: IGI Global. doi:10.4018/978-1-60566-282-4.ch045

Fletcher, P. D. (2004). Portals and policy: Implications of electronic access to U.S. federal government information services. In A. Pavlichev & G. Garson (Eds.), *Digital government: Principles and best practices* (pp. 52–62). Hershey, PA: IGI Global. doi:10.4018/978-1-59140-122-3.ch004

Fletcher, P. D. (2008). Portals and policy: Implications of electronic access to U.S. federal government information services. In A. Anttiroiko (Ed.), *Electronic government: Concepts, methodologies, tools, and applications* (pp. 3970–3979). Hershey, PA: IGI Global. doi:10.4018/978-1-59904-947-2.ch289

Forlano, L. (2004). The emergence of digital government: International perspectives. In A. Pavlichev & G. Garson (Eds.), *Digital government: Principles and best practices* (pp. 34–51). Hershey, PA: IGI Global. doi:10.4018/978-1-59140-122-3.ch003

Franzel, J. M., & Coursey, D. H. (2004). Government web portals: Management issues and the approaches of five states. In A. Pavlichev & G. Garson (Eds.), *Digital government: Principles and best practices* (pp. 63–77). Hershey, PA: IGI Global. doi:10.4018/978-1-59140-122-3.ch005

Gaivéo, J. M. (2013). Security of ICTs supporting healthcare activities. In M. Cruz-Cunha, I. Miranda, & P. Gonçalves (Eds.), *Handbook of research on ICTs for human-centered healthcare and social care services* (pp. 208–228). Hershey, PA: IGI Global. doi:10.4018/978-1-4666-3986-7.ch011

Garson, G. D. (1999). *Information technology and computer applications in public administration: Issues and trends*. Hershey, PA: IGI Global. doi:10.4018/978-1-87828-952-0

Garson, G. D. (2003). Toward an information technology research agenda for public administration. In G. Garson (Ed.), *Public information technology: Policy and management issues* (pp. 331–357). Hershey, PA: IGI Global. doi:10.4018/978-1-59140-060-8.ch014

Garson, G. D. (2004). The promise of digital government. In A. Pavlichev & G. Garson (Eds.), *Digital government: Principles and best practices* (pp. 2–15). Hershey, PA: IGI Global. doi:10.4018/978-1-59140-122-3.ch001

Garson, G. D. (2007). An information technology research agenda for public administration. In G. Garson (Ed.), *Modern public information technology systems: Issues and challenges* (pp. 365–392). Hershey, PA: IGI Global. doi:10.4018/978-1-59904-051-6.ch018

Gasco, M. (2007). Civil servants' resistance towards e-government development. In A. Anttiroiko & M. Malkia (Eds.), *Encyclopedia of digital government* (pp. 190–195). Hershey, PA: IGI Global. doi:10.4018/978-1-59140-789-8.ch028

Gasco, M. (2008). Civil servants' resistance towards e-government development. In A. Anttiroiko (Ed.), *Electronic government: Concepts, methodologies, tools, and applications* (pp. 2580–2588). Hershey, PA: IGI Global. doi:10.4018/978-1-59904-947-2.ch190

Ghere, R. K. (2010). Accountability and information technology enactment: Implications for social empowerment. In E. Ferro, Y. Dwivedi, J. Gil-Garcia, & M. Williams (Eds.), *Handbook of research on overcoming digital divides: Constructing an equitable and competitive information society* (pp. 515–532). Hershey, PA: IGI Global. doi:10.4018/978-1-60566-699-0.ch028

Gibson, I. W. (2012). Simulation modeling of healthcare delivery. In A. Kolker & P. Story (Eds.), *Management engineering for effective healthcare delivery: Principles and applications* (pp. 69–89). Hershey, PA: IGI Global. doi:10.4018/978-1-60960-872-9.ch003

Gil-Garcia, J. R. (2007). Exploring e-government benefits and success factors. In A. Anttiroiko & M. Malkia (Eds.), *Encyclopedia of digital government* (pp. 803–811). Hershey, PA: IGI Global. doi:10.4018/978-1-59140-789-8.ch122

Gil-Garcia, J. R., & González Miranda, F. (2010). E-government and opportunities for participation: The case of the Mexican state web portals. In C. Reddick (Ed.), *Politics, democracy and e-government: Participation and service delivery* (pp. 56–74). Hershey, PA: IGI Global. doi:10.4018/978-1-61520-933-0.ch004

Goldfinch, S. (2012). Public trust in government, trust in e-government, and use of e-government. In Z. Yan (Ed.), *Encyclopedia of cyber behavior* (pp. 987–995). Hershey, PA: IGI Global. doi:10.4018/978-1-4666-0315-8.ch081

Goodyear, M. (2012). Organizational change contributions to e-government project transitions. In S. Aikins (Ed.), *Managing e-government projects: Concepts, issues, and best practices* (pp. 1–21). Hershey, PA: IGI Global. doi:10.4018/978-1-4666-0086-7.ch001

Gordon, S., & Mulligan, P. (2003). Strategic models for the delivery of personal financial services: The role of infocracy. In S. Gordon (Ed.), *Computing information technology: The human side* (pp. 220–232). Hershey, PA: IGI Global. doi:10.4018/978-1-93177-752-0.ch014

Gordon, T. F. (2007). Legal knowledge systems. In A. Anttiroiko & M. Malkia (Eds.), *Encyclopedia of digital government* (pp. 1161–1166). Hershey, PA: IGI Global. doi:10.4018/978-1-59140-789-8.ch175

Graham, J. E., & Semich, G. W. (2008). Integrating technology to transform pedagogy: Revisiting the progress of the three phase TUI model for faculty development. In L. Tomei (Ed.), *Adapting information and communication technologies for effective education* (pp. 1–12). Hershey, PA: IGI Global. doi:10.4018/978-1-59904-922-9.ch001

Grandinetti, L., & Pisacane, O. (2012). Web services for healthcare management. In D. Prakash Vidyarthi (Ed.), *Technologies and protocols for the future of internet design: Reinventing the web* (pp. 60–94). Hershey, PA: IGI Global. doi:10.4018/978-1-4666-0203-8.ch004

Groenewegen, P., & Wagenaar, F. P. (2008). VO as an alternative to hierarchy in the Dutch police sector. In G. Putnik & M. Cruz-Cunha (Eds.), *Encyclopedia of networked and virtual organizations* (pp. 1851–1857). Hershey, PA: IGI Global. doi:10.4018/978-1-59904-885-7.ch245

Gronlund, A. (2001). Building an infrastructure to manage electronic services. In S. Dasgupta (Ed.), *Managing internet and intranet technologies in organizations: Challenges and opportunities* (pp. 71–103). Hershey, PA: IGI Global. doi:10.4018/978-1-878289-95-7.ch006

Gronlund, A. (2002). Introduction to electronic government: Design, applications and management. In Å. Grönlund (Ed.), *Electronic government: Design, applications and management* (pp. 1–21). Hershey, PA: IGI Global. doi:10.4018/978-1-930708-19-8.ch001

Gupta, A., Woosley, R., Crk, I., & Sarnikar, S. (2009). An information technology architecture for drug effectiveness reporting and post-marketing surveillance. In J. Tan (Ed.), *Medical informatics: Concepts, methodologies, tools, and applications* (pp. 631–646). Hershey, PA: IGI Global. doi:10.4018/978-1-60566-050-9.ch047

Hallin, A., & Lundevall, K. (2007). mCity: User focused development of mobile services within the city of Stockholm. In I. Kushchu (Ed.), Mobile government: An emerging direction in e-government (pp. 12-29). Hershey, PA: IGI Global. doi:10.4018/978-1-59140-884-0.ch002

Hallin, A., & Lundevall, K. (2009). mCity: User focused development of mobile services within the city of Stockholm. In S. Clarke (Ed.), Evolutionary concepts in end user productivity and performance: Applications for organizational progress (pp. 268-280). Hershey, PA: IGI Global. doi:10.4018/978-1-60566-136-0.ch017

Hallin, A., & Lundevall, K. (2009). mCity: User focused development of mobile services within the city of Stockholm. In D. Taniar (Ed.), Mobile computing: Concepts, methodologies, tools, and applications (pp. 3455-3467). Hershey, PA: IGI Global. doi:10.4018/978-1-60566-054-7.ch253

Hanson, A. (2005). Overcoming barriers in the planning of a virtual library. In M. Khosrow-Pour (Ed.), *Encyclopedia of information science and technology* (pp. 2255–2259). Hershey, PA: IGI Global. doi:10.4018/978-1-59140-553-5.ch397

Haque, A. (2008). Information technology and surveillance: Implications for public administration in a new word order. In T. Loendorf & G. Garson (Eds.), *Patriotic information systems* (pp. 177–185). Hershey, PA: IGI Global. doi:10.4018/978-1-59904-594-8.ch008

Hauck, R. V., Thatcher, S. M., & Weisband, S. P. (2012). Temporal aspects of information technology use: Increasing shift work effectiveness. In J. Wang (Ed.), *Advancing the service sector with evolving technologies: Techniques and principles* (pp. 87–104). Hershey, PA: IGI Global. doi:10.4018/978-1-4666-0044-7.ch006

Hawk, S., & Witt, T. (2006). Telecommunications courses in information systems programs. *International Journal of Information and Communication Technology Education*, *2*(1), 79–92. doi:10.4018/jicte.2006010107

Helms, M. M., Moore, R., & Ahmadi, M. (2009). Information technology (IT) and the healthcare industry: A SWOT analysis. In J. Tan (Ed.), *Medical informatics: Concepts, methodologies, tools, and applications* (pp. 134–152). Hershey, PA: IGI Global. doi:10.4018/978-1-60566-050-9.ch012

Hendrickson, S. M., & Young, M. E. (2014). Electronic records management at a federally funded research and development center. In J. Krueger (Ed.), *Cases on electronic records and resource management implementation in diverse environments* (pp. 334–350). Hershey, PA: IGI Global. doi:10.4018/978-1-4666-4466-3.ch020

Henman, P. (2010). Social policy and information communication technologies. In J. Martin & L. Hawkins (Eds.), *Information communication technologies for human services education and delivery: Concepts and cases* (pp. 215–229). Hershey, PA: IGI Global. doi:10.4018/978-1-60566-735-5.ch014

Hismanoglu, M. (2011). Important issues in online education: E-pedagogy and marketing. In U. Demiray & S. Sever (Eds.), *Marketing online education programs: Frameworks for promotion and communication* (pp. 184–209). Hershey, PA: IGI Global. doi:10.4018/978-1-60960-074-7.ch012

Ho, K. K. (2008). The e-government development, IT strategies, and portals of the Hong Kong SAR government. In A. Anttiroiko (Ed.), *Electronic government: Concepts, methodologies, tools, and applications* (pp. 715–733). Hershey, PA: IGI Global. doi:10.4018/978-1-59904-947-2.ch060

Holden, S. H. (2003). The evolution of information technology management at the federal level: Implications for public administration. In G. Garson (Ed.), *Public information technology: Policy and management issues* (pp. 53–73). Hershey, PA: IGI Global. doi:10.4018/978-1-59140-060-8.ch003

Holden, S. H. (2007). The evolution of federal information technology management literature: Does IT finally matter? In G. Garson (Ed.), *Modern public information technology systems: Issues and challenges* (pp. 17–34). Hershey, PA: IGI Global. doi:10.4018/978-1-59904-051-6.ch002

Holland, J. W. (2009). Automation of American criminal justice. In M. Khosrow-Pour (Ed.), *Encyclopedia of information science and technology* (2nd ed.; pp. 300–302). Hershey, PA: IGI Global. doi:10.4018/978-1-60566-026-4.ch051

Holloway, K. (2013). Fair use, copyright, and academic integrity in an online academic environment. In *Digital rights management: Concepts, methodologies, tools, and applications* (pp. 917–928). Hershey, PA: IGI Global. doi:10.4018/978-1-4666-2136-7.ch044

Horiuchi, C. (2005). E-government databases. In L. Rivero, J. Doorn, & V. Ferraggine (Eds.), *Encyclopedia of database technologies and applications* (pp. 206–210). Hershey, PA: IGI Global. doi:10.4018/978-1-59140-560-3.ch035

Horiuchi, C. (2006). Creating IS quality in government settings. In E. Duggan & J. Reichgelt (Eds.), *Measuring information systems delivery quality* (pp. 311–327). Hershey, PA: IGI Global. doi:10.4018/978-1-59140-857-4.ch014

Hsiao, N., Chu, P., & Lee, C. (2012). Impact of e-governance on businesses: Model development and case study. In *Digital democracy: Concepts, methodologies, tools, and applications* (pp. 1407–1425). Hershey, PA: IGI Global. doi:10.4018/978-1-4666-1740-7.ch070

Huang, T., & Lee, C. (2010). Evaluating the impact of e-government on citizens: Cost-benefit analysis. In C. Reddick (Ed.), *Citizens and e-government: Evaluating policy and management* (pp. 37–52). Hershey, PA: IGI Global. doi:10.4018/978-1-61520-931-6.ch003

Hunter, M. G., Diochon, M., Pugsley, D., & Wright, B. (2002). Unique challenges for small business adoption of information technology: The case of the Nova Scotia ten. In S. Burgess (Ed.), *Managing information technology in small business: Challenges and solutions* (pp. 98–117). Hershey, PA: IGI Global. doi:10.4018/978-1-930708-35-8.ch006

Hurskainen, J. (2003). Integration of business systems and applications in merger and alliance: Case metso automation. In T. Reponen (Ed.), *Information technology enabled global customer service* (pp. 207–225). Hershey, PA: IGI Global. doi:10.4018/978-1-59140-048-6.ch012

Iazzolino, G., & Pietrantonio, R. (2011). The soveria.it project: A best practice of e-government in southern Italy. In D. Piaggesi, K. Sund, & W. Castelnovo (Eds.), *Global strategy and practice of e-governance: Examples from around the world* (pp. 34–56). Hershey, PA: IGI Global. doi:10.4018/978-1-60960-489-9.ch003

Imran, A., & Gregor, S. (2012). A process model for successful e-government adoption in the least developed countries: A case of Bangladesh. In F. Tan (Ed.), *International comparisons of information communication technologies: Advancing applications* (pp. 321–350). Hershey, PA: IGI Global. doi:10.4018/978-1-61350-480-2.ch014

Inoue, Y., & Bell, S. T. (2005). Electronic/digital government innovation, and publishing trends with IT. In M. Khosrow-Pour (Ed.), *Encyclopedia of information science and technology* (pp. 1018–1023). Hershey, PA: IGI Global. doi:10.4018/978-1-59140-553-5.ch180

Islam, M. M., & Ehsan, M. (2013). Understanding e-governance: A theoretical approach. In M. Islam & M. Ehsan (Eds.), *From government to e-governance: Public administration in the digital age* (pp. 38–49). Hershey, PA: IGI Global. doi:10.4018/978-1-4666-1909-8.ch003

Jaeger, B. (2009). E-government and e-democracy in the making. In M. Khosrow-Pour (Ed.), *Encyclopedia of information science and technology* (2nd ed.; pp. 1318–1322). Hershey, PA: IGI Global. doi:10.4018/978-1-60566-026-4.ch208

Jain, R. B. (2007). Revamping the administrative structure and processes in India for online diplomacy. In A. Anttiroiko & M. Malkia (Eds.), *Encyclopedia of digital government* (pp. 1418–1423). Hershey, PA: IGI Global. doi:10.4018/978-1-59140-789-8.ch217

Jain, R. B. (2008). Revamping the administrative structure and processes in India for online diplomacy. In A. Anttiroiko (Ed.), *Electronic government: Concepts, methodologies, tools, and applications* (pp. 3142–3149). Hershey, PA: IGI Global. doi:10.4018/978-1-59904-947-2.ch233

Jauhiainen, J. S., & Inkinen, T. (2009). E-governance and the information society in periphery. In C. Reddick (Ed.), *Handbook of research on strategies for local e-government adoption and implementation: Comparative studies* (pp. 497–514). Hershey, PA: IGI Global. doi:10.4018/978-1-60566-282-4.ch026

Jensen, M. J. (2009). Electronic democracy and citizen influence in government. In C. Reddick (Ed.), *Handbook of research on strategies for local e-government adoption and implementation: Comparative studies* (pp. 288–305). Hershey, PA: IGI Global. doi:10.4018/978-1-60566-282-4.ch015

Jiao, Y., Hurson, A. R., Potok, T. E., & Beckerman, B. G. (2009). Integrating mobile-based systems with healthcare databases. In J. Erickson (Ed.), *Database technologies: Concepts, methodologies, tools, and applications* (pp. 484–504). Hershey, PA: IGI Global. doi:10.4018/978-1-60566-058-5.ch031

Joia, L. A. (2002). A systematic model to integrate information technology into metabusinesses: A case study in the engineering realms. In F. Tan (Ed.), *Advanced topics in global information management* (Vol. 1, pp. 250–267). Hershey, PA: IGI Global. doi:10.4018/978-1-930708-43-3.ch016

Jones, T. H., & Song, I. (2000). Binary equivalents of ternary relationships in entity-relationship modeling: A logical decomposition approach. *Journal of Database Management, 11*(2), 12–19. doi:10.4018/jdm.2000040102

Juana-Espinosa, S. D. (2007). Empirical study of the municipalitites' motivations for adopting online presence. In L. Al-Hakim (Ed.), *Global e-government: Theory, applications and benchmarking* (pp. 261–279). Hershey, PA: IGI Global. doi:10.4018/978-1-59904-027-1.ch015

Jun, K., & Weare, C. (2012). Bridging from e-government practice to e-government research: Past trends and future directions. In K. Bwalya & S. Zulu (Eds.), *Handbook of research on e-government in emerging economies: Adoption, e-participation, and legal frameworks* (pp. 263–289). Hershey, PA: IGI Global. doi:10.4018/978-1-4666-0324-0.ch013

Junqueira, A., Diniz, E. H., & Fernandez, M. (2010). Electronic government implementation projects with multiple agencies: Analysis of the electronic invoice project under PMBOK framework. In J. Cordoba-Pachon & A. Ochoa-Arias (Eds.), *Systems thinking and e-participation: ICT in the governance of society* (pp. 135–153). Hershey, PA: IGI Global. doi:10.4018/978-1-60566-860-4.ch009

Juntunen, A. (2009). Joint service development with the local authorities. In C. Reddick (Ed.), *Handbook of research on strategies for local e-government adoption and implementation: Comparative studies* (pp. 902–920). Hershey, PA: IGI Global. doi:10.4018/978-1-60566-282-4.ch047

Kamel, S. (2001). *Using DSS for crisis management*. Hershey, PA: IGI Global. doi:10.4018/978-1-87828-961-2.ch020

Kamel, S. (2006). DSS for strategic decision making. In M. Khosrow-Pour (Ed.), *Cases on information technology and organizational politics & culture* (pp. 230–246). Hershey, PA: IGI Global. doi:10.4018/978-1-59904-411-8.ch015

Kamel, S. (2009). The software industry in Egypt as a potential contributor to economic growth. In M. Khosrow-Pour (Ed.), *Encyclopedia of information science and technology* (2nd ed.; pp. 3531–3537). Hershey, PA: IGI Global. doi:10.4018/978-1-60566-026-4.ch562

Kamel, S., & Hussein, M. (2008). Xceed: Pioneering the contact center industry in Egypt. *Journal of Cases on Information Technology, 10*(1), 67–91. doi:10.4018/jcit.2008010105

Kamel, S., & Wahba, K. (2003). The use of a hybrid model in web-based education: "The Global campus project. In A. Aggarwal (Ed.), *Web-based education: Learning from experience* (pp. 331–346). Hershey, PA: IGI Global. doi:10.4018/978-1-59140-102-5.ch020

Kardaras, D. K., & Papathanassiou, E. A. (2008). An exploratory study of the e-government services in Greece. In G. Garson & M. Khosrow-Pour (Eds.), *Handbook of research on public information technology* (pp. 162–174). Hershey, PA: IGI Global. doi:10.4018/978-1-59904-857-4.ch016

Kassahun, A. E., Molla, A., & Sarkar, P. (2012). Government process reengineering: What we know and what we need to know. In *Digital democracy: Concepts, methodologies, tools, and applications* (pp. 1730–1752). Hershey, PA: IGI Global. doi:10.4018/978-1-4666-1740-7.ch086

Khan, B. (2005). Technological issues. In B. Khan (Ed.), *Managing e-learning strategies: Design, delivery, implementation and evaluation* (pp. 154–180). Hershey, PA: IGI Global. doi:10.4018/978-1-59140-634-1.ch004

Khasawneh, A., Bsoul, M., Obeidat, I., & Al Azzam, I. (2012). Technology fears: A study of e-commerce loyalty perception by Jordanian customers. In J. Wang (Ed.), *Advancing the service sector with evolving technologies: Techniques and principles* (pp. 158–165). Hershey, PA: IGI Global. doi:10.4018/978-1-4666-0044-7.ch010

Khatibi, V., & Montazer, G. A. (2012). E-research methodology. In A. Juan, T. Daradoumis, M. Roca, S. Grasman, & J. Faulin (Eds.), *Collaborative and distributed e-research: Innovations in technologies, strategies and applications* (pp. 62–81). Hershey, PA: IGI Global. doi:10.4018/978-1-4666-0125-3.ch003

Kidd, T. (2011). The dragon in the school's backyard: A review of literature on the uses of technology in urban schools. In L. Tomei (Ed.), *Online courses and ICT in education: Emerging practices and applications* (pp. 242–257). Hershey, PA: IGI Global. doi:10.4018/978-1-60960-150-8.ch019

Kidd, T. T. (2010). My experience tells the story: Exploring technology adoption from a qualitative perspective - A pilot study. In H. Song & T. Kidd (Eds.), *Handbook of research on human performance and instructional technology* (pp. 247–262). Hershey, PA: IGI Global. doi:10.4018/978-1-60566-782-9.ch015

Kieley, B., Lane, G., Paquet, G., & Roy, J. (2002). e-Government in Canada: Services online or public service renewal? In Å. Grönlund (Ed.), Electronic government: Design, applications and management (pp. 340-355). Hershey, PA: IGI Global. doi:10.4018/978-1-930708-19-8.ch016

Kim, P. (2012). "Stay out of the way! My kid is video blogging through a phone!": A lesson learned from math tutoring social media for children in underserved communities. In *Wireless technologies: Concepts, methodologies, tools and applications* (pp. 1415–1428). Hershey, PA: IGI Global. doi:10.4018/978-1-61350-101-6.ch517

Kirlidog, M. (2010). Financial aspects of national ICT strategies. In S. Kamel (Ed.), *E-strategies for technological diffusion and adoption: National ICT approaches for socioeconomic development* (pp. 277–292). Hershey, PA: IGI Global. doi:10.4018/978-1-60566-388-3.ch016

Kisielnicki, J. (2006). Transfer of information and knowledge in the project management. In E. Coakes & S. Clarke (Eds.), *Encyclopedia of communities of practice in information and knowledge management* (pp. 544–551). Hershey, PA: IGI Global. doi:10.4018/978-1-59140-556-6.ch091

Kittner, M., & Van Slyke, C. (2006). Reorganizing information technology services in an academic environment. In M. Khosrow-Pour (Ed.), *Cases on the human side of information technology* (pp. 49–66). Hershey, PA: IGI Global. doi:10.4018/978-1-59904-405-7.ch004

Knoell, H. D. (2008). Semi virtual workplaces in German financial service enterprises. In P. Zemliansky & K. St.Amant (Eds.), *Handbook of research on virtual workplaces and the new nature of business practices* (pp. 570–581). Hershey, PA: IGI Global. doi:10.4018/978-1-59904-893-2.ch041

Koh, S. L., & Maguire, S. (2009). Competing in the age of information technology in a developing economy: Experiences of an Indian bank. In S. Koh & S. Maguire (Eds.), *Information and communication technologies management in turbulent business environments* (pp. 326–350). Hershey, PA: IGI Global. doi:10.4018/978-1-60566-424-8.ch018

Kollmann, T., & Häsel, M. (2009). Competence of information technology professionals in internet-based ventures. In I. Lee (Ed.), *Electronic business: Concepts, methodologies, tools, and applications* (pp. 1905–1919). Hershey, PA: IGI Global. doi:10.4018/978-1-60566-056-1.ch118

Kollmann, T., & Häsel, M. (2009). Competence of information technology professionals in internet-based ventures. In A. Cater-Steel (Ed.), *Information technology governance and service management: Frameworks and adaptations* (pp. 239–253). Hershey, PA: IGI Global. doi:10.4018/978-1-60566-008-0.ch013

Kollmann, T., & Häsel, M. (2010). Competence of information technology professionals in internet-based ventures. In *Electronic services: Concepts, methodologies, tools and applications* (pp. 1551–1565). Hershey, PA: IGI Global. doi:10.4018/978-1-61520-967-5.ch094

Kraemer, K., & King, J. L. (2006). Information technology and administrative reform: Will e-government be different? *International Journal of Electronic Government Research, 2*(1), 1–20. doi:10.4018/jegr.2006010101

Kraemer, K., & King, J. L. (2008). Information technology and administrative reform: Will e-government be different? In D. Norris (Ed.), *E-government research: Policy and management* (pp. 1–20). Hershey, PA: IGI Global. doi:10.4018/978-1-59904-913-7.ch001

Lampathaki, F., Tsiakaliaris, C., Stasis, A., & Charalabidis, Y. (2011). National interoperability frameworks: The way forward. In Y. Charalabidis (Ed.), *Interoperability in digital public services and administration: Bridging e-government and e-business* (pp. 1–24). Hershey, PA: IGI Global. doi:10.4018/978-1-61520-887-6.ch001

Lan, Z., & Scott, C. R. (1996). The relative importance of computer-mediated information versus conventional non-computer-mediated information in public managerial decision making. *Information Resources Management Journal, 9*(1), 27–0. doi:10.4018/irmj.1996010103

Law, W. (2004). *Public sector data management in a developing economy*. Hershey, PA: IGI Global. doi:10.4018/978-1-59140-259-6.ch034

Law, W. K. (2005). Information resources development challenges in a cross-cultural environment. In M. Khosrow-Pour (Ed.), *Encyclopedia of information science and technology* (pp. 1476–1481). Hershey, PA: IGI Global. doi:10.4018/978-1-59140-553-5.ch259

Law, W. K. (2009). Cross-cultural challenges for information resources management. In M. Khosrow-Pour (Ed.), *Encyclopedia of information science and technology* (2nd ed.; pp. 840–846). Hershey, PA: IGI Global. doi:10.4018/978-1-60566-026-4.ch136

Law, W. K. (2011). Cross-cultural challenges for information resources management. In *Global business: Concepts, methodologies, tools and applications* (pp. 1924–1932). Hershey, PA: IGI Global. doi:10.4018/978-1-60960-587-2.ch704

Malkia, M., & Savolainen, R. (2004). eTransformation in government, politics and society: Conceptual framework and introduction. In M. Malkia, A. Anttiroiko, & R. Savolainen (Eds.), eTransformation in governance: New directions in government and politics (pp. 1-21). Hershey, PA: IGI Global. doi:10.4018/978-1-59140-130-8.ch001

Mandujano, S. (2011). Network manageability security. In D. Kar & M. Syed (Eds.), *Network security, administration and management: Advancing technology and practice* (pp. 158–181). Hershey, PA: IGI Global. doi:10.4018/978-1-60960-777-7.ch009

Marich, M. J., Schooley, B. L., & Horan, T. A. (2012). A normative enterprise architecture for guiding end-to-end emergency response decision support. In M. Jennex (Ed.), *Managing crises and disasters with emerging technologies: Advancements* (pp. 71–87). Hershey, PA: IGI Global. doi:10.4018/978-1-4666-0167-3.ch006

Markov, R., & Okujava, S. (2008). Costs, benefits, and risks of e-government portals. In G. Putnik & M. Cruz-Cunha (Eds.), *Encyclopedia of networked and virtual organizations* (pp. 354–363). Hershey, PA: IGI Global. doi:10.4018/978-1-59904-885-7.ch047

Martin, N., & Rice, J. (2013). Evaluating and designing electronic government for the future: Observations and insights from Australia. In V. Weerakkody (Ed.), *E-government services design, adoption, and evaluation* (pp. 238–258). Hershey, PA: IGI Global. doi:10.4018/978-1-4666-2458-0.ch014

i. Martinez, A. C. (2008). Accessing administration's information via internet in Spain. In F. Tan (Ed.), *Global information technologies: Concepts, methodologies, tools, and applications* (pp. 2558–2573). Hershey, PA: IGI Global. doi:10.4018/978-1-59904-939-7.ch186

Mbarika, V. W., Meso, P. N., & Musa, P. F. (2006). A disconnect in stakeholders' perceptions from emerging realities of teledensity growth in Africa's least developed countries. In M. Hunter & F. Tan (Eds.), *Advanced topics in global information management* (Vol. 5, pp. 263–282). Hershey, PA: IGI Global. doi:10.4018/978-1-59140-923-6.ch012

Mbarika, V. W., Meso, P. N., & Musa, P. F. (2008). A disconnect in stakeholders' perceptions from emerging realities of teledensity growth in Africa's least developed countries. In F. Tan (Ed.), *Global information technologies: Concepts, methodologies, tools, and applications* (pp. 2948–2962). Hershey, PA: IGI Global. doi:10.4018/978-1-59904-939-7.ch209

Means, T., Olson, E., & Spooner, J. (2013). Discovering ways that don't work on the road to success: Strengths and weaknesses revealed by an active learning studio classroom project. In A. Benson, J. Moore, & S. Williams van Rooij (Eds.), *Cases on educational technology planning, design, and implementation: A project management perspective* (pp. 94–113). Hershey, PA: IGI Global. doi:10.4018/978-1-4666-4237-9.ch006

Melitski, J., Holzer, M., Kim, S., Kim, C., & Rho, S. (2008). Digital government worldwide: An e-government assessment of municipal web sites. In G. Garson & M. Khosrow-Pour (Eds.), *Handbook of research on public information technology* (pp. 790–804). Hershey, PA: IGI Global. doi:10.4018/978-1-59904-857-4.ch069

Memmola, M., Palumbo, G., & Rossini, M. (2009). Web & RFID technology: New frontiers in costing and process management for rehabilitation medicine. In L. Al-Hakim & M. Memmola (Eds.), *Business web strategy: Design, alignment, and application* (pp. 145–169). Hershey, PA: IGI Global. doi:10.4018/978-1-60566-024-0.ch008

Meng, Z., Fahong, Z., & Lei, L. (2008). Information technology and environment. In Y. Kurihara, S. Takaya, H. Harui, & H. Kamae (Eds.), *Information technology and economic development* (pp. 201–212). Hershey, PA: IGI Global. doi:10.4018/978-1-59904-579-5.ch014

Mentzingen de Moraes, A. J., Ferneda, E., Costa, I., & Spinola, M. D. (2011). Practical approach for implementation of governance process in IT: Information technology areas. In N. Shi & G. Silvius (Eds.), *Enterprise IT governance, business value and performance measurement* (pp. 19–40). Hershey, PA: IGI Global. doi:10.4018/978-1-60566-346-3.ch002

Merwin, G. A. Jr, McDonald, J. S., & Odera, L. C. (2008). Economic development: Government's cutting edge in IT. In M. Raisinghani (Ed.), *Handbook of research on global information technology management in the digital economy* (pp. 1–37). Hershey, PA: IGI Global. doi:10.4018/978-1-59904-875-8.ch001

Meso, P., & Duncan, N. (2002). Can national information infrastructures enhance social development in the least developed countries? An empirical investigation. In M. Dadashzadeh (Ed.), *Information technology management in developing countries* (pp. 23–51). Hershey, PA: IGI Global. doi:10.4018/978-1-931777-03-2.ch002

Meso, P. N., & Duncan, N. B. (2002). Can national information infrastructures enhance social development in the least developed countries? In F. Tan (Ed.), *Advanced topics in global information management* (Vol. 1, pp. 207–226). Hershey, PA: IGI Global. doi:10.4018/978-1-930708-43-3.ch014

Middleton, M. (2008). Evaluation of e-government web sites. In G. Garson & M. Khosrow-Pour (Eds.), *Handbook of research on public information technology* (pp. 699–710). Hershey, PA: IGI Global. doi:10.4018/978-1-59904-857-4.ch063

Mingers, J. (2010). Pluralism, realism, and truth: The keys to knowledge in information systems research. In D. Paradice (Ed.), *Emerging systems approaches in information technologies: Concepts, theories, and applications* (pp. 86–98). Hershey, PA: IGI Global. doi:10.4018/978-1-60566-976-2.ch006

Mital, K. M. (2012). ICT, unique identity and inclusive growth: An Indian perspective. In A. Manoharan & M. Holzer (Eds.), *E-governance and civic engagement: Factors and determinants of e-democracy* (pp. 584–612). Hershey, PA: IGI Global. doi:10.4018/978-1-61350-083-5.ch029

Mizell, A. P. (2008). Helping close the digital divide for financially disadvantaged seniors. In F. Tan (Ed.), *Global information technologies: Concepts, methodologies, tools, and applications* (pp. 2396–2402). Hershey, PA: IGI Global. doi:10.4018/978-1-59904-939-7.ch173

Molinari, F., Wills, C., Koumpis, A., & Moumtzi, V. (2011). A citizen-centric platform to support networking in the area of e-democracy. In H. Rahman (Ed.), *Cases on adoption, diffusion and evaluation of global e-governance systems: Impact at the grass roots* (pp. 282–302). Hershey, PA: IGI Global. doi:10.4018/978-1-61692-814-8.ch014

Molinari, F., Wills, C., Koumpis, A., & Moumtzi, V. (2013). A citizen-centric platform to support networking in the area of e-democracy. In H. Rahman (Ed.), *Cases on progressions and challenges in ICT utilization for citizen-centric governance* (pp. 265–297). Hershey, PA: IGI Global. doi:10.4018/978-1-4666-2071-1.ch013

Monteverde, F. (2010). The process of e-government public policy inclusion in the governmental agenda: A framework for assessment and case study. In J. Cordoba-Pachon & A. Ochoa-Arias (Eds.), *Systems thinking and e-participation: ICT in the governance of society* (pp. 233–245). Hershey, PA: IGI Global. doi:10.4018/978-1-60566-860-4.ch015

Moodley, S. (2008). Deconstructing the South African government's ICT for development discourse. In A. Anttiroiko (Ed.), *Electronic government: Concepts, methodologies, tools, and applications* (pp. 622–631). Hershey, PA: IGI Global. doi:10.4018/978-1-59904-947-2.ch053

Moodley, S. (2008). Deconstructing the South African government's ICT for development discourse. In C. Van Slyke (Ed.), *Information communication technologies: Concepts, methodologies, tools, and applications* (pp. 816–825). Hershey, PA: IGI Global. doi:10.4018/978-1-59904-949-6.ch052

Mora, M., Cervantes-Perez, F., Gelman-Muravchik, O., Forgionne, G. A., & Mejia-Olvera, M. (2003). DMSS implementation research: A conceptual analysis of the contributions and limitations of the factor-based and stage-based streams. In G. Forgionne, J. Gupta, & M. Mora (Eds.), *Decision-making support systems: Achievements and challenges for the new decade* (pp. 331–356). Hershey, PA: IGI Global. doi:10.4018/978-1-59140-045-5.ch020

Mörtberg, C., & Elovaara, P. (2010). Attaching people and technology: Between e and government. In S. Booth, S. Goodman, & G. Kirkup (Eds.), *Gender issues in learning and working with information technology: Social constructs and cultural contexts* (pp. 83–98). Hershey, PA: IGI Global. doi:10.4018/978-1-61520-813-5.ch005

Murphy, J., Harper, E., Devine, E. C., Burke, L. J., & Hook, M. L. (2011). Case study: Lessons learned when embedding evidence-based knowledge in a nurse care planning and documentation system. In A. Cashin & R. Cook (Eds.), *Evidence-based practice in nursing informatics: Concepts and applications* (pp. 174–190). Hershey, PA: IGI Global. doi:10.4018/978-1-60960-034-1.ch014

Mutula, S. M. (2013). E-government's role in poverty alleviation: Case study of South Africa. In H. Rahman (Ed.), *Cases on progressions and challenges in ICT utilization for citizen-centric governance* (pp. 44–68). Hershey, PA: IGI Global. doi:10.4018/978-1-4666-2071-1.ch003

Nath, R., & Angeles, R. (2005). Relationships between supply characteristics and buyer-supplier coupling in e-procurement: An empirical analysis. *International Journal of E-Business Research, 1*(2), 40–55. doi:10.4018/jebr.2005040103

Nissen, M. E. (2006). Application cases in government. In M. Nissen (Ed.), *Harnessing knowledge dynamics: Principled organizational knowing & learning* (pp. 152–181). Hershey, PA: IGI Global. doi:10.4018/978-1-59140-773-7.ch008

Norris, D. F. (2003). Leading-edge information technologies and American local governments. In G. Garson (Ed.), *Public information technology: Policy and management issues* (pp. 139–169). Hershey, PA: IGI Global. doi:10.4018/978-1-59140-060-8.ch007

Norris, D. F. (2008). Information technology among U.S. local governments. In G. Garson & M. Khosrow-Pour (Eds.), *Handbook of research on public information technology* (pp. 132–144). Hershey, PA: IGI Global. doi:10.4018/978-1-59904-857-4.ch013

Northrop, A. (1999). The challenge of teaching information technology in public administration graduate programs. In G. Garson (Ed.), *Information technology and computer applications in public administration: Issues and trends* (pp. 1–22). Hershey, PA: IGI Global. doi:10.4018/978-1-87828-952-0.ch001

Northrop, A. (2003). Information technology and public administration: The view from the profession. In G. Garson (Ed.), *Public information technology: Policy and management issues* (pp. 1–19). Hershey, PA: IGI Global. doi:10.4018/978-1-59140-060-8.ch001

Northrop, A. (2007). Lip service? How PA journals and textbooks view information technology. In G. Garson (Ed.), *Modern public information technology systems: Issues and challenges* (pp. 1–16). Hershey, PA: IGI Global. doi:10.4018/978-1-59904-051-6.ch001

Null, E. (2013). Legal and political barriers to municipal networks in the United States. In A. Abdelaal (Ed.), *Social and economic effects of community wireless networks and infrastructures* (pp. 27–56). Hershey, PA: IGI Global. doi:10.4018/978-1-4666-2997-4.ch003

Okunoye, A., Frolick, M., & Crable, E. (2006). ERP implementation in higher education: An account of pre-implementation and implementation phases. *Journal of Cases on Information Technology*, *8*(2), 110–132. doi:10.4018/jcit.2006040106

Olasina, G. (2012). A review of egovernment services in Nigeria. In A. Tella & A. Issa (Eds.), *Library and information science in developing countries: Contemporary issues* (pp. 205–221). Hershey, PA: IGI Global. doi:10.4018/978-1-61350-335-5.ch015

Orgeron, C. P. (2008). A model for reengineering IT job classes in state government. In G. Garson & M. Khosrow-Pour (Eds.), *Handbook of research on public information technology* (pp. 735–746). Hershey, PA: IGI Global. doi:10.4018/978-1-59904-857-4.ch066

Owsinski, J. W., & Pielak, A. M. (2011). Local authority websites in rural areas: Measuring quality and functionality, and assessing the role. In Z. Andreopoulou, B. Manos, N. Polman, & D. Viaggi (Eds.), *Agricultural and environmental informatics, governance and management: Emerging research applications* (pp. 39–60). Hershey, PA: IGI Global. doi:10.4018/978-1-60960-621-3.ch003

Owsiński, J. W., Pielak, A. M., Sęp, K., & Stańczak, J. (2014). Local web-based networks in rural municipalities: Extension, density, and meaning. In Z. Andreopoulou, V. Samathrakis, S. Louca, & M. Vlachopoulou (Eds.), *E-innovation for sustainable development of rural resources during global economic crisis* (pp. 126–151). Hershey, PA: IGI Global. doi:10.4018/978-1-4666-4550-9.ch011

Pagani, M., & Pasinetti, C. (2008). Technical and functional quality in the development of t-government services. In A. Anttiroiko (Ed.), *Electronic government: Concepts, methodologies, tools, and applications* (pp. 2943–2965). Hershey, PA: IGI Global. doi:10.4018/978-1-59904-947-2.ch220

Pani, A. K., & Agrahari, A. (2005). On e-markets in emerging economy: An Indian experience. In M. Khosrow-Pour (Ed.), *Advanced topics in electronic commerce* (Vol. 1, pp. 287–299). Hershey, PA: IGI Global. doi:10.4018/978-1-59140-819-2.ch015

Papadopoulos, T., Angelopoulos, S., & Kitsios, F. (2011). A strategic approach to e-health interoperability using e-government frameworks. In A. Lazakidou, K. Siassiakos, & K. Ioannou (Eds.), *Wireless technologies for ambient assisted living and healthcare: Systems and applications* (pp. 213–229). Hershey, PA: IGI Global. doi:10.4018/978-1-61520-805-0.ch012

Papadopoulos, T., Angelopoulos, S., & Kitsios, F. (2013). A strategic approach to e-health interoperability using e-government frameworks. In *User-driven healthcare: Concepts, methodologies, tools, and applications* (pp. 791–807). Hershey, PA: IGI Global. doi:10.4018/978-1-4666-2770-3.ch039

Papaleo, G., Chiarella, D., Aiello, M., & Caviglione, L. (2012). Analysis, development and deployment of statistical anomaly detection techniques for real e-mail traffic. In T. Chou (Ed.), *Information assurance and security technologies for risk assessment and threat management: Advances* (pp. 47–71). Hershey, PA: IGI Global. doi:10.4018/978-1-61350-507-6.ch003

Papp, R. (2003). Information technology & FDA compliance in the pharmaceutical industry. In M. Khosrow-Pour (Ed.), *Annals of cases on information technology* (Vol. 5, pp. 262–273). Hershey, PA: IGI Global. doi:10.4018/978-1-59140-061-5.ch017

Parsons, T. W. (2007). Developing a knowledge management portal. In A. Tatnall (Ed.), *Encyclopedia of portal technologies and applications* (pp. 223–227). Hershey, PA: IGI Global. doi:10.4018/978-1-59140-989-2.ch039

Passaris, C. E. (2007). Immigration and digital government. In A. Anttiroiko & M. Malkia (Eds.), *Encyclopedia of digital government* (pp. 988–994). Hershey, PA: IGI Global. doi:10.4018/978-1-59140-789-8.ch148

Pavlichev, A. (2004). The e-government challenge for public administration. In A. Pavlichev & G. Garson (Eds.), *Digital government: Principles and best practices* (pp. 276–290). Hershey, PA: IGI Global. doi:10.4018/978-1-59140-122-3.ch018

Penrod, J. I., & Harbor, A. F. (2000). Designing and implementing a learning organization-oriented information technology planning and management process. In L. Petrides (Ed.), *Case studies on information technology in higher education: Implications for policy and practice* (pp. 7–19). Hershey, PA: IGI Global. doi:10.4018/978-1-878289-74-2.ch001

Planas-Silva, M. D., & Joseph, R. C. (2011). Perspectives on the adoption of electronic resources for use in clinical trials. In M. Guah (Ed.), *Healthcare delivery reform and new technologies: Organizational initiatives* (pp. 19–28). Hershey, PA: IGI Global. doi:10.4018/978-1-60960-183-6.ch002

Pomazalová, N., & Rejman, S. (2013). The rationale behind implementation of new electronic tools for electronic public procurement. In N. Pomazalová (Ed.), *Public sector transformation processes and internet public procurement: Decision support systems* (pp. 85–117). Hershey, PA: IGI Global. doi:10.4018/978-1-4666-2665-2.ch006

Postorino, M. N. (2012). City competitiveness and airport: Information science perspective. In M. Bulu (Ed.), *City competitiveness and improving urban subsystems: Technologies and applications* (pp. 61–83). Hershey, PA: IGI Global. doi:10.4018/978-1-61350-174-0.ch004

Poupa, C. (2002). Electronic government in Switzerland: Priorities for 2001-2005 - Electronic voting and federal portal. In Å. Grönlund (Ed.), *Electronic government: Design, applications and management* (pp. 356–369). Hershey, PA: IGI Global. doi:10.4018/978-1-930708-19-8.ch017

Powell, S. R. (2010). Interdisciplinarity in telecommunications and networking. In *Networking and telecommunications: Concepts, methodologies, tools and applications* (pp. 33–40). Hershey, PA: IGI Global. doi:10.4018/978-1-60566-986-1.ch004

Priya, P. S., & Mathiyalagan, N. (2011). A study of the implementation status of two e-governance projects in land revenue administration in India. In M. Shareef, V. Kumar, U. Kumar, & Y. Dwivedi (Eds.), *Stakeholder adoption of e-government services: Driving and resisting factors* (pp. 214–230). Hershey, PA: IGI Global. doi:10.4018/978-1-60960-601-5.ch011

Prysby, C., & Prysby, N. (2000). Electronic mail, employee privacy and the workplace. In L. Janczewski (Ed.), *Internet and intranet security management: Risks and solutions* (pp. 251–270). Hershey, PA: IGI Global. doi:10.4018/978-1-878289-71-1.ch009

Prysby, C. L., & Prysby, N. D. (2003). Electronic mail in the public workplace: Issues of privacy and public disclosure. In G. Garson (Ed.), *Public information technology: Policy and management issues* (pp. 271–298). Hershey, PA: IGI Global. doi:10.4018/978-1-59140-060-8.ch012

Prysby, C. L., & Prysby, N. D. (2007). You have mail, but who is reading it? Issues of e-mail in the public workplace. In G. Garson (Ed.), *Modern public information technology systems: Issues and challenges* (pp. 312–336). Hershey, PA: IGI Global. doi:10.4018/978-1-59904-051-6.ch016

Radl, A., & Chen, Y. (2005). Computer security in electronic government: A state-local education information system. *International Journal of Electronic Government Research, 1*(1), 79–99. doi:10.4018/jegr.2005010105

Rahman, H. (2008). Information dynamics in developing countries. In C. Van Slyke (Ed.), *Information communication technologies: Concepts, methodologies, tools, and applications* (pp. 104–114). Hershey, PA: IGI Global. doi:10.4018/978-1-59904-949-6.ch008

Ramanathan, J. (2009). Adaptive IT architecture as a catalyst for network capability in government. In P. Saha (Ed.), *Advances in government enterprise architecture* (pp. 149–172). Hershey, PA: IGI Global. doi:10.4018/978-1-60566-068-4.ch007

Ramos, I., & Berry, D. M. (2006). Social construction of information technology supporting work. In M. Khosrow-Pour (Ed.), *Cases on information technology: Lessons learned* (Vol. 7, pp. 36–52). Hershey, PA: IGI Global. doi:10.4018/978-1-59140-673-0.ch003

Ray, D., Gulla, U., Gupta, M. P., & Dash, S. S. (2009). Interoperability and constituents of interoperable systems in public sector. In V. Weerakkody, M. Janssen, & Y. Dwivedi (Eds.), *Handbook of research on ICT-enabled transformational government: A global perspective* (pp. 175–195). Hershey, PA: IGI Global. doi:10.4018/978-1-60566-390-6.ch010

Reddick, C. G. (2007). E-government and creating a citizen-centric government: A study of federal government CIOs. In G. Garson (Ed.), *Modern public information technology systems: Issues and challenges* (pp. 143–165). Hershey, PA: IGI Global. doi:10.4018/978-1-59904-051-6.ch008

Reddick, C. G. (2010). Citizen-centric e-government. In C. Reddick (Ed.), *Homeland security preparedness and information systems: Strategies for managing public policy* (pp. 45–75). Hershey, PA: IGI Global. doi:10.4018/978-1-60566-834-5.ch002

Reddick, C. G. (2010). E-government and creating a citizen-centric government: A study of federal government CIOs. In C. Reddick (Ed.), *Homeland security preparedness and information systems: Strategies for managing public policy* (pp. 230–250). Hershey, PA: IGI Global. doi:10.4018/978-1-60566-834-5.ch012

Reddick, C. G. (2010). Perceived effectiveness of e-government and its usage in city governments: Survey evidence from information technology directors. In C. Reddick (Ed.), *Homeland security preparedness and information systems: Strategies for managing public policy* (pp. 213–229). Hershey, PA: IGI Global. doi:10.4018/978-1-60566-834-5.ch011

Reddick, C. G. (2012). Customer relationship management adoption in local governments in the United States. In S. Chhabra & M. Kumar (Eds.), *Strategic enterprise resource planning models for e-government: Applications and methodologies* (pp. 111–124). Hershey, PA: IGI Global. doi:10.4018/978-1-60960-863-7.ch008

Reeder, F. S., & Pandy, S. M. (2008). Identifying effective funding models for e-government. In A. Anttiroiko (Ed.), *Electronic government: Concepts, methodologies, tools, and applications* (pp. 1108–1138). Hershey, PA: IGI Global. doi:10.4018/978-1-59904-947-2.ch083

Riesco, D., Acosta, E., & Montejano, G. (2003). An extension to a UML activity graph from workflow. In L. Favre (Ed.), *UML and the unified process* (pp. 294–314). Hershey, PA: IGI Global. doi:10.4018/978-1-93177-744-5.ch015

Ritzhaupt, A. D., & Gill, T. G. (2008). A hybrid and novel approach to teaching computer programming in MIS curriculum. In S. Negash, M. Whitman, A. Woszczynski, K. Hoganson, & H. Mattord (Eds.), *Handbook of distance learning for real-time and asynchronous information technology education* (pp. 259–281). Hershey, PA: IGI Global. doi:10.4018/978-1-59904-964-9.ch014

Roche, E. M. (1993). International computing and the international regime. *Journal of Global Information Management, 1*(2), 33–44. doi:10.4018/jgim.1993040103

Rocheleau, B. (2007). Politics, accountability, and information management. In G. Garson (Ed.), *Modern public information technology systems: Issues and challenges* (pp. 35–71). Hershey, PA: IGI Global. doi:10.4018/978-1-59904-051-6.ch003

Rodrigues Filho, J. (2010). E-government in Brazil: Reinforcing dominant institutions or reducing citizenship? In C. Reddick (Ed.), *Politics, democracy and e-government: Participation and service delivery* (pp. 347–362). Hershey, PA: IGI Global. doi:10.4018/978-1-61520-933-0.ch021

Rodriguez, S. R., & Thorp, D. A. (2013). eLearning for industry: A case study of the project management process. In A. Benson, J. Moore, & S. Williams van Rooij (Eds.), Cases on educational technology planning, design, and implementation: A project management perspective (pp. 319-342). Hershey, PA: IGI Global. doi:10.4018/978-1-4666-4237-9.ch017

Roman, A. V. (2013). Delineating three dimensions of e-government success: Security, functionality, and transformation. In J. Gil-Garcia (Ed.), *E-government success factors and measures: Theories, concepts, and methodologies* (pp. 171–192). Hershey, PA: IGI Global. doi:10.4018/978-1-4666-4058-0.ch010

Ross, S. C., Tyran, C. K., & Auer, D. J. (2008). Up in smoke: Rebuilding after an IT disaster. In H. Nemati (Ed.), *Information security and ethics: Concepts, methodologies, tools, and applications* (pp. 3659–3675). Hershey, PA: IGI Global. doi:10.4018/978-1-59904-937-3.ch248

Ross, S. C., Tyran, C. K., Auer, D. J., Junell, J. M., & Williams, T. G. (2005). Up in smoke: Rebuilding after an IT disaster. *Journal of Cases on Information Technology*, *7*(2), 31–49. doi:10.4018/jcit.2005040103

Roy, J. (2008). Security, sovereignty, and continental interoperability: Canada's elusive balance. In T. Loendorf & G. Garson (Eds.), *Patriotic information systems* (pp. 153–176). Hershey, PA: IGI Global. doi:10.4018/978-1-59904-594-8.ch007

Rubeck, R. F., & Miller, G. A. (2009). vGOV: Remote video access to government services. In A. Scupola (Ed.), Cases on managing e-services (pp. 253-268). Hershey, PA: IGI Global. doi:10.4018/978-1-60566-064-6.ch017

Saekow, A., & Boonmee, C. (2011). The challenges of implementing e-government interoperability in Thailand: Case of official electronic correspondence letters exchange across government departments. In Y. Charalabidis (Ed.), *Interoperability in digital public services and administration: Bridging e-government and e-business* (pp. 40–61). Hershey, PA: IGI Global. doi:10.4018/978-1-61520-887-6.ch003

Saekow, A., & Boonmee, C. (2012). The challenges of implementing e-government interoperability in Thailand: Case of official electronic correspondence letters exchange across government departments. In *Digital democracy: Concepts, methodologies, tools, and applications* (pp. 1883–1905). Hershey, PA: IGI Global. doi:10.4018/978-1-4666-1740-7.ch094

Sagsan, M., & Medeni, T. (2012). Understanding "knowledge management (KM) paradigms" from social media perspective: An empirical study on discussion group for KM at professional networking site. In M. Cruz-Cunha, P. Gonçalves, N. Lopes, E. Miranda, & G. Putnik (Eds.), *Handbook of research on business social networking: Organizational, managerial, and technological dimensions* (pp. 738–755). Hershey, PA: IGI Global. doi:10.4018/978-1-61350-168-9.ch039

Sahi, G., & Madan, S. (2013). Information security threats in ERP enabled e-governance: Challenges and solutions. In *Enterprise resource planning: Concepts, methodologies, tools, and applications* (pp. 825–837). Hershey, PA: IGI Global. doi:10.4018/978-1-4666-4153-2.ch048

Sanford, C., & Bhattacherjee, A. (2008). IT implementation in a developing country municipality: A sociocognitive analysis. *International Journal of Technology and Human Interaction*, *4*(3), 68–93. doi:10.4018/jthi.2008070104

Schelin, S. H. (2003). E-government: An overview. In G. Garson (Ed.), *Public information technology: Policy and management issues* (pp. 120–138). Hershey, PA: IGI Global. doi:10.4018/978-1-59140-060-8.ch006

Schelin, S. H. (2004). Training for digital government. In A. Pavlichev & G. Garson (Eds.), *Digital government: Principles and best practices* (pp. 263–275). Hershey, PA: IGI Global. doi:10.4018/978-1-59140-122-3.ch017

Schelin, S. H. (2007). E-government: An overview. In G. Garson (Ed.), *Modern public information technology systems: Issues and challenges* (pp. 110–126). Hershey, PA: IGI Global. doi:10.4018/978-1-59904-051-6.ch006

Schelin, S. H., & Garson, G. (2004). Theoretical justification of critical success factors. In G. Garson & S. Schelin (Eds.), *IT solutions series: Humanizing information technology: Advice from experts* (pp. 4–15). Hershey, PA: IGI Global. doi:10.4018/978-1-59140-245-9.ch002

Scime, A. (2002). Information systems and computer science model curricula: A comparative look. In M. Dadashzadeh, A. Saber, & S. Saber (Eds.), *Information technology education in the new millennium* (pp. 146–158). Hershey, PA: IGI Global. doi:10.4018/978-1-931777-05-6.ch018

Scime, A. (2009). Computing curriculum analysis and development. In M. Khosrow-Pour (Ed.), *Encyclopedia of information science and technology* (2nd ed.; pp. 667–671). Hershey, PA: IGI Global. doi:10.4018/978-1-60566-026-4.ch108

Scime, A., & Wania, C. (2008). Computing curricula: A comparison of models. In C. Van Slyke (Ed.), *Information communication technologies: Concepts, methodologies, tools, and applications* (pp. 1270–1283). Hershey, PA: IGI Global. doi:10.4018/978-1-59904-949-6.ch088

Seidman, S. B. (2009). An international perspective on professional software engineering credentials. In H. Ellis, S. Demurjian, & J. Naveda (Eds.), *Software engineering: Effective teaching and learning approaches and practices* (pp. 351–361). Hershey, PA: IGI Global. doi:10.4018/978-1-60566-102-5.ch018

Seifert, J. W. (2007). E-government act of 2002 in the United States. In A. Anttiroiko & M. Malkia (Eds.), *Encyclopedia of digital government* (pp. 476–481). Hershey, PA: IGI Global. doi:10.4018/978-1-59140-789-8.ch072

Seifert, J. W., & Relyea, H. C. (2008). E-government act of 2002 in the United States. In A. Anttiroiko (Ed.), *Electronic government: Concepts, methodologies, tools, and applications* (pp. 154–161). Hershey, PA: IGI Global. doi:10.4018/978-1-59904-947-2.ch013

Seufert, S. (2002). E-learning business models: Framework and best practice examples. In M. Raisinghani (Ed.), *Cases on worldwide e-commerce: Theory in action* (pp. 70–94). Hershey, PA: IGI Global. doi:10.4018/978-1-930708-27-3.ch004

Shareef, M. A., & Archer, N. (2012). E-government service development. In M. Shareef, N. Archer, & S. Dutta (Eds.), *E-government service maturity and development: Cultural, organizational and technological perspectives* (pp. 1–14). Hershey, PA: IGI Global. doi:10.4018/978-1-60960-848-4.ch001

Shareef, M. A., & Archer, N. (2012). E-government initiatives: Review studies on different countries. In M. Shareef, N. Archer, & S. Dutta (Eds.), *E-government service maturity and development: Cultural, organizational and technological perspectives* (pp. 40–76). Hershey, PA: IGI Global. doi:10.4018/978-1-60960-848-4.ch003

Shareef, M. A., Kumar, U., & Kumar, V. (2011). E-government development: Performance evaluation parameters. In M. Shareef, V. Kumar, U. Kumar, & Y. Dwivedi (Eds.), *Stakeholder adoption of e-government services: Driving and resisting factors* (pp. 197–213). Hershey, PA: IGI Global. doi:10.4018/978-1-60960-601-5.ch010

Shareef, M. A., Kumar, U., Kumar, V., & Niktash, M. (2012). Electronic-government vision: Case studies for objectives, strategies, and initiatives. In M. Shareef, N. Archer, & S. Dutta (Eds.), *E-government service maturity and development: Cultural, organizational and technological perspectives* (pp. 15–39). Hershey, PA: IGI Global. doi:10.4018/978-1-60960-848-4.ch002

Shukla, P., Kumar, A., & Anu Kumar, P. B. (2013). Impact of national culture on business continuity management system implementation. *International Journal of Risk and Contingency Management, 2*(3), 23–36. doi:10.4018/ijrcm.2013070102

Shulman, S. W. (2007). The federal docket management system and the prospect for digital democracy in U S rulemaking. In G. Garson (Ed.), *Modern public information technology systems: Issues and challenges* (pp. 166–184). Hershey, PA: IGI Global. doi:10.4018/978-1-59904-051-6.ch009

Simonovic, S. (2007). Problems of offline government in e-Serbia. In A. Anttiroiko & M. Malkia (Eds.), *Encyclopedia of digital government* (pp. 1342–1351). Hershey, PA: IGI Global. doi:10.4018/978-1-59140-789-8.ch205

Simonovic, S. (2008). Problems of offline government in e-Serbia. In A. Anttiroiko (Ed.), *Electronic government: Concepts, methodologies, tools, and applications* (pp. 2929–2942). Hershey, PA: IGI Global. doi:10.4018/978-1-59904-947-2.ch219

Singh, A. M. (2005). Information systems and technology in South Africa. In M. Khosrow-Pour (Ed.), *Encyclopedia of information science and technology* (pp. 1497–1502). Hershey, PA: IGI Global. doi:10.4018/978-1-59140-553-5.ch263

Singh, S., & Naidoo, G. (2005). Towards an e-government solution: A South African perspective. In W. Huang, K. Siau, & K. Wei (Eds.), *Electronic government strategies and implementation* (pp. 325–353). Hershey, PA: IGI Global. doi:10.4018/978-1-59140-348-7.ch014

Snoke, R., & Underwood, A. (2002). Generic attributes of IS graduates: An analysis of Australian views. In F. Tan (Ed.), *Advanced topics in global information management* (Vol. 1, pp. 370–384). Hershey, PA: IGI Global. doi:10.4018/978-1-930708-43-3.ch023

Sommer, L. (2006). Revealing unseen organizations in higher education: A study framework and application example. In A. Metcalfe (Ed.), *Knowledge management and higher education: A critical analysis* (pp. 115–146). Hershey, PA: IGI Global. doi:10.4018/978-1-59140-509-2.ch007

Song, H., Kidd, T., & Owens, E. (2011). Examining technological disparities and instructional practices in English language arts classroom: Implications for school leadership and teacher training. In L. Tomei (Ed.), *Online courses and ICT in education: Emerging practices and applications* (pp. 258–274). Hershey, PA: IGI Global. doi:10.4018/978-1-60960-150-8.ch020

Speaker, P. J., & Kleist, V. F. (2003). Using information technology to meet electronic commerce and MIS education demands. In A. Aggarwal (Ed.), *Web-based education: Learning from experience* (pp. 280–291). Hershey, PA: IGI Global. doi:10.4018/978-1-59140-102-5.ch017

Spitler, V. K. (2007). Learning to use IT in the workplace: Mechanisms and masters. In M. Mahmood (Ed.), *Contemporary issues in end user computing* (pp. 292–323). Hershey, PA: IGI Global. doi:10.4018/978-1-59140-926-7.ch013

Stellefson, M. (2011). Considerations for marketing distance education courses in health education: Five important questions to examine before development. In U. Demiray & S. Sever (Eds.), *Marketing online education programs: Frameworks for promotion and communication* (pp. 222–234). Hershey, PA: IGI Global. doi:10.4018/978-1-60960-074-7.ch014

Straub, D. W., & Loch, K. D. (2006). Creating and developing a program of global research. *Journal of Global Information Management, 14*(2), 1–28. doi:10.4018/jgim.2006040101

Straub, D. W., Loch, K. D., & Hill, C. E. (2002). Transfer of information technology to the Arab world: A test of cultural influence modeling. In M. Dadashzadeh (Ed.), *Information technology management in developing countries* (pp. 92–134). Hershey, PA: IGI Global. doi:10.4018/978-1-931777-03-2.ch005

Straub, D. W., Loch, K. D., & Hill, C. E. (2003). Transfer of information technology to the Arab world: A test of cultural influence modeling. In F. Tan (Ed.), *Advanced topics in global information management* (Vol. 2, pp. 141–172). Hershey, PA: IGI Global. doi:10.4018/978-1-59140-064-6.ch009

Suki, N. M., Ramayah, T., Ming, M. K., & Suki, N. M. (2013). Factors enhancing employed job seekers intentions to use social networking sites as a job search tool. In A. Mesquita (Ed.), *User perception and influencing factors of technology in everyday life* (pp. 265–281). Hershey, PA: IGI Global. doi:10.4018/978-1-4666-1954-8.ch018

Suomi, R. (2006). Introducing electronic patient records to hospitals: Innovation adoption paths. In T. Spil & R. Schuring (Eds.), *E-health systems diffusion and use: The innovation, the user and the use IT model* (pp. 128–146). Hershey, PA: IGI Global. doi:10.4018/978-1-59140-423-1.ch008

Swim, J., & Barker, L. (2012). Pathways into a gendered occupation: Brazilian women in IT. *International Journal of Social and Organizational Dynamics in IT, 2*(4), 34–51. doi:10.4018/ijsodit.2012100103

Tarafdar, M., & Vaidya, S. D. (2006). Adoption and implementation of IT in developing nations: Experiences from two public sector enterprises in India. In M. Khosrow-Pour (Ed.), *Cases on information technology planning, design and implementation* (pp. 208–233). Hershey, PA: IGI Global. doi:10.4018/978-1-59904-408-8.ch013

Tarafdar, M., & Vaidya, S. D. (2008). Adoption and implementation of IT in developing nations: Experiences from two public sector enterprises in India. In G. Garson & M. Khosrow-Pour (Eds.), *Handbook of research on public information technology* (pp. 905–924). Hershey, PA: IGI Global. doi:10.4018/978-1-59904-857-4.ch076

Thesing, Z. (2007). Zarina thesing, pumpkin patch. In M. Hunter (Ed.), *Contemporary chief information officers: Management experiences* (pp. 83–94). Hershey, PA: IGI Global. doi:10.4018/978-1-59904-078-3.ch007

Thomas, J. C. (2004). Public involvement in public administration in the information age: Speculations on the effects of technology. In M. Malkia, A. Anttiroiko, & R. Savolainen (Eds.), *eTransformation in governance: New directions in government and politics* (pp. 67–84). Hershey, PA: IGI Global. doi:10.4018/978-1-59140-130-8.ch004

Treiblmaier, H., & Chong, S. (2013). Trust and perceived risk of personal information as antecedents of online information disclosure: Results from three countries. In F. Tan (Ed.), *Global diffusion and adoption of technologies for knowledge and information sharing* (pp. 341–361). Hershey, PA: IGI Global. doi:10.4018/978-1-4666-2142-8.ch015

van Grembergen, W., & de Haes, S. (2008). IT governance in practice: Six case studies. In W. van Grembergen & S. De Haes (Eds.), *Implementing information technology governance: Models, practices and cases* (pp. 125–237). Hershey, PA: IGI Global. doi:10.4018/978-1-59904-924-3.ch004

van Os, G., Homburg, V., & Bekkers, V. (2013). Contingencies and convergence in European social security: ICT coordination in the back office of the welfare state. In M. Cruz-Cunha, I. Miranda, & P. Gonçalves (Eds.), *Handbook of research on ICTs and management systems for improving efficiency in healthcare and social care* (pp. 268–287). Hershey, PA: IGI Global. doi:10.4018/978-1-4666-3990-4.ch013

Velloso, A. B., Gassenferth, W., & Machado, M. A. (2012). Evaluating IBMEC-RJ's intranet usability using fuzzy logic. In M. Cruz-Cunha, P. Gonçalves, N. Lopes, E. Miranda, & G. Putnik (Eds.), *Handbook of research on business social networking: Organizational, managerial, and technological dimensions* (pp. 185–205). Hershey, PA: IGI Global. doi:10.4018/978-1-61350-168-9.ch010

Villablanca, A. C., Baxi, H., & Anderson, K. (2009). Novel data interface for evaluating cardiovascular outcomes in women. In A. Dwivedi (Ed.), *Handbook of research on information technology management and clinical data administration in healthcare* (pp. 34–53). Hershey, PA: IGI Global. doi:10.4018/978-1-60566-356-2.ch003

Villablanca, A. C., Baxi, H., & Anderson, K. (2011). Novel data interface for evaluating cardiovascular outcomes in women. In *Clinical technologies: Concepts, methodologies, tools and applications* (pp. 2094–2113). Hershey, PA: IGI Global. doi:10.4018/978-1-60960-561-2.ch806

Virkar, S. (2011). Information and communication technologies in administrative reform for development: Exploring the case of property tax systems in Karnataka, India. In J. Steyn, J. Van Belle, & E. Mansilla (Eds.), *ICTs for global development and sustainability: Practice and applications* (pp. 127–149). Hershey, PA: IGI Global. doi:10.4018/978-1-61520-997-2.ch006

Virkar, S. (2013). Designing and implementing e-government projects: Actors, influences, and fields of play. In S. Saeed & C. Reddick (Eds.), *Human-centered system design for electronic governance* (pp. 88–110). Hershey, PA: IGI Global. doi:10.4018/978-1-4666-3640-8.ch007

Wallace, A. (2009). E-justice: An Australian perspective. In A. Martínez & P. Abat (Eds.), *E-justice: Using information communication technologies in the court system* (pp. 204–228). Hershey, PA: IGI Global. doi:10.4018/978-1-59904-998-4.ch014

Wang, G. (2012). E-democratic administration and bureaucratic responsiveness: A primary study of bureaucrats' perceptions of the civil service e-mail box in Taiwan. In K. Kloby & M. D'Agostino (Eds.), *Citizen 2.0: Public and governmental interaction through web 2.0 technologies* (pp. 146–173). Hershey, PA: IGI Global. doi:10.4018/978-1-4666-0318-9.ch009

Wangpipatwong, S., Chutimaskul, W., & Papasratorn, B. (2011). Quality enhancing the continued use of e-government web sites: Evidence from e-citizens of Thailand. In V. Weerakkody (Ed.), *Applied technology integration in governmental organizations: New e-government research* (pp. 20–36). Hershey, PA: IGI Global. doi:10.4018/978-1-60960-162-1.ch002

Wedemeijer, L. (2006). Long-term evolution of a conceptual schema at a life insurance company. In M. Khosrow-Pour (Ed.), *Cases on database technologies and applications* (pp. 202–226). Hershey, PA: IGI Global. doi:10.4018/978-1-59904-399-9.ch012

Whybrow, E. (2008). Digital access, ICT fluency, and the economically disadvantages: Approaches to minimize the digital divide. In F. Tan (Ed.), *Global information technologies: Concepts, methodologies, tools, and applications* (pp. 1409–1422). Hershey, PA: IGI Global. doi:10.4018/978-1-59904-939-7.ch102

Whybrow, E. (2008). Digital access, ICT fluency, and the economically disadvantages: Approaches to minimize the digital divide. In C. Van Slyke (Ed.), *Information communication technologies: Concepts, methodologies, tools, and applications* (pp. 764–777). Hershey, PA: IGI Global. doi:10.4018/978-1-59904-949-6.ch049

Wickramasinghe, N., & Geisler, E. (2010). Key considerations for the adoption and implementation of knowledge management in healthcare operations. In M. Saito, N. Wickramasinghe, M. Fuji, & E. Geisler (Eds.), *Redesigning innovative healthcare operation and the role of knowledge management* (pp. 125–142). Hershey, PA: IGI Global. doi:10.4018/978-1-60566-284-8.ch009

Wickramasinghe, N., & Geisler, E. (2012). Key considerations for the adoption and implementation of knowledge management in healthcare operations. In *Organizational learning and knowledge: Concepts, methodologies, tools and applications* (pp. 1316–1328). Hershey, PA: IGI Global. doi:10.4018/978-1-60960-783-8.ch405

Wickramasinghe, N., & Goldberg, S. (2007). A framework for delivering m-health excellence. In L. Al-Hakim (Ed.), *Web mobile-based applications for healthcare management* (pp. 36–61). Hershey, PA: IGI Global. doi:10.4018/978-1-59140-658-7.ch002

Wickramasinghe, N., & Goldberg, S. (2008). Critical success factors for delivering m-health excellence. In N. Wickramasinghe & E. Geisler (Eds.), *Encyclopedia of healthcare information systems* (pp. 339–351). Hershey, PA: IGI Global. doi:10.4018/978-1-59904-889-5.ch045

Wyld, D. (2009). Radio frequency identification (RFID) technology. In J. Symonds, J. Ayoade, & D. Parry (Eds.), *Auto-identification and ubiquitous computing applications* (pp. 279–293). Hershey, PA: IGI Global. doi:10.4018/978-1-60566-298-5.ch017

Yaghmaei, F. (2010). Understanding computerised information systems usage in community health. In J. Rodrigues (Ed.), *Health information systems: Concepts, methodologies, tools, and applications* (pp. 1388–1399). Hershey, PA: IGI Global. doi:10.4018/978-1-60566-988-5.ch088

Yee, G., El-Khatib, K., Korba, L., Patrick, A. S., Song, R., & Xu, Y. (2005). Privacy and trust in e-government. In W. Huang, K. Siau, & K. Wei (Eds.), *Electronic government strategies and implementation* (pp. 145–190). Hershey, PA: IGI Global. doi:10.4018/978-1-59140-348-7.ch007

Yeh, S., & Chu, P. (2010). Evaluation of e-government services: A citizen-centric approach to citizen e-complaint services. In C. Reddick (Ed.), *Citizens and e-government: Evaluating policy and management* (pp. 400–417). Hershey, PA: IGI Global. doi:10.4018/978-1-61520-931-6.ch022

Young-Jin, S., & Seang-tae, K. (2008). E-government concepts, measures, and best practices. In A. Anttiroiko (Ed.), *Electronic government: Concepts, methodologies, tools, and applications* (pp. 32–57). Hershey, PA: IGI Global. doi:10.4018/978-1-59904-947-2.ch004

Yun, H. J., & Opheim, C. (2012). New technology communication in American state governments: The impact on citizen participation. In K. Bwalya & S. Zulu (Eds.), *Handbook of research on e-government in emerging economies: Adoption, e-participation, and legal frameworks* (pp. 573–590). Hershey, PA: IGI Global. doi:10.4018/978-1-4666-0324-0.ch029

Zhang, N., Guo, X., Chen, G., & Chau, P. Y. (2011). User evaluation of e-government systems: A Chinese cultural perspective. In F. Tan (Ed.), *International enterprises and global information technologies: Advancing management practices* (pp. 63–84). Hershey, PA: IGI Global. doi:10.4018/978-1-60960-605-3.ch004

Zuo, Y., & Hu, W. (2011). Trust-based information risk management in a supply chain network. In J. Wang (Ed.), *Supply chain optimization, management and integration: Emerging applications* (pp. 181–196). Hershey, PA: IGI Global. doi:10.4018/978-1-60960-135-5.ch013

Compilation of References

Abdulwahed, M., Nagy, Z. K., & Blanchard, R. (2008). The Tri Lab, a Novel View of Laboratory Education. In *Proceedings of the EE2008 - International conference on innovation, good practice and research in engineering education*. Loughborough, UK: HEA Engineering Subject Centre.

Abella, J., Jericó, S., & Martínez-Hansen, M. (2012). *Validación de una metodología de trabajo para diseño industrial y desarrollo de producto mediante el diseño de 3 versiones de un producto: moto ecológica. Trabajo Fin de Grado. EINA, Graduado en Ingeniería en Diseño Industrial y Desarrollo de Producto. TAZ-TFG-2012-682*. Universidad de Zaragoza.

Abelló Gamazo, A., Rodríguez González, M. E., Urpí Tubella, A., Burgués Illa, X., Casany Guerrero, M. J., Martín Escofet, C., & Quer Bosor, M. C. (2008). LEARN-SQL: automatic assesment of SQL Based on IMS QTI specification. In *8th IEEE International Conference on Advanced Learning Technologies* (pp. 592-593). IEEE.

Adner, R., & Kapoor, R. (2010). Value creation in innovation ecosystems: How the structure of technological interdependence affects firm performance in new technology generations. *Strategic Management Journal, 31*(3), 306–333. doi:10.1002/smj.821

Agherdien, N. (2011). Twitter and Edulink: balancing passive consumption with knowledge creation. In P. Balcaen (Ed.), *Proceedings of the 6th international conference on e-learning* (pp. 489–492). Reading: Academic Conferences, Ltd.

Aguilar-Melchor, C., Fau, S., Fontaine, C., Gogniat, G., & Sirdey, R. (2013). Recent advances in homomorphic encryption: A possible future for signal processing in the encrypted domain. *IEEE Signal Processing Magazine, 30*(2), 108–117. doi:10.1109/MSP.2012.2230219

Ahmed, A. (2013, June 26). The Abu Dhabi Group plans to sell its 100% stake in WARID Telecom Private Limited (WTPL) – AHL Research. *Investor Guide 360*. Retrieved from http://investorguide360.com/latest-economic-news/abu-dhabi-group-plans-to-sell-its-100-stake-in-WARID-telecom-private-limited-wtpl-ahl-research/

Ajtai, M. (1996). Generating hard instances of lattice problems. In *Proceedings of the twenty-eighth annual ACM symposium on Theory of computing* (pp. 99-108). ACM. doi:10.1145/237814.237838

Akkermans, H., & Vos, B. (2003). Amplification in service supply chains: An exploratory case study from the telecom industry. *Production and Operations Management, 12*(2), 204–223. doi:10.1111/j.1937-5956.2003.tb00501.x

Albion, P. R. (2008). Web 2.0 in Teacher Education: Two Imperatives for Action. *Computers in the Schools, 25*(3-4), 181–198. doi:10.1080/07380560802368173

Ali, N. (2017). Pre-service Teachers ' Perception of Quick Response (QR) Code integration in Classroom Activities. *TOJET: The Turkish Online Journal of Educational Technology, 16*(1), 93–100.

Al-Jabri, I. M., & Roztocki, N. (2015). Adoption of ERP systems: Does information transparency matter? *Telematics and Informatics, 32*(2), 300–310. doi:10.1016/j.tele.2014.09.005

Alkraiji, A. I. (2017). The Efficiency of the Top Management and the Reality of the Decision Support Systems in Saudi Government Organizations. In New Contribution in Information Systems and Technologies, Advances in Intelligent Systems and Computing (vol. 569). Springer International Publishing.

Alkraiji, A. I. (2016). Cross-Case Analysis of Top Management Characteristics and Enterprise Information System Success. *Journal of Cases on Information Technology, 18*(3), 40–58. doi:10.4018/JCIT.2016070104

Allan, B. (2007). *Blended Learning tools for teaching and training*. London: Facet Publishing.

Aloini, D., Dulmin, R., & Mininno, V. (2007). Risk management in ERP project introduction: Review of the literature. *Information & Management, 44*(6), 547–567. doi:10.1016/j.im.2007.05.004

Alqahtani, N. D., Al-Jewair, T., AL-Moammar, K., Albarakati, S. F., & ALkofide, E. A. (2015). Live demonstration versus procedural video: A comparison of two methods for teaching an orthodontic laboratory procedure. *BMC Medical Education, 15*(1), 199. doi:10.1186/s12909-015-0479-y PMID:26537393

Ambrose, H. (2005). *Layout (2nd ed.)*. Barcelona, Spain: Parramón.

Ames, S., Venkitasubramaniam, M., Page, A., Kocabas, O., & Soyata, T. (2015). Secure health monitoring in the cloud using homomorphic encryption: A branching-program formulation. In *Enabling Real-Time Mobile Cloud Computing through Emerging Technologies* (pp. 116–152). IGI Global. doi:10.4018/978-1-4666-8662-5.ch004

Amid, A. (2014). Organizational Levels Model for Measuring the Effectiveness of Enterprise Resource Planning System (Case Study TUGA Company, Iran). *Universal Journal of Industrial and Business Management, 2*, 25–30.

Anderson, L. W., Krathwohl, D. R., & Bloom, B. S. (2005). *A taxonomy for learning, teaching, and assessing*. Longman. Retrieved from http://www.lavoisier.fr/livre/notice.asp?ouvrage=1603118

Anthopoulos, L., Reddick, C. G., Giannakidou, I., & Mavridis, N. (2016). Why e-government projects fail? An analysis of the Healthcare.gov website. *Government Information Quarterly, 33*(1), 161–173. doi:10.1016/j.giq.2015.07.003

Anzai, Y., & Simon, H. A. (1979). The theory of learning by doing. *Psychological Review, 86*(2), 124–140. doi:10.1037/0033-295X.86.2.124 PMID:493441

Appelbaum, S. H. (1997). Socio-technical systems theory: An intervention strategy for organizational development. *Management Decision, 35*(6), 452–463. doi:10.1108/00251749710173823

Arkorful, V., & Abaidoo, N. (2015). The role of e-learning, advantages and disadvantages of its adoption in higher education. *International Journal of Instructional Technology and Distance Learning, 12*(1), 29–42.

Armknecht, F., & Sadeghi, A. R. (2008). A New Approach for Algebraically Homomorphic Encryption. *IACR Cryptology ePrint Archive, 2008*, 422.

Armstrong, C. P., & Sambamurthy, V. (1999). Information technology assimilation in firms: The influence of senior leadership and IT infrastructures. *Information Systems Research, 10*(4), 304–327. doi:10.1287/isre.10.4.304

Arteaga, R., & Duarte, A. (2010). Motivational factors that influence the acceptance of Moodle using TAM. *Computers in Human Behavior, 26*(6), 1632–1640. doi:10.1016/j.chb.2010.06.011

Attaa, A. (2016). *PTA okays Mobilink-Warid Merger with Conditions*. Retrieved from https://propakistani.pk/2016/05/23/pta-okays-mobilink-warid-merger-with-conditions/

Attaa, A. (2017a). *Post Merger: Jazz Posts 44% Growth in Revenues During Q4 2016*. Retrieved from https://propakistani.pk/2017/02/27/post-merger-jazz-posts-44-growth-revenues-q4-2016/

Attaa, A. (2017b). *Post Merger: Jazz Posts 44% Growth in Revenues During Q4 2016*. Retrieved from https://propakistani.pk/2017/05/11/jazz-posts-5-4-revenue-growth-q1-2017/

Avolio, B. J., Sosik, J. J., Kahai, S. S., & Baker, B. (2014). E-leadership: Re-examining transformations in leadership source and transmission. *The Leadership Quarterly, 25*(1), 105–131. doi:10.1016/j.leaqua.2013.11.003

Bacchiocchi, E., & Montobbio, F. (2010). International Knowledge Diffusion and Home-bias Effect: Do USPTO and EPO Patent Citations Tell the Same Story? *The Scandinavian Journal of Economics, 112*, 441–470.

Bachmann, S., & Moroney, W. F. (2005, September). A Consumer's Guide to Free Internet Questionnaire Development Tools. *Proceedings of the Human Factors and Ergonomics Society Annual Meeting, 49*(24), 2085–2089. doi:10.1177/154193120504902404

Baloch, F. (2015, January 23). One year later: Warid Telecom comes back with a bang. *The Express Tribune*. Retrieved from http://tribune.com.pk/story/826035

Bansal, V., & Agarwal, A. (2015). Enterprise resource planning: Identifying relationships among critical success factors. *Business Process Management Journal, 21*(6), 1337–1352. doi:10.1108/BPMJ-12-2014-0128

Barnett, E., & Casper, M. (2001). A Definition of Social Environment. *American Journal of Public Health, 91*(3), 465. doi:10.2105/AJPH.91.3.465a PMID:11249033

Barron, B. J., Schwartz, D. L., Vye, N. J., Moore, A., Petrosino, A., Zech, L., & Bransford, J. D. (1998). Doing with understanding: Lessons from research on problem-and project-based learning. *Journal of the Learning Sciences, 7*(3-4), 271–311. doi:10.1080/10508406.1998.9672056

Bekavac, I., & Praničević, D. G. (2015). Web analytics tools and web metrics tools: An overview and comparative analysis. *Croatian Operational Research Review, 6*(2), 373–386. doi:10.17535/crorr.2015.0029

Benaloh, J. D. C. (1987). *Verifiable secret-ballot elections.* Yale University.

Bennett, S., Maton, K., & Kervin, L. (2008). The 'digital natives' debate: A critical review of the evidence. *British Journal of Educational Technology, 39*(5), 775–786. doi:10.1111/j.1467-8535.2007.00793.x

Benyus, J. M. (2002). *Biomimicry: Innovation Inspired by Nature.* New York: Perennial.

Bilgram, V., Brem, A., & Voigt, K. I. (2008). User-centric innovations in new product development—Systematic identification of lead users harnessing interactive and collaborative online-tools. *International Journal of Innovation Management, 12*(03), 419–458. doi:10.1142/S1363919608002096

Bligh, P., & Turk, D. (2004). *CRM unplugged: releasing CRM's strategic value.* John Wiley & Sons.

Blood, R. (2004). How blogging software reshapes the online community. *Communications of the ACM, 47*(12), 53–55. doi:10.1145/1035134.1035165

Bohl, O., Scheuhase, J., Sengler, R., & Winand, U. (2002, December). The sharable content object reference model (SCORM)-a critical review. In Computers in education, 2002. proceedings. international conference on (pp. 950-951). IEEE.

Bolorizadeh, A., Brannen, M., Gibbs, R., & Mack, T. (2012). Making Instruction Mobile. *The Reference Librarian, 53*(4), 373–383. doi:10.1080/02763877.2012.707488

Boneh, D., & Franklin, M. (2001). Identity-based encryption from the Weil pairing. In Advances in Cryptology—CRYPTO 2001 (pp. 213-229). Springer Berlin Heidelberg. doi:10.1007/3-540-44647-8_13

Boneh, D., Goh, E. J., & Nissim, K. (2005). Evaluating 2-DNF formulas on ciphertexts. In Theory of cryptography (pp. 325-341). Springer Berlin Heidelberg.

Bonifácio, V. D. B. (2012). QR-Coded Audio Periodic Table of the Elements: A Mobile-Learning Tool. *Journal of Chemical Education, 89*(4), 552–554. doi:10.1021/ed200541e

Borau, K., Ullrich, C., Feng, J., & Shen, R. (2009). Microblogging for language learning: Using twitter to train communicative and cultural competence. In *Advances in Web Based Learning–ICWL 2009* (pp. 78–87). Springer. doi:10.1007/978-3-642-03426-8_10

Boschma, R. (2005). Proximity and innovation: A critical assessment. *Regional Studies, 39*(1), 61–74. doi:10.1080/0034340052000320887

Bower, M., Hedberg, J. G., & Kuswara, A. (2010). A framework for Web 2.0 learning design. *Educational Media International, 47*(3), 177–198. doi:10.1080/09523987.2010.518811

Bradford, M., & Florin, J. (2003). Examining the role of innovation diffusion factors on the implementation success of enterprise resource planning systems. *International Journal of Accounting Information Systems, 4*(3), 205–225. doi:10.1016/S1467-0895(03)00026-5

Brakerski, Z., & Vaikuntanathan, V. (2014). Efficient fully homomorphic encryption from (standard) LWE. *SIAM Journal on Computing, 43*(2), 831–871. doi:10.1137/120868669

Bramhall, M., Radley, K., & Metcalf, J. (2008). Users as producers: students using video to develop learner autonomy. In *Proceedings of the EE2008 - International conference on innovation, good practice and research in engineering education.* Loughborough, UK: HEA Engineering Subject Centre.

Braun, V., & Clarke, V. (2006). Using thematic analysis in psychology. *Qualitative Research in Psychology, 3*(2), 77–101. doi:10.1191/1478088706qp063oa

Brem, A., Nylund, P. A., & Hitchen, E. (2017). Open Innovation and Intellectual Property Rights: How do SMEs benefit from patents, industrial designs, trademarks and copyrights? *Management Decision, 55*(6), 1285–1306. doi:10.1108/MD-04-2016-0223

Brohi, B. (2013, June 27). WARID Telecom up for sale. *Dawn News*. Retrieved from http://www.dawn.com/news/1021149

Brown, J. S., Collins, A., & Duguid, P. (1989). Situated Cognition and the Culture of Learning. *Educational Researcher, 18*(1), 32–42. doi:10.3102/0013189X018001032

Brown, T., & Wyatt, J. (2010). *Design thinking for social innovation.* IDEO. *Development Outreach, 12*(1), 29–31. doi:10.1596/1020-797X_12_1_29

Bryman, A. (2015). *Social research methods* (2nd ed.). New York, NY: Oxford University Press.

BuiltWith. (2016). *Google Analytics Usage Statistics*. Retrieved 3 November 2016, from http://trends.builtwith.com/analytics/Google-Analytics

Buonanno, G., Faverio, P., Pigni, F., Ravarini, A., Sciuto, D., & Tagliavini, M. (2005). Factors affecting ERP systems adoption: A comparative analysis between SMEs and large companies. *Journal of Enterprise Information Management, 18*(4), 384–426. doi:10.1108/17410390510609572

Carlsson, S. A., Henningsson, S., Hrastinski, S., & Keller, C. (2011). Socio-technical IS design science research: Developing design theory for IS integration management. *Information Systems and e-Business Management, 9*(1), 109–131. doi:10.1007/s10257-010-0140-6

Carr, N. (2003). IT Doesn't Matter. *Harvard Business Review*. PMID:12747161

Carroll, J. M. (2009). *Learning in Communities. Interdisciplinary Perspectives on Human Centered Information Technology*. London: Springer.

Caspi, A., Gorsky, P., & Privman, M. (2005). Viewing comprehension: Students' learning preferences and strategies when studying from video. *Instructional Science, 33*(1), 31–47. doi:10.1007/s11251-004-2576-x

Chan, E. (2008). *5 Goals For Online Businesses and The Strategies To Reach Them*. Retrieved 10 August 2017, from http://businesscatalyst.com/bc-blog/5-goals-for-online-businesses-and-the-strategies-to-reach-them

Chang, V. (2016). Review and discussion: E-learning for academia and industry. *International Journal of Information Management, 36*(3), 476–485. doi:10.1016/j.ijinfomgt.2015.12.007

Chan, Y. M. (2010). Video instructions as support for beyond classroom learning. *Procedia: Social and Behavioral Sciences, 9*, 1313–1318. doi:10.1016/j.sbspro.2010.12.326

Chatterjee, D., Grewal, R., & Sambamurthy, V. (2002). Shaping up for e-commerce: Institutional enablers of the organizational assimilation of web technologies. *Management Information Systems Quarterly, 26*(2), 65–89. doi:10.2307/4132321

Chatti, M. A., Jarke, M., & Frosch-Wilke, D. (2007). The future of e-learning: A shift to knowledge networking and social software. *International Journal of Knowledge and Learning, 3*(4), 404–420. doi:10.1504/IJKL.2007.016702

Chen, I. J. (2001). Planning for ERP systems: Analysis and future trends. *Business Process Management Journal, 7*(5), 374–386. doi:10.1108/14637150110406768

Chesbrough, H. W. (2003). *Open Innovation: The New Imperative for Creating and Profiting from Technology*. Cambridge, MA: Harvard Business School Publishing.

Christensen, C. M. (2016). *The innovator's dilemma: when new technologies cause great firms to fail*. Harvard Business Review Press.

Churches, A. (2007). Educational Origami. *Bloom's and ICT Tools*. Retrieved from http://edorigami.wikispaces.com/Bloom%27s+and+ICT+tools

Click, A., & Petit, J. (2010). Social networking and Web 2.0 in information literacy. *The International Information & Library Review, 42*(2), 137–142. doi:10.1080/10572317.2010.10762855

Cloutier, J. (1975). *A era de Emerec ou a comunicação audio-scripto-visual na era dos self-media* [The age of EMEREC or the audio-script-visual communication in the era of self-media]. Ministério da Educação e Investigação Científica - Instituto de Tecnologia Educativa.

Coaten, N. (2003). Blended e-learning. *Educaweb, 69*. Retrieved July 20, 2017, from http://www.educaweb.com/esp/servicios/monografico/formacionvirtual/1181076.asp

Coleman, H., & Unrau, Y. A. (2010). Qualitative Data Analysis. In R. M. Grinnell & Y. A. Unrau (Eds.), *Social Work: Research and Evaluation. Foundations of Evidence-Based Practice* (pp. 447–464). New York: Oxford University Press.

Collis, B. (1998). WWW-Based Environments for Collaborative Group Work. *Education and Information Technologies, 3*(3/4), 231–245. doi:10.1023/A:1009697415029

Collis, D. J., & Montgomery, C. A. (2008). Competing on resources. *Harvard Business Review*, (June-August): 2008.

Conole, G., & Alevizou, P. (2010). *A literature review of the use of Web 2.0 tools in Higher Education.* A report commissioned by the Higher Education Academy. Retrieved from http://www.jisctechdis.ac.uk/assets/EvidenceNet/Conole_Alevizou_2010.pdf

Corbeil, J. R., & Corbeil, M. E. (2011). The birth of a social networking phenomenon. *Cutting-edge Technologies in Higher Education, 1*, 13–32. doi:10.1108/S2044-9968(2011)0000001004

Coron, J. S., Mandal, A., Naccache, D., & Tibouchi, M. (2011). Fully Homomorphic Encryption over the Integers with Shorter Public Keys. In Crypto (Vol. 6841, pp. 487-504). Academic Press. doi:10.1007/978-3-642-22792-9_28

Corral-Marfil, J. A., Arimany-Serrat, N., Ferràs, X., & Nylund, P. A. (2015). PlanetFactory: Competing under rapid technological change. *Journal of Cases on Information Technology, 17*(2), 1–14. doi:10.4018/JCIT.2015040101

Cortiella, C., & Horowitz, S. H. (2014). *The State of learning disabilities: facts, trends and emerging issues*. New York: National Center for Learning Disabilities.

Cramer, R., Damgård, I., & Nielsen, J. (2001). Multiparty computation from threshold homomorphic encryption. *Advances in cryptology—EUROCRYPT 2001*, 280-300.

Creswell, J. (2005). *Educational Research: Planning, Conducting, and Evaluating Quantitative and Qualitative Research*. Upper Saddle River, NJ: Pearson Education.

Croft, N., Dalton, A., & Grant, M. (2015). Overcoming Isolation in Distance Learning: Building a Learning Community through Time and Space. *Journal for Education in the Built Environment, 5*(1), 27–64. doi:10.11120/jebe.2010.05010027

Cruse, E. (2006). *Using Educational Video in the Classroom: Theory, Research and Practice*. Retrieved 6 June 2017, from http://www.libraryvideo.com/articles/article26.asp

D'Aveni, R. A., Dagnino, G. B., & Smith, K. G. (2010). The age of temporary advantage. *Strategic Management Journal, 31*(13), 1371–1385. doi:10.1002/smj.897

Damgård, I., & Jurik, M. (2003). A length-flexible threshold cryptosystem with applications. In *Information Security and Privacy* (pp. 350–364). Springer Berlin Heidelberg. doi:10.1007/3-540-45067-X_30

Davenport, T. H., Harris, J. G., & Cantrell, S. (2004). Enterprise systems and ongoing process change. *Business Process Management Journal*, *10*(1), 16–26. doi:10.1108/14637150410518301

Davis, F. D., Bagozzi, R. P., & Warshaw, P. R. (1989). User acceptance of computer technology: A comparison of two theoretical models. *Management Science*, *35*(8), 982–1003. doi:10.1287/mnsc.35.8.982

De Europa, C. (2002). *Marco común europeo de referencia para las lenguas. Strasburgo: Consejo de Europa*. Ministerio de Educación, Cultura y Deporte/Instituto Cervantes.

De la Fuente Sánchez, D., Hernández Solís, M., & Pra Martos, I. (2013). El mini video como recurso didáctico na aprendizagem de materias cuantitativas [Mini videos as didactic resources in quantitative subjects]. *Revista Iberoamericana de Educación a Distancia*, *16*(2), 177–192. doi:10.5944/ried.16.2.9911

Denso Wave Incorporated. (n.d.a). *Answers to your questions about the QR code*Retrieved 14 November 2017, from http://www.qrcode.com/en/about/version.html

Denso Wave Incorporated. (n.d.b). *Information capacity and versions of the QR code*. Retrieved 15 March 2016, from http://www.qrcode.com/en/

Dervan, P. (2014). Increasing in-class student engagement using Socrative (an online Student Response System). *AISHE-J: The All Ireland Journal of Teaching and Learning in Higher Education*, *6*(3).

Di Bartolo, C. (2000). Naturaleza Como Modelo, Naturaleza Como Sistema. *Experimenta: Ediciones de Diseño,* (31), 39-45.

Díaz, F., & Hernández, G. (2002). *Estrategias docentes para un aprendizaje significativo*. McGraw-Hill.

Diffie, W., & Hellman, M. (1976). New directions in cryptography. *IEEE Transactions on Information Theory*, *22*(6), 644–654. doi:10.1109/TIT.1976.1055638

Dong, L., Neufeld, D., & Higgins, C. (2009). Top management support of enterprise systems implementations. *Journal of Information Technology*, *24*(1), 55–80. doi:10.1057/jit.2008.21

Downes, S. (2005). Feature: E-learning 2.0. *Elearn magazine, 2005*(10), 1.

Doyle, C., Sammon, D., & Neville, K. (2016). A design science research (DSR) case study: building an evaluation framework for social media enabled collaborative learning environments (SMECLEs). *Journal of Decision Systems, 25*(sup1), 125-144.

Duan, L., Huang, J., & Walrand, J. (2015). Economic Analysis of 4G Upgrade Timing. *Mobile Computing. IEEE Transactions on*, *14*(5), 975–989.

Dubois, L. (2010). *11 Best Web Analytics Tools*. Retrieved 12 December 2016, from http://www.inc.com/guides/12/2010/11-best-web-analytics-tools.html

Duffy, P. D., & Bruns, A. (2006). *The use of blogs, wikis and RSS in education: A conversation of possibilities*. Retrieved from http://eprints.qut.edu.au/5398

Duffy, T. M., & Cunningham, D. J. (1996). Constructivism: Implications for the Design and Delivery of Instruction. In D. H. Jonassen (Ed.), *Handbook of Research for Educational Communications and Technology* (pp. 170–198). New York: Macmillan Library Reference.

Dunlap, J. C., & Lowenthal, P. R. (2009). Tweeting the night away: Using Twitter to enhance social presence. *Journal of Information Systems Education*, *20*(2), 129–135.

Durak, G., Ozkeskin, E. E., & Ataizi, M. (2016). QR codes in education and communication. *Turkish Online Journal of Distance Education*, *17*(1), 42–58. doi:10.17718/tojde.89156

Dyer, J. H., & Singh, H. (1998). The relational view: Cooperative strategy and sources of interorganizational competitive advantage. *Academy of Management Review*, *23*(4), 660–679.

Dykes, B. (2014). *Web Analytics Kick Start Guide A Primer on the Fundamentals of Digital Analytics*. Peachpit. Retrieved from http://ptgmedia.pearsoncmg.com/imprint_downloads/adobepress/promotions/analytics_ksg/WebAnalyticsKickStartGuide.pdf

Ebner, M., & Schiefner, M. (2008). Microblogging-more than fun. *Proceedings of IADIS mobile learning conference*, 155-159.

Ebner, M. (2009). Interactive lecturing by integrating mobile devices and micro-blogging in higher education. *CIT. Journal of Computing and Information Technology*, *17*(4), 371–381. doi:10.2498/cit.1001382

Ebner, M., Lienhardt, C., Rohs, M., & Meyer, I. (2010). Microblogs in Higher Education–A chance to facilitate informal and process-oriented learning? *Computers & Education*, *55*(1), 92–100. doi:10.1016/j.compedu.2009.12.006

Ebner, M., & Maurer, H. (2008). Can Microblogs and Weblogs change traditional scientific writing? In *World Conference on E-Learning in Corporate* (Vol. 2008, pp. 768–776). Government, Healthcare, and Higher Education. Retrieved from http://www.editlib.org/p/29699

Edel-Navarro, R. (2004). Educación a distancia y eficiencia terminal exitosa: El caso de la sede Tejupilco en la Universidad Virtual del Tecnológico de Monterrey. *Revista de Educación a Distancia*, *III*(12), 1–23.

Edmodo – The Secure, Social Learning Network for Teachers and Students. (n.d.). Retrieved May 8, 2013, from http://trendingeducation.com/?p=69

Edmodo: What students think. (n.d.). *Box of Tricks*. Retrieved May 8, 2013, from http://www.boxoftricks.net/2008/09/edmodo-what-students-think/

Edwards, S. (2013). *10 Common LMS Questions You Can Answer with Google Analytics*. Retrieved 10 August 2017, from https://www.opensesame.com/blog/10-common-lms-questions-you-can-answer-google-analytics

Ehlers, U.-D., & Hilera, J. R. (2012). Special Issue on quality in e-learning [Editorial]. *Journal of Computer Assisted Learning*, *28*(1), 1–3. doi:10.1111/j.1365-2729.2011.00448.x

Eisenhardt, K. M. (1989). Building theories from case study research. *Academy of Management Review*, *4*(4), 532–550.

Eisenmann, T., Parker, G., & Van Alstyne, M. (2011). Platform envelopment. *Strategic Management Journal*, *32*(12), 1270–1285. doi:10.1002/smj.935

Elbanna, A. (2010). Rethinking IS project boundaries in practice: A multiple-projects perspective. *The Journal of Strategic Information Systems*, *19*(1), 39–51. doi:10.1016/j.jsis.2010.02.005

Elbanna, A. (2013). Top management support in multiple-project environments: An in-practice view. *European Journal of Information Systems*, *22*(3), 278–294. doi:10.1057/ejis.2012.16

Elbeck, M. (2014). Selecting a free web-hosted survey tool for student use. *E-Journal of Business Education & Scholarship of Teaching*, *8*(2), 54.

ElGamal, T. (1984). A public key cryptosystem and a signature scheme based on discrete logarithms. In *Advances in cryptology* (pp. 10–18). Springer Berlin Heidelberg.

Ellis, D. (2012). *Google Analytics as a tool in development of e-learning artefacts: A case study*. Presented at the ascilite 2012, New Zealand. Retrieved from http://www.ascilite.org/conferences/Wellington12/2012/images/custom/ellis,_damon_-_google_analytics.pdf

Ellison, N. B. (2007). Social network sites: Definition, history, and scholarship. *Journal of Computer-Mediated Communication*, *13*(1), 210–230. doi:10.1111/j.1083-6101.2007.00393.x

Emmeche, C., & Hoffmeyer, J. N. (1991). From language to nature: The semiotic metaphor in biology. *Semiotica*, *84*(1/2), 1–42. doi:10.1515/semi.1991.84.1-2.1

Erkin, Z., Veugen, T., Toft, T., & Lagendijk, R. L. (2012). Generating private recommendations efficiently using homomorphic encryption and data packing. *IEEE Transactions on Information Forensics and Security*, *7*(3), 1053–1066. doi:10.1109/TIFS.2012.2190726

European Union. (2006). *Recommendation of the European Parliament and of the Council on Key Competences for Lifelong Learning*. Official Journal of the European Union L394 of 30.12.2006, pp.0010-0018.

Evangelidou, F. (2017). Kavala's 6th Graders towards Mastering Mathematics: A Blended Learning Proposal at Disabled Students' Side. In A. Palalas & P. Pawluk (Eds.), *Blended Learning Theory and Practice Proceedings of the 2nd World Conference on Blended Learning (IABL2017)* (pp. 79-84). Toronto: International Association for Blended Learning and George Brown College.

Evans, P. B., & Wurster, T. S. (1997, September). Strategy and the new economics of information. *Harvard Business Review*, 65–82. PMID:10170332

Fang, W. (2007). *Using Google Analytics for improving library website content and design: a case study*. Academic Press. 10.7282/T3MK6B6N

Feisel, L. D., & Rosa, A. J. (2005). The Role of the Laboratory in Undergraduate Engineering Education. *Journal of Engineering Education*, *94*(1), 121–130. doi:10.1002/j.2168-9830.2005.tb00833.x

Fernandes, J. (2017). *Estudo da Google indica que adopção de smartphones triplica em Portugal* [Google report indicates that smartphones use in Portugal triples]. Retrieved 6 July 2017, from http://wintech.pt/53-noticias/google/23040-estudo-da-google-indica-que-adopcao-de-smartphones-triplica-em-portugal

Fernandez, V., Simo, P., Ines, A., Maria, A.-S., Nuria, S., Amante, B., & Federico, G. et al. (2011). 'Low-Cost Educational Videos' for Engineering Students: A New Concept Based on Video Stream and Youtube Channels. *International Journal of Engineering Education*, *27*(3), 1–10.

Ferriman, N. (2013). The impact of blended e-learning on undergraduate academic essay writing in English (L2). *Computers & Education*, *60*(1), 243–253. doi:10.1016/j.compedu.2012.07.008

FEUP-LEA. (n.d.). *Vincere: Ciências e Engenharia em português* [Vincere: Science and Engineering in Portuguese]. Retrieved 15 March 2016, from https://paginas.fe.up.pt/~lea/projvincere/

Fichman, R. G., Dos Santos, B. L., & Zheng, Z. (2014). Digital innovation as a fundamental and powerful concept in the information Systems curriculum. *Management Information Systems Quarterly*, *38*(2), 329–353. doi:10.25300/MISQ/2014/38.2.01

Finocchiaro, M., & Sako, S. (1983). Foreign Language Testing: A Practical Approach. Regents Publishing Company, Inc.

Fishenden, C., & Markland, E. (2006). A systems approach to elementary laboratory instruction. *European Journal of Engineering Education*, *4*(3-4), 293–302. doi:10.1080/0304379800040306

French, K., & Poterba, J. (1991). Investor diversification and international equity markets. *The American Economic Review*, *81*(2), 222–226.

Frost, T. (2001). The geographic sources of foreign subsidiaries' innovation. *Strategic Management Journal*, *22*(2), 101–124. doi:10.1002/1097-0266(200101)22:2<101::AID-SMJ155>3.0.CO;2-G

Fulton, K. P. (2014). *Time for Learning: Top 10 Reasons Why Flipping the Classroom Can Change Education*. Corwin Press. doi:10.4135/9781483332772

Gan, B., Menkhoff, T., & Smith, R. (2015). Enhancing students' learning process through interactive digital media: New opportunities for collaborative learning. *Computers in Human Behavior*, *51*, 652–663. doi:10.1016/j.chb.2014.12.048

Gao, F., Luo, T., & Zhang, K. (2012a). Tweeting for learning: A critical analysis of research on microblogging in education published in 2008–2011. *British Journal of Educational Technology*, *43*(5), 783–801. doi:10.1111/j.1467-8535.2012.01357.x

Gaonkar, S., Li, J., Choudhury, R. R., Cox, L., & Schmidt, A. (2008). Micro-blog: sharing and querying content through mobile phones and social participation. In *Proceedings of the 6th international conference on Mobile systems, applications, and services* (pp. 174–186). Retrieved from http://dl.acm.org/citation.cfm?id=1378620

García, F. J., & García, J. (2005). Educational Hypermedia Resources Facilitator. *Computers & Education*, *44*(3), 301–325. doi:10.1016/j.compedu.2004.02.004

García-Holgado, A., & García-Peñalvo, F. J. (2013). The evolution of the technological ecosystems: An architectural proposal to enhancing learning processes. In F. J. García-Peñalvo (Ed.), *Proceedings of the First International Conference on Technological Ecosystems for Enhancing Multiculturality (TEEM'13)* (pp. 565-571). New York, NY: ACM. doi:10.1145/2536536.2536623

García-Holgado, A., & García-Peñalvo, F. J. (2016). Architectural pattern to improve the definition and implementation of eLearning ecosystems. *Science of Computer Programming*, *129*, 20–34. doi:10.1016/j.scico.2016.03.010

García-Peñalvo, F. J. (2008). *Advances in E-Learning: Experiences and Methodologies*. Hershey, PA: Information Science Reference. doi:10.4018/978-1-59904-756-0

García-Peñalvo, F. J., Hernández-García, Á., Conde-González, M. Á., Fidalgo-Blanco, Á., Sein-Echaluce Lacleta, M. L., Alier-Forment, M., & Iglesias-Pradas, S. et al. (2015). Learning services-based technological ecosystems. In G. R. Alves, & M. C. Felgueiras (Eds.), *Proceedings of the Third International Conference on Technological Ecosystems for Enhancing Multiculturality (TEEM'15)* (pp. 467-472). New York: ACM. doi:10.1145/2808580.2808650

García-Peñalvo, F. J., & Seoane-Pardo, A. M. (2015). *Una revisión actualizada del concepto de eLearning*. Décimo Aniversario. *Education in the Knowledge Society*, *16*(1), 119–144. doi:10.14201/eks2015161119144

Garrison, D. R., & Kanuka, H. (2004). Blended learning: Uncovering its transformative potential in higher education. *The Internet and Higher Education*, *7*(2), 95–105. doi:10.1016/j.iheduc.2004.02.001

Gaspar, A. (2014). Metade da população portuguesa já utiliza smartphones [Half of the Portuguese population uses smartphones]. *Jornal de Notícias*. Retrieved from http://www.jn.pt/PaginaInicial/Tecnologia/Interior.aspx?content_id=4114579

Gasson, S. (2006). A genealogical study of boundary-spanning IS design. *European Journal of Information Systems*, *15*(1), 26–41. doi:10.1057/palgrave.ejis.3000594

Gentry, C. (2009). *A fully homomorphic encryption scheme* (Doctoral dissertation). Stanford University.

Gentry, C., Sahai, A., & Waters, B. (2013). Homomorphic encryption from learning with errors: Conceptually-simpler, asymptotically-faster, attribute-based. In *Advances in Cryptology–CRYPTO 2013* (pp. 75–92). Berlin: Springer. doi:10.1007/978-3-642-40041-4_5

Giannakos, M. N. (2013). Exploring the video-based learning research: A review of the literature. *British Journal of Educational Technology*, *44*(6), E191–E195. doi:10.1111/bjet.12070

Giddens, A., & Sutton, P. W. (2013). *Sociology* (7th ed.). Cambridge, UK: Polity Press.

Gilbert, A. H., & Ward, S. G. (2012). Does" IT Doesn't Matter" Matter?: A Study Of Innovation And Information Systems Issues. *The Review of Business Information Systems (Online)*, *16*(4), 177. doi:10.19030/rbis.v16i4.7437

Gladwell, M. (2006). *The tipping point: How little things can make a big difference*. Little, Brown.

Glaser, B. G., & Strauss, A. L. (1967). *The Discovery of Grounded Theory: Strategies for Qualitative Research*. New York: Aldine Publishing Company.

González, A. B., Rodríguez, M., Olmos, S., Borham, M., & García, F. (2013). Experimental evaluation of the impact of b-learning methodologies on engineering students in Spain. *Computers in Human Behavior*, *29*(2), 370–377. doi:10.1016/j.chb.2012.02.003

González, M. Á., Martín, M. E., Llamas, C., Martínez, Ó., Vegas, J., Herguedas, M., & Hernández, C. (2015). Teaching and learning physics with smartphones. *Journal of Cases on Information Technology*, *17*(1), 31–50. doi:10.4018/JCIT.2015010103

Goodhue, D. L., Wixom, B. H., & Watson, H. J. (2002). Realizing business benefits through CRM: Hitting the right target in the right way. *MIS Quarterly Executive*, *1*(2), 79–94.

Google Analytics Solutions. (2017). *Success Stories*. Retrieved from https://www.google.com/analytics/success-stories/#?modal_active=none

Google Analytics. (2013). *Fairmont Hotels harness the power of Google Analytics to optimize social media marketing*. Retrieved from https://static.googleusercontent.com/media/www.google.com/en/us/analytics/customers/pdfs/fairmont-raffles-hotels-international.pdf

Google Analytics. (2015a). *AlfaStrakhovanie Doubles Transaction Rates With Google Analytics and Enhanced Ecommerce*. Retrieved from https://static.googleusercontent.com/media/www.google.com/en/us/analytics/customers/pdfs/alfastrakhovanie.pdf

Google Analytics. (2015b). *Cancer.org Donations Rise 5.4% With Help From Google Analytics*. Retrieved from https://static.googleusercontent.com/media/www.google.com/en/us/analytics/customers/pdfs/americancancersociety.pdf

Google Analytics. (2015c). *Mumzworld Reaches 300% ROAS With Enhanced Ecommerce and Data Import From Google Analytics*. Retrieved from https://services.google.com/fh/files/misc/mumzworld-enhanced-ecommerce.pdf

Gordon, W. (1963). *Sinéctica. El desarrollo de la capacidad creadora*. México DF, Mexico: Editorial Herrero.

Goset, J., & Espinoza, P. (2014). Perception of the use of teaching videos in preclinical training of Chilean dentistry students. *Journal Oral Of Research*, *3*(2), 90–94. doi:10.17126/joralres.2014.022

Gottliebson, D., Layton, N., & Wilson, E. (2010). Comparative effectiveness report: Online survey tools. *Disability and Rehabilitation. Assistive Technology*, *5*(6), 401–410. doi:10.3109/17483101003793404 PMID:20441440

Gracia, M. J. F., García, R. T. M., Papí, J. A. F., & Aycart, G. J. F. (2010). Pruebas objetivas en Moodle, Campus Virtual y sistema tradicional. Análisis comparativo de su gestión. In VIII Jornades de Xarxes d'Investigació en Docència Universitària [Recurso electrónico]: noves titulacions i canvi universitari= VIII Jornadas de Redes de Investigación en Docencia Universitaria: nuevas titulaciones y cambio universitario (pp. 1659-1673). Universidad de Alicante.

Gradel, K., & Edson, A. J. (2013). QR Codes in Higher Ed: Fad or Functional Tool? Journal of Educational Technology Systems, 41(1), 45–67. doi:10.2190/ET.41.1.e

Graham, C. R. (2006). Blended Learning Systems. Definition, Current Trends, and Future Directions. In C. R. Graham & C. J. Bonk (Eds.), *The Handbook of Blended Learning. Global Perspectives, Local Design* (pp. 3–21). San Francisco: Eds. Pfeiffer.

Greasley, A., & Wang, Y. C. (2016). Building the hybrid organization through ERP and enterprise social software. *Computers in Industry*, *82*, 69–81. doi:10.1016/j.compind.2016.05.007

Greenhow, C., Robelia, B., & Hughes, J. E. (2009). Response to comments: Research on learning and teaching with Web 2.0: Bridging conversations. *Educational Researcher*, *38*(4), 280–283. doi:10.3102/0013189X09336675

Gregor, S., & Jones, D. (2007). The anatomy of a design theory. *Journal of AIS*, *8*(5), 312–335.

Gros, B., García, I., & Lara, P. (2009). El desarrollo de herramientas de apoyo para el trabajo colaborativo en entornos virtuales de aprendizaje. *RIED*, *12*(2), 115–138.

GrupoMarktest. (2016). *Smartphone continua a crescer em Portugal* [Smartphone continues to grow in Portugal]. Retrieved 5 July 2017, from http://www.marktest.com/wap/a/n/id~2046.aspx

Gunther McGrath, R. (2013, June). Transient advantage. *Harvard Business Review*.

Gupta, D. S., & Biswas, G. P. (2015c). Identity-Based/Attribute-Based cryptosystem using threshold value without Shamir's Secret Sharing. In *Signal Processing, Computing and Control (ISPCC), 2015 International Conference on* (pp. 307-311). IEEE.

Gupta, D. S., & Biswas, G. P. (2015a). Securing Voice Call Transmission over Cellular Communication. *Procedia Computer Science*, *57*, 752–756. doi:10.1016/j.procs.2015.07.469

Gupta, D. S., & Biswas, G. P. (2015b). Secure Computation on Cloud Storage: A Homomorphic Approach. *Journal of Cases on Information Technology*, *17*(3), 22–29. doi:10.4018/JCIT.2015070103

Gupta, D. S., & Biswas, G. P. (2016). Cryptanalysis of Wang et al.'s lattice-based key exchange protocol. *Perspectives on Science*, *8*, 228–230. doi:10.1016/j.pisc.2016.04.034

Gupta, D. S., & Biswas, G. P. (2017a). A Secure Cloud Storage using ECC-Based Homomorphic Encryption. *International Journal of Information Security and Privacy*, *11*(3), 54–62. doi:10.4018/IJISP.2017070105

Gupta, D. S., & Biswas, G. P. (2017b). On Securing Bi-and Tri-partite Session Key Agreement Protocol Using IBE Framework. *Wireless Personal Communications*, 1–20. doi:10.1007/s11277-017-4399-5

Gupta, D. S., & Biswas, G. P. (2017c). An ECC-based Authenticated Group Key Exchange Protocol in IBE framework. *International Journal of Communication Systems*, e3363. doi:10.1002/dac.3363

Güzer, B., & Caner, H. (2014). The Past, Present and Future of Blended Learning: An in Depth Analysis of Literature. *Procedia: Social and Behavioral Sciences*, *116*, 4596–4603. doi:10.1016/j.sbspro.2014.01.992

Hadley, A. O., & Reiken, E. (1993). Teaching Language in Context, and Teaching Language in Context--Workbook. Heinle & Heinle Publishers.

Haintz, C., Pichler, K., & Ebner, M. (2014). Developing a Web-Based Question-Driven Audience Response System Supporting BYOD. *J. UCS*, *20*(1), 39–56.

Halupa, C. M., & Caldwell, B. W. (2015). A Comparison of Two Engineering Statics Courses: Traditional Lecture-based and Lecture-based with Online Supplemental Video. *International Journal of Higher Education*, *4*(1), 232–240. doi:10.5430/ijhe.v4n1p232

Hambrick, D. C., & Mason, P. A. (1984). Upper echelons: The organization as a reflection of its top managers. *Academy of Management Review*, *9*(2), 193–206.

Hasan, L., Morris, A., & Probets, S. (2009). Using Google Analytics to Evaluate the Usability of E-Commerce Sites. In M. Kurosu (Ed.), *Human Centered Design* (pp. 697–706). Springer Berlin Heidelberg; doi:10.1007/978-3-642-02806-9_81

Hawkes, M., & Cambre, M. (2000). The cost factor. *Technological Horizons in Education*, *28*(1), 26–32.

Heikkilä, J. (2002). From supply to demand chain management: Efficiency and customer satisfaction. *Journal of Operations Management*, *20*(6), 747–767. doi:10.1016/S0272-6963(02)00038-4

Helander, M. G., & Tham, M. P. (2003). Hedonomics – affective human factors design. *Ergonomics*, *46*(13-14), 1269–1272. doi:10.1080/00140130310001610810 PMID:14612318

Helm, R., & Gritsch, S. (2014). Examining the influence of uncertainty on marketing-mix strategy elements in emerging business to business-export markets. *International Business Review, 23*(2), 418–428. doi:10.1016/j.ibusrev.2013.06.007

Helms, M., Vattam, S. S., & Goel, A. K. (2009). Biologically Inspired Design: Process and Products. *Design Studies, 30*(5), 606–622. doi:10.1016/j.destud.2009.04.003

Henderson, A. J. (2003). *The e-learning question and answer book: a survival guide for trainers and business managers*. New York: American Management Association.

Hernández, R., Fernández, C., & Baptista, P. (2010). *Metodología de la investigación*. McGraw Hill.

Herrmann, P., & Datta, D. K. (2005). Relationships between top management team characteristics and international diversification: An empirical investigation. *British Journal of Management, 16*(1), 69–78. doi:10.1111/j.1467-8551.2005.00429.x

Hewitt, J. (2015). E-Learning. In R. Gunstone (Ed.), *Encyclopedia of Science Education* (pp. 362–363). Dordrecht, The Netherlands: Springer Netherlands; doi:10.1007/978-94-007-2150-0_36

Holloway, G. (2014). *The Business Benefits of Google Analytics*. Retrieved 10 August 2017, from https://www.koozai.com/blog/analytics/google-analytics-business-benefits/

Holloway, I., & Todres, L. (2003). The status of method: Flexibility, consistency and coherence. *Qualitative Research, 3*(3), 345–357. doi:10.1177/1468794103033004

Holotescu, C., & Grosseck, G. (2009). Using microblogging in education. Case Study: Cirip. ro. In *6th International Conference on e-Learning Applications*. Retrieved from http://www.cblt.soton.ac.uk/multimedia/PDFs%20Twitter/8551345-Using-microblogging-in-education-Case-Study-Ciripro.pdf

Honebein, P. C. (1996). Seven Goals for the Design of Constuctivist Learning Environment. In B. G. Wilson (Ed.), *Constructivist learning environments: Case studies in instructional design* (pp. 11–24). Englewood Cliffs, NJ: Educational Technology Publications.

Hošková-Mayerová, Š., & Rosická, Z. (2015). E-Learning Pros and Cons: Active Learning Culture? *Procedia: Social and Behavioral Sciences, 191*, 958–962. doi:10.1016/j.sbspro.2015.04.702

Howcroft, D., Newell, S., & Wagner, E. (2004). Understanding the contextual influences on enterprise system design, implementation, use and evaluation. *The Journal of Strategic Information Systems, 13*(4), 271–277. doi:10.1016/j.jsis.2004.11.010

Hsin, W.-J., & Cigas, J. (2013). Short videos improve student learning in online education. *Journal of Computing Sciences in Colleges, 28*(5), 253–259.

Hsu, Y.-C., & Ching, Y.-H. (2011). Microblogging for strengthening a virtual learning community in an online course. *Knowledge Management & E-Learning: An International Journal, 3*(4), 585–598.

Iansiti, M., & Levien, R. (2004). Strategy as Ecology. *Harvard Business Review*, *82*(3), 68–78. PMID:15029791

Ibtihal, M., & Hassan, N. (2017). Homomorphic Encryption as a Service for Outsourced Images in Mobile Cloud Computing Environment. *International Journal of Cloud Applications and Computing*, *7*(2), 27–40. doi:10.4018/IJCAC.2017040103

Ihanainen, P., & Moravec, J. (2011). Pointillist, cyclical, and overlapping: Multidimensional facets of time in online education. *International Review of Research in Open and Distance Learning*, *12*(7), 27–39. doi:10.19173/irrodl.v12i7.1023

Irani, Z., Ezingeard, J., Grieve, R., & Race, P. (1999). Case study approach to carrying out information systems research: A critique. *International Journal of Computer Applications in Technology*, *12*(2), 190–198. doi:10.1504/IJCAT.1999.000203

Jabri, P. (2017). *Mobile phone users cross 140.5mn with 72.52pc tele-density*. Retrieved from http://www.brecorder.com/2017/07/31/362290/mobile-phone-users-cross-140-5mn-with-72-52pc-tele-density/

Jacobs, S. (2008). *College Degree.com | The Facebook Classroom: 25 Facebook Apps That Are Perfect for Online Education*. Retrieved 11 June 2013, from http://www.collegedegree.com

Jaffe, A. B., & Trajtenberg, M. (1999). International Knowledge Flows: Evidence from Patent Citations. *Economics of Innovation and New Technology*, *8*(1-2), 105–136. doi:10.1080/10438599900000006

Jammernegg, W., & Reiner, G. (2007). Performance improvement of supply chain processes by coordinated inventory and capacity management. *International Journal of Production Economics*, *108*(1), 183–190. doi:10.1016/j.ijpe.2006.12.047

Jamu, J. T., Lowi-Jones, H., & Mitchell, C. (2016). Just in time? Using QR codes for multi-professional learning in clinical practice. [doi:10.1016/j.nepr.2016.03.007 PMID:27428702]. *Nurse Education in Practice*, *19*, 107–112.

Jarc, J. (2010). *Edmodo – a free, web 2.0 classroom management tool*. Retrieved 11 June 2013, from http://trendingeducation.com/?p=190

Jarvenpaa, S. L., & Ives, B. (1991). Executive involvement and participation in the management of information technology. *Management Information Systems Quarterly*, *15*(2), 205–227. doi:10.2307/249382

Java, A., Song, X., Finin, T., & Tseng, B. (2007). Why we twitter: understanding microblogging usage and communities. In *Proceedings of the 9th WebKDD and 1st SNA-KDD 2007 workshop on Web mining and social network analysis* (pp. 56–65). Retrieved from http://dl.acm.org/citation.cfm?id=1348556

Jenkins, H., Clinton, K., Purushotma, R., Robison, A. J., & Weigel, M. (2006). *Confronting the challenges of participatory culture: Media education for the 21st century. MacArthur Foundation.* Retrieved 10/29/2009 from http://www. digitallearning. macfound. org/atf/cf/%7B7E45C7E0-A3E0-4B89-AC9C-E807E1B0AE4E% 7D/JENKINS_WHITE_PAPER.pdf

Ji, Z., García Seoane, J., Vázquez Osorio, T., Flores Herráez, R., & Álvarez Vázquez, M. (2015). *Empleo de Google Drive en evaluación universitaria.* Academic Press.

Johnson, L., Adams Becker, S., Estrada, V., & Freeman, A. (2014). *NMC Horizon Report: 2014 Higher Education Edition.* Austin, TX: The New Media Consortium.

Kakasevski, G., Mihajlov, M., Arsenovski, S., & Chungurski, S. (2008, June). Evaluating usability in learning management system Moodle. In *Information Technology Interfaces, 2008. ITI 2008. 30th International Conference on* (pp. 613-618). IEEE. doi:10.1109/ITI.2008.4588480

Kamrujjaman, Md., & Alam, Md. N. (2014). *Feasibility of Migration from 3G to 4G including Comparative Study.* Northern University Bangladesh. Retrieved from https://www.researchgate.net/publication/271647287

Khalid, H. M. (2006). Embracing diversity in user needs for affective design. *Applied Ergonomics, 37*(4), 409–418. doi:10.1016/j.apergo.2006.04.005 PMID:16764817

Khan, W. (2016). *Mobilink-Warid Merger: Mobilink to Warid calls will now count as on-net.* Retrieved from http://techrave.pk/2016/10/mobilink-warid-merger-mobilink-warid-calls-will-now-count-net/

Khan, G., Moon, J., Park, H., Swar, B., & Rho, J. (2010). A socio-technical perspective on e-government issues in developing countries: A scientometrics approach. *Scientometrics, 78*(2), 267–286.

King, S. F., & Burgess, T. F. (2006). Beyond critical success factors: A dynamic model of enterprise system innovation. *International Journal of Information Management, 26*(1), 59–69. doi:10.1016/j.ijinfomgt.2005.10.005

Koblitz, N. (1987). Elliptic curve cryptosystems. *Mathematics of Computation, 48*(177), 203–209. doi:10.1090/S0025-5718-1987-0866109-5

Kocabas, O., & Soyata, T. (2015). Towards privacy-preserving medical cloud computing using homomorphic encryption. In Enabling Real-Time Mobile Cloud Computing through Emerging Technologies (pp. 213-246). IGI Global. doi:10.4018/978-1-4666-8662-5.ch007

Kocoglu, Z., Ozek, Y., & Kesli, Y. (2011). Blended learning: Investigating its potential in an English language teacher training program. *Australasian Journal of Educational Technology, 27*(7), 1124–1134. doi:10.14742/ajet.908

Kravtsov, H., & Kravtsov, D. (2008). Knowledge Control Model of Distance Learning System on IMS Standard. In Innovative Techniques in Instruction Technology, E-learning, E-assessment, and Education (pp. 195-198). Springer Netherlands. doi:10.1007/978-1-4020-8739-4_35

Kravtsova, O. A., & Galiguzova, A. V. (2015). Blended Learning: In Search of the Right Blend. In D. A. Kryachkov, E. B. Yastrebova, & O. A. Kravtsova (Eds.), *The Magic of Innovation. New Techniques and Technologies in Teaching Foreign Languages* (pp. 67–83). Newcastle, UK: Cambridge Scholars Publishing.

Krivickas, R. V., & Krivickas, J. (2007). Laboratory instruction in engineering education. *Global Journal of Engineering. Education, 11*(2), 191–196.

Kumaravadivelu, B. (1994). The postmethod condition: (E) merging strategies for second/foreign language teaching. *TESOL Quarterly, 28*(1), 27–48. doi:10.2307/3587197

Lado, R. (1961). *Language Testing: The Construction and Use of Foreign Language Tests. A Teacher's Book*. Academic Press.

Law, C., & So, S. (2010). QR codes in education. *Journal of Educational Technology Development and Exchange, 3*(1), 85–100. doi:10.18785/jetde.0301.07

Lee, L. (2010). Exploring wiki-mediated collaborative writing: A case-study in an elementary Spanish course. *CALICO Journal, 27*(2), 260–276. doi:10.11139/cj.27.2.260-276

Lenhart, A., Purcell, K., Smith, A., & Zickuhr, K. (2010). *Social media & mobile internet use among teens and young adults*. Pew Internet & American Life Project. Retrieved from http://web.pewinternet.org/~/media/Files/Reports/2010/PIP_Social_Media_and_Young_Adults_Report_Final_with_toplines.pdf

Lento, T., Welser, H. T., Gu, L., & Smith, M. (2006). *The ties that blog: Examining the relationship between social ties and continued participation in the wallop weblogging system. In 3rd Annual Workshop on the Weblogging Ecosystem* (p. 13). Retrieved from http://citeseerx.ist.psu.edu/viewdoc/download?doi=10.1.1.61.6907&rep=rep1&type=pdf

Liang, H., Saraf, N., Hu, Q., & Xue, Y. (2007). Assimilation of enterprise systems: The effect of institutional pressures and the mediating role of top management. *Management Information Systems Quarterly, 31*(1), 59–87.

Li, S. (2011). A matrix-based clustering approach for the decomposition of design problems. *Research in Engineering Design, 22*(4), 263–278. doi:10.1007/s00163-011-0111-z

Litzinger, T., Lattuca, L. R., Hadgraft, R., & Newstetter, W. (2011). Engineering Education and the Development of Expertise. *Journal of Engineering Education, 100*(1), 123–150. doi:10.1002/j.2168-9830.2011.tb00006.x

Liu, G. H. W., Wang, E. T. G., & Chua, C. E. H. (2015). Leveraging social capital to obtain top management support in IT projects. *Journal of the Association for Information Systems, 16*(8), 707–737.

Liu, S., Zhang, J., Keil, M., & Chen, T. (2010). Comparing senior executive and project manager perceptions of IT project risk: A Chinese Delphi stud. *Information Systems Journal, 20*(4), 319–355. doi:10.1111/j.1365-2575.2009.00333.x

Lodato, F. (2000). Biónica: La Naturaleza como herramienta de innovación. *Experimenta: Ediciones de Diseño,* (31), 46-51.

Lombardo, N. T., Morrow, A., & Le Ber, J. (2012). Rethinking mobile delivery: Using Quick Response codes to access information at the point of need. *Medical Reference Services Quarterly, 31*(1), 14–24. doi:10.1080/02763869.2012.641817 PMID:22289092

Loonam, J. A., & McDonagh, J. (2005). Exploring top management support for the introduction of enterprise information systems: A literature review. *Irish Journal of Management, 26*(1), 163–178.

López-Alt, A., Tromer, E., & Vaikuntanathan, V. (2012). On-the-fly multiparty computation on the cloud via multikey fully homomorphic encryption. In *Proceedings of the forty-fourth annual ACM symposium on Theory of computing* (pp. 1219-1234). ACM. doi:10.1145/2213977.2214086

López, C., & Matesanz, M. (2009). *Las plataformas de aprendizaje. Del mito a la realidad.* Madrid: Editorial Biblioteca Nueva.

López-Forniés, I., & Berges Muro, L. (2012). A Top-down Biomimetic Design Process For Product Concept Generation. *International Journal of Design & Nature and Ecodynamics, 7*(1), 27–48. doi:10.2495/DNE-V7-N1-27-48

Lorenzo, O., Bjorn, C., Oksana, K., & Angel, D. (2017). Integration through orchestration: the interplay between enterprise systems & inventory management capabilities. *Journal of Enterprise Information Management.*

Lowe, B., & Laffey, D. (2011). Is Twitter for the birds? Using Twitter to enhance student learning in a marketing course. *Journal of Marketing Education, 33*(2), 183–192. doi:10.1177/0273475311410851

Luna, E. (2002). Requerimientos básicos de los planes de estudio para la educación a distancia. In *Proceedings of the II Congreso Internacional de Educación Abierta y a Distancia: CICESE, ANUIES.* Universidad Autónoma de Baja California.

Lyytinen, K., Newman, M., & Al-Muharfi, A. (2009). Institutionalizing enterprise resource planning in the Saudi steel industry: A punctuated socio-technical analysis. *Journal of Information Technology, 24*(4), 286–304. doi:10.1057/jit.2009.14

Malaga, R. A., & Koppel, N. B. (2016). A Comparison Of Video Formats For Online Teaching. *Contemporary Issues in Education Research, 10*(1), 7–12. doi:10.19030/cier.v10i1.9876

Malik, K. (2012). Inception to dissolution: E-learning initiative at an Indian institute. *Journal of Cases on Information Technology, 14*(3), 35–56. doi:10.4018/jcit.2012070103

Maloney, E. (2007). What Web 2.0 can teach us about learning. *The Chronicle of Higher Education, 25*(18), B26.

Manchado, E.; Romero, C.; Blanco, T.; Casas, R.; López, J.M. (2015): Una experiencia de aprendizaje colaborativo, basada en la adaptación de metodologías de Design Thinking. *La sociedad del aprendizaje. Actas del iii congreso internacional sobre aprendizaje, innovación y competitividad. Cinaic 2015, 1,* 434-439.

Manchado, E., & Berges, L. (2013). Sistemas de retículas: Un método para diseñar nuevos conceptos de producto hacia el usuario. *DYNA. Año, 80*(181), 16–24.

Manchado, E., & López, I. (2012). Coordinación por módulos de asignaturas en el Grado de Ingeniería de Diseño Industrial y Desarrollo de Producto de la UZ. *REDU, 10*(3), 195–207.

Manzini, E. (1991). *Artefacts: vers une nouvelle écologie de l'environnement artificial.* Paris, France: Editions du Centre Pompidou.

Manzini, E., & Coad, R. (2015). *Design, when everybody designs: An introduction to design for social innovation.* MIT Press.

Marín, R., & De la Torre, S. (1991). *Manual de la creatividad.* Barcelona: Vicens Vives.

Marketingland. (2015). *How Many Websites Use Google Analytics? Estimates Suggest 30-50 Million.* Retrieved 3 November 2016, from http://marketingland.com/as-google-analytics-turns-10-we-ask-how-many-websites-use-it-151892

Markman, G. D., & Waldron, T. L. (2014). Small entrants and large incumbents: A framework of micro entry. *The Academy of Management Perspectives, 28*(2), 179–197. doi:10.5465/amp.2011.0112

Mayère, A., Grabot, B., & Bazet, I. (2008).The mutual influence of the tool and the organization. In B. Grabot, A, Mayère, & I Bazet (Eds.), ERP systems and organizational change (pp. 1–11). London: Springer-Verlag.

Mc Guckin, C., & Crowley, N. (2012). Using Google Analytics to Evaluate the Impact of the CyberTraining Project. *Cyberpsychology, Behavior, and Social Networking, 15*(11), 625–629. doi:10.1089/cyber.2011.0460 PMID:23075266

McAdam, R., & Galloway, A. (2005). Enterprise resource planning and organizational innovation: A management perspective. *Industrial Management & Data Systems, 105*(3), 280–290. doi:10.1108/02635570510590110

McFedries, P. (2007). Technically speaking: All a-twitter. *Spectrum, IEEE, 44*(10), 84–84.

McKay, S. L. (2002). *Teaching English as an international language: Rethinking goals and perspectives.* OUP.

McLeod, L., & Doolin, B. (2012). Information systems development as situated socio-technical change: A process approach. *European Journal of Information Systems, 21*(2), 176–191. doi:10.1057/ejis.2011.43

Medved, J. P. (2015). *LMS Industry User Research Report*. Retrieved 12 December 2016, from http://www.capterra.com/learning-management-system-software/user-research

Melchor, C. A., Castagnos, G., & Gaborit, P. (2008). Lattice-based homomorphic encryption of vector spaces. In *Information Theory, 2008. ISIT 2008. IEEE International Symposium on* (pp. 1858-1862). IEEE. doi:10.1109/ISIT.2008.4595310

Meseguer-Martinez, A., Ros-Galvez, A., & Rosa-Garcia, A. (2017). Satisfaction with online teaching videos: A quantitative approach. *Innovations in Education and Teaching International*, *54*(1), 62–67. doi:10.1080/14703297.2016.1143859

Mihalca, R., Uta, A., Andreescu, A., & Intorsureanu, I. (2008). Knowledge management in e-learning systems. *Revista Informatica Economică*, *46*(2), 60–65.

Miles, M. B., & Huberman, A. (1994). *Qualitative Data Analysis: An Expanded Sourcebook*. Newbury Park, CA: SAGE.

Miller, V. S. (1985). Use of elliptic curves in cryptography. In Advances in Cryptology—CRYPTO'85 Proceedings (pp. 417-426). Springer Berlin Heidelberg.

Milwich, M., Speck, T., Speck, O., Stegmaier, T., & Planck, H. (2006). Biomimetics and Technical Textiles: Solving Engineering Problems with the Help of Nature's Wisdom. *American Journal of Botany*, *93*(10), 1455–1465. doi:10.3732/ajb.93.10.1455 PMID:21642093

Mobile-QR-Codes.org. (n.d.a). *History of QR Codes*. Retrieved 5 July 2017, from http://www.mobile-qr-codes.org/history-of-qr-codes.html

Mobile-QR-Codes.org. (n.d.b). *How Are QR Codes Better Than Barcodes*. Retrieved 5 July 2017, from http://www.mobile-qr-codes.org/qr-codes-vs-barcodes.html

Moles, A. (1972). *Théorie des objets*. Paris, France: Ed. Universitaires.

Molinsky, A. (2007). Cross-cultural code-switching: The psychological challenges of adapting behavior in foreign cultural interactions. *Academy of Management Review*, *32*(2), 622–640. doi:10.5465/AMR.2007.24351878

Momoh, A., Roy, R., & Shehab, E. (2010). Challenges in enterprise resource planning implementation: State-of-the-art. *Business Process Management Journal*, *16*(4), 537–565. doi:10.1108/14637151011065919

Montoya, M. S. R., & Hernández, D. (2016). Inverted Learning Environments with Technology, Innovation and Flexibility. *Journal of Information Technology Research*, *9*(1), 18–33. doi:10.4018/JITR.2016010102

Moore, J. (1993). Predators and prey: A new ecology of competition. *Harvard Business Review*, *71*(3), 75–86. PMID:10126156

Moradi, M., Liu, L., & Luchies, C. (2016). Identifying Important Aspects in Developing Interactive Instructional Videos. In *ASEE Midwest Section Conference*. Retrieved from https://conferences.k-state.edu/aseemidwest2016/files/2016/06/Paper-Moradi-Liu-Luchies-29kd4x4.pdf

Morrison, M. A., Haley, E., Sheehan, K. B., & Taylor, R. E. (2011). *Using qualitative research in advertising: strategies, techniques, and applications*. SAGE Publications, Incorporated.

Mostaghel, R., Oghazi, P., Beheshti, H. M., & Hultman, M. (2015). Strategic use of enterprise systems among service firms: Antecedents and consequences. *Journal of Business Research*, *68*(7), 1544–1549. doi:10.1016/j.jbusres.2015.01.049

Motiwalla, L. F., & Thompson, J. (2012). *Enterprise systems for management*. Upper Saddle River, NJ: Pearson Education.

Munari, B. (1983). *Cómo nacen los objetos? Apuntes para una metodología proyectual*. Barcelona, Spain: Gustavo Gili.

Munir, K., & Mohammed, L. A. (2012). Policy for Mobile Devices to Access Grid Infrastructure. In Grid and Cloud Computing: Concepts, Methodologies, Tools and Applications (pp. 1593-1608). IGI Global. doi:10.4018/978-1-4666-0879-5.ch704

Naccache, D., & Stern, J. (1998). A new public key cryptosystem based on higher residues. In *Proceedings of the 5th ACM conference on Computer and communications security* (pp. 59-66). ACM. doi:10.1145/288090.288106

Nardi, B. A., Schiano, D. J., Gumbrecht, M., & Swartz, L. (2004). Why we blog. *Communications of the ACM*, *47*(12), 41–46. doi:10.1145/1035134.1035163

Neufeld, D. J., Dong, L., & Higgins, C. (2007). Charismatic leadership and user acceptance of information technology. *European Journal of Information Systems*, *16*(4), 494–510. doi:10.1057/palgrave.ejis.3000682

Nobelis, N., Boudaoud, K., Delettre, C., & Riveill, M. (2012). Designing Security Properties-Centric Communication Protocols using a Component-Based Approach. *International Journal of Distributed Systems and Technologies*, *3*(1), 1–16. doi:10.4018/jdst.2012010101

Norman, D. A. (2005). *El diseño emocional; por qué nos gustan (o no) los objetos cotidianos*. Barcelona, Spain: Paidós Ibérica.

Nuttall, C. (1996). *Teaching reading skills in a foreign language*. Heinemann.

Observatory of Educational Innovation of the Tecnológico de Monterrey. (2014). *Flipped Learning. Monterrey*. Retrieved from http://observatorio.itesm.mx/edutrendsaprendizajeinvertido

Okamoto, T., & Uchiyama, S. (1998). A new public-key cryptosystem as secure as factoring. In Advances in Cryptology—EUROCRYPT'98 (pp. 308-318). Springer Berlin Heidelberg. doi:10.1007/BFb0054135

Oliver, R. (2000). When Teaching Meets Learning: Design, Principles and Strategies for Web-based Learning Environments that Support Knowledge Construction. In R. Sims, M., O'Reilly & S. Sawkins (Eds.), *Learning to Choose: Choosing to Learn. Proceedings of the 17th Annual ASCILITE Conference* (pp. 17-28). Lismore, Australia: Southern Cross University Press.

Orlikowski, W. J., & Iacono, C. S. (2001). Research commentary: Desperately seeking the "IT" in IT research—A call to theorizing the IT artifact. *Information Systems Research*, *12*(2), 121–134. doi:10.1287/isre.12.2.121.9700

Osguthorpe, R. T., & Graham, C. R. (2003). Blended Learning Environments: Definitions and Directions. *Quarterly Review of Distance Education*, *4*(3), 227–233.

Osterwalder, A., & Pigneur, Y. (2010). *Business model generation: a handbook for visionaries, game changers, and challengers*. Hoboken, NJ: John Wiley & Sons.

Oyelere, S. S., Paliktzoglou, V., & Suhonen, J. (2016). M-learning in Nigerian higher education: An experimental study with Edmodo. *International Journal of Social Media and Interactive Learning Environments*, *4*(1), 43–62. doi:10.1504/IJSMILE.2016.075055

Özen, Z., Bakioğlu, F. Ö. K., & Beden, Ş. (2014). The Examination of User Habits through the Google Analytic Data of Academic Education Platforms. *International Journal of E-Adoption*, *6*(2), 31–45. doi:10.4018/ijea.2014070103

Ozkan, B., & McKenzie, B. (2008). *Social networking tools for teacher education*. Society for Information Technology & Teacher Education International Conference; Retrieved from http://www.editlib.org/p/27640

Ozkan, S., & Koseler, R. (2009). Multi-dimensional students' evaluation of e-learning systems in the higher education context: An empirical investigation. *Computers & Education*, *53*(4), 1285–1296. doi:10.1016/j.compedu.2009.06.011

Paavola, S., Lipponen, L., & Hakkarainen, K. (2004). Models of innovative knowledge communities and three metaphors of learning. *Review of Educational Research*, *74*(4), 557–576. doi:10.3102/00346543074004557

Paillier, P. (1999). Public-key cryptosystems based on composite degree residuosity classes. In Advances in cryptology—EUROCRYPT'99 (pp. 223-238). Springer Berlin Heidelberg doi:10.1007/3-540-48910-X_16

Pakkala, H., Presser, K., & Christensen, T. (2012). Using Google Analytics to measure visitor statistics: The case of food composition websites. *International Journal of Information Management*, *32*(6), 504–512. doi:10.1016/j.ijinfomgt.2012.04.008

Paliktzoglou, V., Giousmpasoglou, C., & Marinakou, E. (2016). Assessment of Students' Familiarity, Adoption, and Use of Social Media in Bahrain. *International Journal of Technology Diffusion*, *7*(4), 82–90. doi:10.4018/IJTD.2016100105

Pandermarakis, Z. G., Sotiropoulou, A. B., Passa, D. S., & Mitsopoulos, G. D. (2012). The Role of Engineering Educational Laboratories at a Thesis Level. *SEFI (European Society for Engineering Education) 40th Annual Conference Proceedings.*

Pappas, C. (2015a). *The Top 8 Open Source Learning Management Systems.* Retrieved 12 December 2016, from https://elearningindustry.com/top-open-source-learning-management-systems

Pappas, C. (2015b). *The Top LMS Statistics and Facts for 2015 You Need to Know.* Retrieved 12 December 2016, from https://elearningindustry.com/top-lms-statistics-and-facts-for-2015

Pappas, C. (2016). *8 Types of eLearning Analytics to Obtain from Google and Your LMS.* Retrieved 3 October 2016, from https://elearningindustry.com/types-elearning-analytics-obtain-google-lms

Patterson, C., Crooks, D., & Lunyk-Child, O. (2002). A New Perspective on Competencies for Self-Directed Learning. *The Journal of Nursing Education, 41*(1), 25–31. PMID:11843104

Peres, P., & Mesquita, A. (2016). Flipping the classroom to gain time: A pedagogical innovative model. *Journal of Cases on Information Technology, 18*(4), 36–52. doi:10.4018/JCIT.2016100103

Perifanou, M. A. (2009). Language micro-gaming: Fun and informal microblogging activities for language learning. In *Best Practices for the Knowledge Society. Knowledge, Learning, Development and Technology for All* (pp. 1–14). Springer. Retrieved from http://link.springer.com/chapter/10.1007/978-3-642-04757-2_1

Pinto Llorente, A. M., Sánchez Gómez, M. C., & García-Peñalvo, F. J. (2014). Assessing the effectiveness of a technological model to improve written skills in English in higher education. In F. J. García-Peñalvo (Ed.), *Proceedings of the Second International Conference on Technological Ecosystems for Enhancing Multiculturality (TEEM'14)* (pp. 69-74). New York: ACM. doi:10.1145/2669711.2669881

Pinto Llorente, A. M., Sánchez Gómez, M. C., & García-Peñalvo, F. J. (2014). Students' essential characteristics for learning English in a hypermedia modular model. In F. J. García-Peñalvo (Ed.), *Proceedings of the Second International Conference on Technological Ecosystems for Enhancing Multiculturality (TEEM'14)* (pp. 107-111). New York: ACM. doi:10.1145/2669711.2669887

Pinto Llorente, A. M., Sánchez Gómez, M. C., & García-Peñalvo, F. J. (2016). Assessing the Effectiveness of Interactive and Collaborative Resources to Improve Reading and Writing in English. *International Journal of Human Capital and Information Technology Professionals, 7*(1), 66–85. doi:10.4018/IJHCITP.2016010105

Pinto, A. M., Sánchez, M. C., & García, F. J. (2015b). Developing a VLE to Enable the Innovative Learning of English Pronunciation. In *Proceedings of the 3rd International Conference on Technological Ecosystems for Enhancing Multiculturality* (pp. 83-89). New York: ACM. doi:10.1145/2808580.2808594

Pinto-Llorente, A. M., Sánchez-Gómez, M. C., & García-Peñalvo, F. J. (2015a). To Be or Not to Be Successful? That Does Not Only Depend on Technology, But Also on Human Factors. *Journal of Cases on Information Technology, 17*(1), 51–69. doi:10.4018/JCIT.2015010104

Pinto-Llorente, A. M., Sánchez-Gómez, M. C., & García-Peñalvo, F. J. (2017). A Mixed Methods Research of Pre-service Teachers' Perceptions about the Benefits of Wiki-Based Tasks and Discussion Boards. In A. P. Costa, L. P. Reis, F. N. De Souza, & A. Moreire (Eds.), *Computer Supported Qualitative Research, Advances in Intelligent Systems and Computing* (pp. 260–276). Springer International Publishing.

Pinto-Llorente, A., Sánchez-Gómez, M. C., & García-Peñalvo, F. J. (2015). Student Perception of the Use of a Blended-Learning Model to Improve Grammatical Competence. In G. R. Alves, & M. C. Felgueiras (Eds.), *Proceedings of the Third International Conference on Technological Ecosystems for Enhancing Multiculturality (TEEM'15)* (pp. 91-98). New York: ACM. doi:10.1145/2808580.2808595

Plaza, B. (2009). Monitoring web traffic source effectiveness with Google Analytics: An experiment with time series. *Aslib Proceedings*, *61*(5), 474–482. doi:10.1108/00012530910989625

Plaza, B. (2011). Google Analytics for measuring website performance. *Tourism Management*, *32*(3), 477–481. doi:10.1016/j.tourman.2010.03.015

Pollock, N., & Hyysalo, S. (2014). The business of being a user: The role of the reference actor in shaping packaged enterprise system acquisition and development. *Management Information Systems Quarterly*, *38*(2), 473–496. doi:10.25300/MISQ/2014/38.2.07

Polonsky, M. J., & Waller, D. S. (2010). *Designing and managing a research project: A business student's guide*. Sage Publications, Incorporated. Retrieved from http://books.google.com/books?hl=en&lr=&id=D3aXAgQpkJUC&oi=fnd&pg=PR1&dq=Polonsky,+M.,+%26+Waller,+D.+(2005).+Designing+and+managing+a+research+project%3B+A+business+student%27s+guide+(1st+ed.).+Thousand+Oaks,+CA:+Sage+Publications.+&ots=1udeGVc2_8&sig=qcYfugQ0JdN_4fkHoGnIJe31mh0

Porter, M. E. (2008, January). The five competitive forces that shape strategy. *Harvard Business Review*. PMID:18271320

Prensky, M. (2001). Digital Natives, Digital Immigrants Part 1. *On the Horizon*, *9*(5), 1–6. doi:10.1108/10748120110424816

Rafay, A., & Khan, A. (2016). Shift towards Next Generation Networks (NGNs) for sustainability: Evidence from an emerging Economy. *Journal of Cases on Information Technology*, *18*(3), 1–12. doi:10.4018/JCIT.2016070101

Raitman, R., Augar, N., & Zhou, W. (2005). Employing wikis for online collaboration in the e-learning environment: Case study. In *Information Technology and Applications, 2005. ICITA 2005. Third International Conference on* (Vol. 2, pp. 142–146). Retrieved from http://ieeexplore.ieee.org/xpls/abs_all.jsp?arnumber=1488944

Rajagopal, P. (2002). An innovation—diffusion view of implementation of enterprise resource planning (ERP) systems and development of a research model. *Information & Management*, *40*(2), 87–114. doi:10.1016/S0378-7206(01)00135-5

Ramsden, A. (2008). *How and why are people using Twitter: A small group study*. Retrieved from http://opus.bath.ac.uk/id/eprint/471

Rangan, S. (2000). Search and deliberation in international exchange: Microfoundations to some macro patterns. *Journal of International Business Studies*, *31*(2), 205–222. doi:10.1057/palgrave.jibs.8490902

Redecker, C., Ala-Mutka, K., Bacigalupo, M., Ferrari, A., & Punie, Y. (2009). *Learning 2.0: The impact of Web 2.0 innovations on education and training in Europe*. Final Report. European Commission-Joint Research Center-Institute for Porspective Technological Studies, Seville. Retrieved from ftp://ftp.jrc.es/pub/EURdoc/EURdoc/JRC55629.pdf

Reed, E., & Sharp, D. (2003). Minimum Technical Competencies for Distance Learning Students. *Journal of Research on Technology in Education*, *34*(3), 319–325.

Ricoy, M. C., & Couto, M. (2014). As boas práticas com TIC e a utilidade atribuída pelos alunos recém-integrados à universidade [Good practices with ICT and the use attributed by higher education freshmen]. *Educação e Pesquisa*, *40*(4), 897–912. doi:10.1590/S1517-97022014005000005

Rigby, D. K., & Ledingham, D. (2004). CRM done right. *Harvard Business Review*, *82*(11), 118–130. PMID:15559450

Rinaldo, S. B., Tapp, S., & Laverie, D. A. (2011). Learning by Tweeting Using Twitter as a Pedagogical Tool. *Journal of Marketing Education*, *33*(2), 193–203. doi:10.1177/0273475311410852

Rivers, W. M. (1981). *Teaching foreign-language skills. University of Chicago Press.*

Rivest, R. L., Adleman, L., & Dertouzos, M. L. (1978). On data banks and privacy homomorphisms. *Foundations of Secure Computation, 4*(11), 169-180.

Rodriguez, C. O. (2012). MOOCs and the AI-Stanford like Courses: Two Successful and Distinct Course Formats for Massive Open Online Courses. *European Journal of Open, Distance and E-Learning, 15*(2). Retrieved from http://www.eurodl.org/index.php?p=archives&year=2013&halfyear=2&article=516

Rodríguez, G., Gil, J., & García, E. (1999). *Metodología de la investigación cualitativa*. Málaga: Aljibe.

Rodríguez, G., & Gómez, M. A. (2010). Análisis de contenido y textual de datos cualitativos. In S. Nieto (Ed.), *Principios, métodos y técnicas esenciales para la investigación educativa* (pp. 447–469). Madrid: Dykinson.

Rogado, A. B. G., Quintana, A. M. V., Conde, M. J. R., Martín, I. R., Abad, F. M., Miguelañez, S. O., & Diego, P. B. (2015, March 11). Adquisición de la competencia de prevención de riesgos en el laboratorio mediante tecnología móvil. *Revista del Congrés Internacional de Docència Universitària i Innovació* (CIDUI). Retrieved from http://www.cidui.org/revistacidui/index.php/cidui/article/view/663

Rogers, E. M. (2003). *Diffusion of innovations*. Free Press.

Rogers, G., & Bouey, E. (2005). Participant Observation. In R. M. Grinnell & Y. A. Unrau (Eds.), *Social Work: Research and Evaluation. Quantitative and Qualitative Approaches* (pp. 231–244). New York: Oxford University Press.

Rom, A., & Rohde, C. (2006). Enterprise resource planning systems, strategic enterprise management systems and management accounting: A Danish study. *Journal of Enterprise Information Management*, *19*(1), 50–66. doi:10.1108/17410390610636878

Roschelle, J., & Teasley, S. D. (1995). The Construction of Shared Knowledge in Collaborative Problem Solving. In C. O'Malley (Ed.), *Computer-supported collaborative learning* (pp. 69–97). Berlin: Springer Verlag. doi:10.1007/978-3-642-85098-1_5

Roshko, T. (2010). The Pedagogy of Bio-Design: Methodology Development. *WIT Transactions on Ecology and the Environment.*, *138*, 545–558. doi:10.2495/DN100491

Ross, C., Terras, M., Warwick, C., & Welsh, A. (2011). Enabled backchannel: Conference Twitter use by digital humanists. *The Journal of Documentation*, *67*(2), 214–237. doi:10.1108/00220411111109449

Rouse, M. (2005). *What is Web analytics? - Definition from WhatIs.com.* Retrieved 10 August 2017, from http://searchcrm.techtarget.com/definition/Web-analytics

Rowe, M., Bozalek, V., & Frantz, J. (2013). Using Google Drive to facilitate a blended approach to authentic learning. *British Journal of Educational Technology*, *44*(4), 594–606. doi:10.1111/bjet.12063

Rubin, H. J., & Rubin, I. S. (2011). *Qualitative interviewing: The art of hearing data.* SAGE Publications, Incorporated.

Ruivo, P., & Neto, M. (2011). Sustainable enterprise KPIs and ERP post adoption. In *Proceeding of the 6th Iberian Conference on Information Systems and Technologies* (pp. 1–7). Piscataway, NJ: IEEE.

Ruivo, P., Oliveira, T., & Neto, M. (2014). Examine ERP post-implementation stages of use and value: Empirical evidence from Portuguese SMEs. *International Journal of Accounting Information Systems*, *15*(2), 166–184. doi:10.1016/j.accinf.2014.01.002

Rumelt, R. (2011). *Good strategy, bad strategy, the difference and why it matters.* Deckle Edge.

Ryan, G., & Bernard, H. (2000). Data management and analysis methods. In N. K. Denzin & Y. S. Lincoln (Eds.), *Handbook of qualitative research* (2nd ed.; pp. 769–802). Thousand Oaks, CA: Sage.

Safran, C., Guetl, C., & Helic, D. (2007). The Impact of Web 2.0 on Learning at a Technical University-A usage survey. In *World Conference on E-Learning in Corporate* (Vol. 2007, pp. 436–443). Government, Healthcare, and Higher Education. Retrieved from http://www.editlib.org/p/26364/

Safran, R. J., Doerr, V. A., Sherman, P. W., Doerr, E. D., Flaxman, S. M., & Winkler, D. W. (2007). Group breeding in vertebrates: Linking individual-and population-level approaches. *Evolutionary Ecology Research*, *9*(7), 1163.

Salinas, J. (2000). El aprendizaje colaborativo con los nuevos canales de comunicación. In J. Cabero, J. Salinas, A. Duarte, A. & J. Domingo (Eds), Nuevas Tecnologías aplicadas a la Educación (pp. 199-228). Madrid: Síntesis.

Sánchez-Gómez, M. C., Pinto-Llorente, A. M., & García-Peñalvo, F. J. (2010). Blended learning University students' perception of digital competence. In *Proceedings of the Conference on European Applied Business Research (EABR) and European College Teaching & Learning (ETLC)* (pp. 1-11). Dublin, Ireland: The Clute Institute for Academic Research.

Sánchez, M. C., Delgado, M. C., & Santos, M. C. (2012). *El proceso de la investigación cualitativa. Manual de procedimiento: ejemplificación con una tesis doctoral*. Valladolid: Fundación Intras.

Sánchez, M. C., & Revuelta, I. F. (2004). El proceso de transcripción en el marco de la metodología cualitativa actual. *Enseñanza*, *23*, 367–386.

Sander, T., Young, A., & Yung, M. (1999). Non-interactive cryptocomputing for NC 1. In *Foundations of Computer Science, 1999. 40th Annual Symposium on* (pp. 554-566). IEEE.

Sanders, P. (1982). Phenomenology: A new way of viewing organizational research. *Academy of Management Review*, *7*(3), 353–360.

Sankey, M., & Hunt, L. (2014). Flipped university classrooms: Using technology to enable sound pedagogy. *Journal of Cases on Information Technology*, *16*(2), 26–38. doi:10.4018/jcit.2014040103

Sarker, S., & Lee, A. (2003). Using a case study to test the role of three key social enablers in ERP implementation. *Information & Management*, *40*(8), 813–829. doi:10.1016/S0378-7206(02)00103-9

Sawyer, S., & Jarrahi, M. H. (2014). Sociotechnical approaches to the study of Information Systems. In A. Tucker & H. Topi (Eds.), Computing Handbook (3rd ed.; pp. 5.1-5.27). New York: Chapman and Hall. doi:10.1201/b16768-7

Schmidt, R., Lyytinen, K., Keil, M., & Cule, P. (2001). Identifying software project risks: An international Delphi study. *Journal of Management Information Systems*, *17*(4), 5–36. doi:10.1080/07421222.2001.11045662

Schnackenberg, H. L., Vega, E. S., & Simard, D. A. (2014). Paradigm shift: Introduction of a social media network and web 2.0 technology into a college classroom environment. *Journal of Cases on Information Technology*, *16*(2), 1–12. doi:10.4018/jcit.2014040101

Sedera, D., Lokuge, S., Grover, V., Sarker, S., & Sarker, S. (2016). Innovating with enterprise systems and digital platforms: A contingent resource-based theory view. *Information & Management*, *53*(3), 366–379. doi:10.1016/j.im.2016.01.001

Selwyn, N. (2009). Faceworking: Exploring students' education-related use of Facebook. *Learning, Media and Technology*, *34*(2), 157–174. doi:10.1080/17439880902923622

Sexton, R. L. (2006). Using Short Movie and Television Clips in the Economics Principles Class. *Journal of Economic Education*, *37*(4), 406–417. doi:10.3200/JECE.37.4.406-417

Shairani, K. T. (2016). Mobilink and Warid are now one happy family — but what's in it for customers? *Dawn News*. Retrieved from https://www.dawn.com/news/1272471

Shang, S., & Seddon, P. B. (2002). Assessing and managing the benefits of enterprise systems: The business manager's perspective. *Information Systems Journal*, *12*(4), 271–299. doi:10.1046/j.1365-2575.2002.00132.x

Shen, L. P., & Shen, R. M. (2004). Learning content recommendation service based-on simple sequencing specification. In *Advances in Web-Based Learning–ICWL 2004* (pp. 363–370). Springer Berlin Heidelberg. doi:10.1007/978-3-540-27859-7_47

Shi, L., & Xie, Z. (2010). Designing product to improve affective communication. *Intelligent Systems and Applications (ISA), 2010 2nd International Workshop on*, 1-4.

Shin, D.-H., Jung, J., & Chang, B.-H. (2012). The psychology behind QR codes: User experience perspective. *Computers in Human Behavior*, *28*(4), 1417–1426. doi:10.1016/j.chb.2012.03.004

Shohamy, E. (1988). A proposed framework for testing the oral language of second/foreign language learners. *Studies in Second Language Acquisition*, *10*(02), 165–179. doi:10.1017/S0272263100007294

Siddiqui, S. (2017). *Mobile phone subscribers reach close to 140m*. Retrieved from https://tribune.com.pk/story/1471997/mobile-phone-subscribers-reach-close-140m/

Simonson, M., Smaldino, S., Albright, M., & Zvacek, S. (2003). *Teaching and learning at a distance foundations of distance education*. Person.

Simo, P., Fernandez, V., Algaba, I., Salan, N., Enache, M., Albareda-Sambola, M., & Rajadell, M. et al. (2010). Video stream and teaching channels: Quantitative analysis of the use of low-cost educational videos on the web. *Procedia: Social and Behavioral Sciences*, *2*(2), 2937–2941. doi:10.1016/j.sbspro.2010.03.444

Slater, S. F., Mohr, J. J., & Sengupta, S. (2014). Radical product innovation capability: Literature review, synthesis, and illustrative research propositions. *Journal of Product Innovation Management*, *31*(3), 552–566. doi:10.1111/jpim.12113

Somers, T., & Nelson, K. (2004). A taxonomy of players and activities across the ERP project life cycle. *Information & Management*, *41*(3), 257–278. doi:10.1016/S0378-7206(03)00023-5

Spathis, C., & Constantinides, S. (2003). The usefulness of ERP systems for effective management. *Industrial Management & Data Systems*, *103*(9), 677–685. doi:10.1108/02635570310506098

Srivardhana, T., & Pawlowski, S. D. (2007). ERP systems as an enabler of sustained business process innovation: A knowledge-based view. *The Journal of Strategic Information Systems, 16*(1), 51–69. doi:10.1016/j.jsis.2007.01.003

Staehr, L., Shanks, G., & Seddon, P. B. (2012). An explanatory framework for achieving business benefits from ERP systems. *Journal of the Association for Information Systems, 13*(6), 424.

Stallard, C. K., & Cocker, J. (2014). *Education technology and the failure of American schools*. Lanham, MD: Rowman & Littlefield.

Stangos, N. (1987). Conceptos de arte moderno. Madrid, Spain: Ed.Alianza, Col. Alianza y Forma nº 53.

StatCounter. (2016). *Mobile and tablet internet usage exceeds desktop for first time worldwide | StatCounter Global Stats*. Retrieved 14 December 2016, from http://gs.statcounter.com/press/mobile-and-tablet-internet-usage-exceeds-desktop-for-first-time-worldwide

Statista. (2016a). *Global smartphone shipments market forecast 2010-2020 | Statistic*. Retrieved 14 December 2016, from https://www.statista.com/statistics/263441/global-smartphone-shipments-forecast/

Statista. (2016b). *Laptop, PC, tablet sales 2010-2020 | Statistic*. Retrieved 14 December 2016, from https://www.statista.com/statistics/272595/global-shipments-forecast-for-tablets-laptops-and-desktop-pcs/

Statista. (2017). *Smartphone users worldwide 2014-2020*. Retrieved 6 July 2017, from https://www.statista.com/statistics/330695/number-of-smartphone-users-worldwide/

Strauss, A., & Corbin, J. (1994). Grounded Theory Methodology. An Overview. In N. K. Denzin & Y. S. Lincoln (Eds.), *Handbook of Qualitative Research* (pp. 273–285). Thousand Oaks, CA: SAGE.

Strauss, A., & Corbin, J. (1998). *Basics of Qualitative Research: Techniques and Procedures for Developing Grounded Theory*. Thousand Oaks, CA: Sage.

Streatfield, P. (2001). *The paradox of control in organizations*. London: Routledge. doi:10.4324/9780203361504

Suarez, F., & Lanzolla, G. (2005, April). The half-truth of first-mover advantage. *Harvard Business Review*. PMID:15807045

Sumner, M. (1999, April). Critical success factors in enterprise wide information management systems projects. In *Proceedings of the 1999 ACM SIGCPR conference on Computer personnel research* (pp. 297-303). ACM. doi:10.1145/299513.299722

Surgenor, B., & Firth, K. (2011). The role of the laboratory in design engineering education. *Proceedings of the 2nd Canadian Engineering Education Association*. doi:10.24908/pceea.v0i0.3848

Swan, A. (1990). *Cómo diseñar retículas*. México DF, Mexico: Gustavo Gili.

Swanson, E. B., & Wang, P. (2005). Knowing why and how to innovate with packaged business software. *Journal of Information Technology*, *20*(1), 20–31. doi:10.1057/palgrave.jit.2000033

Takagi, M., & Teshigawara, Y. (2006). A WBT system enabling to create new or similar quizzes collaboratively by students. In *Proceedings of the 3rd International Conference on Educational Technology* (pp. 263-268). ACTA Press.

Tang. C. M., & Chaw, L. Y. (2016). Digital Literacy: A Prerequisite for Effective Learning in a Blended Learning Environment? *The Electronic Journal of e-Learning, 14*(1), 54-65.

Technologytimes.pk. (2016). *CCP approves Mobilink merger with Warid*. Retrieved from http://technologytimes.pk/post.php?title=CCP+approves+Mobilink+merger+with+Warid

Telg, R. (1999). *Producing an Educational Video*. AEC343/WC024. Retrieved from http://edis.ifas.ufl.edu/wc024

The Marketing Bureau UK Ltd. (n.d.). *FAQs*. Retrieved 15 March 2016, from http://www.freeqrcodetracker.com/pages/faqs

Thenews.com.pk. (2016). *Islamabad High Court allows merger of Mobilink, Warid*. Retrieved from https://www.thenews.com.pk/print/173740-Islamabad-High-Court-allows-merger-of-Mobilink-Warid

Thompson, C. (2011, July 15). How Khan Academy is changing the rules of education. *Wired Digital*.

Thorne, K. (2003). *Blended Learning: How to Integrate Online and Traditional Learning*. London: Kogan.

Tihanyi, L., Ellstrand, A., Daily, C., & Dalton, D. (2000). Composition of the top management team and firm international diversification. *Journal of Management*, *26*(6), 1157–1177. doi:10.1177/014920630002600605

Tîrziu, A.-M., & Vrabie, C. (2015). Education 2.0: E-Learning Methods. *Procedia: Social and Behavioral Sciences*, *186*, 376–380. doi:10.1016/j.sbspro.2015.04.213

Traser, C. J., Hoffman, L. A., Seifert, M. F., & Wilson, A. B. (2015). Investigating the use of quick response codes in the gross anatomy laboratory. *Anatomical Sciences Education*, *8*(5), 421–429. doi:10.1002/ase.1499 PMID:25288343

Travis, L. (2010). One of many free survey tools: Google docs. *Journal of Electronic Resources in Medical Libraries*, *7*(2), 105–114. doi:10.1080/15424065.2010.482902

Trinidad, A., Carretero, V., & Soriano, R. M. (2006). *Teoría fundamentada "Grounded Theory" La construcción de la teoría a través del análisis interpretacional*. Madrid: CIS.

Tutorials Point. (2015). *Web Analytics*. Retrieved from https://www.tutorialspoint.com/web_analytics/web_analytics_tutorial.pdf

Twu, H. (2010). A predictive study of Wiki interaction: Can attitude toward Wiki predict Wiki interaction in High-Context Cultures groups? *Journal of Educational Technology Development and Exchange*, *3*(1), 57–68. doi:10.18785/jetde.0301.05

Uhl-Bien, M., & Marion, R. (2009). Complexity leadership in bureaucratic forms of organizing: A meso model. *The Leadership Quarterly*, *20*(4), 631–650. doi:10.1016/j.leaqua.2009.04.007

Van Dijk, M., Gentry, C., Halevi, S., & Vaikuntanathan, V. (2010). Fully homomorphic encryption over the integers. In *Annual International Conference on the Theory and Applications of Cryptographic Techniques* (pp. 24-43). Springer Berlin Heidelberg.

Van Wart, M. (2015). Evaluating transformational leaders: The challenging case of Eric Shinseki and the US Department of Veterans Affairs. *Public Administration Review*, *75*(5), 760–769. doi:10.1111/puar.12393

Van Wart, M., Roman, A., Wang, X., & Liu, C. (2017). Integrating ICT adoption issues into (e-) leadership theory. *Telematics and Informatics*, *34*(5), 527–537. doi:10.1016/j.tele.2016.11.003

Vat, K. H. (2010). Virtual Organizaing Professional Learning Communities through a Servant-leader Model of Appreciative Coaching. In Y. Inoue (Ed.), Cases on Online and Blended Learning Technologies in Higher Education: Concepts and Practices (pp.183-206). Hershey, PA: Information Science Reference, IGI Global.

Viladás, X. (2008). *Diseño rentable: diez temas a debate*. Barcelona, Spain: Index Book.

Villagrasa, S., Fonseca, D., Redondo, E., & Duran, J. (2014). Teaching case of gamification and visual technologies for education. *Journal of Cases on Information Technology*, *16*(4), 38–57. doi:10.4018/jcit.2014100104

Vincent, J. (2007). *Biomimetics: Strategies for Product Design Inspired by Nature - A Mission to the Netherlands and Germany. Report of A DTI Global Watch Mission*. Pera.

Viñolas i Marlet, J. (2005). *Diseño ecológico: Hacia un diseño y una producción en armonía con la naturaleza*. Barcelona, Spain: Blume.

Vinyets, J. (2009). Showing the Value of Ethnography in Business. *Ethnographic Praxis in Industry Conference Proceedings, (1),* 162-169.

Vogten, H., Martens, H., Nadolski, R., Tattersall, C., Van Rosmalen, P., & Koper, R. (2006, July). CopperCore Service Integration-Integrating IMS Learning Design and IMS Question and Test Interoperability. In *Advanced Learning Technologies, 2006. Sixth International Conference on* (pp. 378-382). IEEE. doi:10.1109/ICALT.2006.1652450

Von Hippel, E. (1986). Lead users: A source of novel product concepts. *Management Science*, *32*(7), 791–805. doi:10.1287/mnsc.32.7.791

Voorheis, C. M., Singleton, H. M., Silvers, T. J., Snead, A. E., & Moroney, W. F. (2004, September). A consumer's guide to internet questionnaire development tools: II. *Proceedings of the Human Factors and Ergonomics Society Annual Meeting*, *48*(21), 2446–2449. doi:10.1177/154193120404802101

Vygotsky, L. S. (1978). *Mind in Society: The Development of Higher Psychological Processes*. Cambridge, MA: Harvard University Press.

Waisberg, D., & Kaushik, A. (2009). Web Analytics 2.0: Empowering customer centricity. *The Original Search Engine Marketing Journal*, *2*(1), 5–11.

Walsh, J. (1988). Selectivity and selective perception: An investigation of managers' belief structures and information processing. *Academy of Management Journal*, *31*(4), 873–896. doi:10.2307/256343

Walsh, L., & Kahn, P. (2010). *Collaborative Working in Higher Education. The Social Academy*. New York: Routledge.

Waridtel.com. (2016). *VimpelCom and Dhabi Group announce completion of Mobilink and Warid transaction*. Retrieved from http://www.waridtel.com/media-center/press-release/archive/381

Watson, E. E. & Schneider, H. (1999). Using ERP systems in education. *Communications of the AIS*, *1*(2), Article No.3.

Watson, W. R., & Watson, S. L. (2007). What are learning management systems, what are they not, and what should they become. *TechTrends*, *51*(2), 29.

Web Analytics Association. (2008). *Web Analytics Definitions*. Retrieved from https://www.digitalanalyticsassociation.org/Files/PDF_standards/WebAnalyticsDefinitions.pdf

Weill, P., & Ross, J. W. (2009). *IT savvy: what top executives must know to go from pain to gain*. Harvard Business Press.

Weiss, C. (2011). *E-Learning End Users Have Spoken – time for the industry to listen*. Retrieved June 12, 2015, from http://elearninfo247.com

Wei, Z. (2016). A pairing-based homomorphic encryption scheme for multi-user settings. *International Journal of Technology and Human Interaction*, *12*(2), 72–82. doi:10.4018/IJTHI.2016040106

Wheeler, S. (2012). e-Learning and Digital Learning. In Encyclopedia of the Sciences of Learning (pp. 1109–1111). Boston, MA: Springer US.

Whittington, K., Owen-Smith, J., & Powell, W. (2009). Networks, propinquity and innovation in knowledge-intensive industries. *Administrative Science Quarterly*, *54*(1), 90–122. doi:10.2189/asqu.2009.54.1.90

Wikipedia. (2017). List of countries by smartphone penetration. Retrieved 14 November 2017, from https://en.wikipedia.org/wiki/List_of_countries_by_smartphone_penetration

Wirtz, B. W., Mathieu, A., & Shilke, O. (2007). Strategy in high-velocity environments. *Long Range Planning*, *40*(3), 295–313. doi:10.1016/j.lrp.2007.06.002

Wladawsky-Berger, I. (2013, May). *'IT Doesn't Matter:' 10 Years Later*. Retrieved from: https://blogs.wsj.com/cio/2013/05/24/it-doesnt-matter-10-years-later/

Wohlers, S. D., & Jamieson, J. (2014). "What in me is Dark, Illumine": developing a semantic URL learning analytics solution for Moodle. In B. Hegarty, J. McDonald, & S. K. Loke (Eds.), Rhetoric and reality: Critical perspectives on educational technology – Conference proceedings (pp. 110–118). Academic Press. Retrieved from http://ascilite.org/conferences/dunedin2014/files/fullpapers/105-Wohlers.pdf

Wright, N. (2010). Twittering in teacher education: Reflecting on practicum experiences. *Open Learning*, *25*(3), 259–265. doi:10.1080/02680513.2010.512102

Xia, J. (2011). The third-generation-mobile (3G) policy and deployment in China: Current status, challenges, and prospects. *Telecommunications Policy*, *35*(1), 51–63. doi:10.1016/j.telpol.2010.12.008

Xia, J. (2012). Reprint of: Competition and regulation in China's 3G/4G mobile communications industry—Institutions, governance, and telecom SOEs. *Telecommunications Policy*, *36*(10), 798–816. doi:10.1016/j.telpol.2012.10.002

xyleme.com. (2014). *Taking a Google Analytics Approach to Measuring Learning*. Retrieved 3 October 2016, from http://resources.xyleme.com/h/i/63857530-taking-a-google-analytics-approach-to-measuring-learning

Yamamoto, G. T., Özen, Z., & Beden, Ş. (2012). *Examination of LMS Logs in Academic Education Platforms: A Case Study*. Presented at the 4th International Future-Learning Conference on Innovations in Learning for the Future 2012: e-Learning, İstanbul, Turkey.

Yang, Q., Li, J. J., & Weiss, D. M. (2009). A survey of coverage-based testing tools. *The Computer Journal*, *52*(5), 589–597. doi:10.1093/comjnl/bxm021

Yasin, A. (2017). Pakistan to be the first country in South Asia to test 5G services. *Dawn News*. Retrieved from https://www.dawn.com/news/1326401

Yeh, E., & Swinehart, N. (2017). A Learner-Centered Approach to Technology Integration: Online Geographical Tools in the ESL Classroom. In J. Keengwe & G. Onchwari (Eds.), Handbook of Research on Learner-Centered Pedagogy in Teacher Education and Professional Development (pp. 1-22). Hershey, PA: Information Science Reference, IGI Global.

Yen Teoh, S., & Pan, S. (2008). Understanding the influences of social integration in enterprise systems use. *Journal of Enterprise Information Management*, *21*(5), 493–511. doi:10.1108/17410390810904256

Yen, J. C., & Lee, C. Y. (2011). Exploring problem solving patterns and their impact on learning achievement in a blended learning environment. *Computers & Education*, *56*(1), 138–145. doi:10.1016/j.compedu.2010.08.012

Yigitbasioglu, O. (2015). The role of institutional pressures and top management support in the intention to adopt cloud computing solutions. *Journal of Enterprise Information Management*, *28*(4), 579–594. doi:10.1108/JEIM-09-2014-0087

Yılmaz, Y. (2012a). Knowledge management in e-learning practices. *TOJET: The Turkish Online Journal of Educational Technology, 11*(2).

Yılmaz, Y. (2012b). Transition to knowledge society in Turkey: Current state and future perspectives. *Turkish Studies, 13*(3), 509–522. doi:10.1080/14683849.2012.717439

Yin, R. (2003). *Case study research design and methods* (3rd ed.). Thousand Oaks, CA: Sage.

Yoffie, D. B., & Cusumano, M. A. (1999). The competitive dynamics of Internet time. *Harvard Business Review, 77*(1), 70–80. PMID:10345393

YouTube for Press. (n. d.). Retrieved 14 November 2017, from https://www.youtube.com/yt/press/statistics.html

Yuuichi, S., Toshihiro, K., Seisuke, Y., & Hiroshi, N. (2006, July). Web-based rapid authoring tool for LMS quiz creation. In *Information Technology Based Higher Education and Training, 2006. ITHET'06. 7th International Conference on* (pp. 617-620). IEEE. doi:10.1109/ITHET.2006.339676

Zhang, Y. (2007). A teaching case for a distance learning course: Teaching digital image processing. *Journal of Cases on Information Technology, 9*(4), 30–39. doi:10.4018/jcit.2007100103

Zhu, Y., Li, Y., Wang, W., & Chen, J. (2010). What leads to post-implementation success of ERP? An empirical study of the Chinese retail industry. *International Journal of Information Management, 30*(3), 265–276. doi:10.1016/j.ijinfomgt.2009.09.007

Zwicky, F. (1969). *Discovery, invention, research through the morphological analysis*. New York: McMillan.

About the Contributors

Andrew Borchers is currently Professor of Management and Associate Dean of the College of Business at Lipscomb University in Nashville, TN. Before starting this position in 2011, he served for 10 years as an Associate Professor and department head at Kettering University in Flint, MI. Prior to his full-time academic life, Andy spend over 20 years in the automotive industry working for General Motors and Electronic Data Systems. Andy's academic background includes a bachelors in industrial administration from Kettering University (formerly General Motors Institute), an MBA from Vanderbilt and a DBA from Nova Southeastern His research and teaching interests are varied - having taught in five academic units at Lipscomb (College of Business, College of Computing, College of Pharmacy and Health Sciences, Institute for Sustainable Practice and Institute for Conflict Management). Andy's publications include a number of case studies and empirical studies in entrepreneurship, information technology, project management, supply chain management and other areas.

* * *

Abdullah Ibrahim Alkraiji is an Associated Professor in the Department of Information Technology and Communications, King Fahd Security College (KFSC), Riyadh. He holds a BSc in computer science in 1996 from King Saud University (Riyadh, Saudi Arabia), a MSs in computer science in 2002 from Florida Institute of Technology (Melbourne, Florida, US), a PhD in Information Systems (IS) from Loughborough University (Loughborough, UK). During his career, Dr. Alkraiji has participated in various educational and information systems projects and developments. Dr. From a methodological point of view, Dr. Alkraiji's research has generally adopted a qualitative and quantitative approaches to study such subjects of interest in the area of IS including Organizational aspects of information systems, Information Systems Evaluation and Integration, Information Systems for Development and Sociotechnical of IS Adoption and implementation. In doing so, Dr. Alkraiji has contributed to research in the field of information systems with numerous international publications.

Núria Arimany Serrat is a faculty member of the University of Vic - Central University of Catalonia. She received her Ph.D. from the University of Barcelona and her research focuses on the financing of R&D.

Andresa Baptista is a technician in the Mechanical Engineering Department, School of Engineering, Polytechnic of Porto. She holds a Bachelor's degree in Instrumentation Engineering and Industrial Quality, a Bachelor's degree in Mechanical Engineering and a master's degree in Mechanical Engineering -Thermal Energy. She is a member of the Centre for Research and Development in Mechanical Engineering (CIDEM) of the Polytechnic of Porto.

Luis Berges-Muro, PhD Industrial Engineer, has been Vice-Rector of the University of Zaragoza (Spain), and Director of its Engineering Design and Manufacturing Department for more than ten years; consultor to several different companies and/or institutions, currently continues his teaching and research in various fields of Industrial Design and Manufacturing Engineering at the University of Zaragoza, and participates in studies of Degree, Postgraduate and Doctorate at its School of Engineering and Architecture.

Prithvi Bhattacharya is a practitioner-researcher in the field of Strategic Alignment of Business and Information Technology. With a Bachelor of Commerce (Hons) Degree in Accounting and Masters Degree in Information Technology, he has worked as a business analyst and consultant for large, reputed organizations like Accenture and the University of Melbourne. He has also completed a PhD from the University of Melbourne and had almost a decade of University teaching experience, currently holding a faculty position at the Higher Colleges of Technology, UAE. Prithvi has published in several leading conferences like the International Conference of Information Systems and has several leading journal papers.

G. P. Biswas received B.Sc (Engg.) and M.Sc (Engg.) degrees in Electrical & Electronics Engineering and Computer Science & Engineering, respectively. He completed his Ph.D. degree in Computer Science & Engineering from Indian Institute of Technology, Kharagpur, India. He is currently working as a Professor in the Department of Computer Science & Engineering, Indian Institute of Technology (ISM), Dhanbad, Jharkhand, India. He has around 20 years of teaching and research experiences and published more than 100 research articles in Journals, Conferences and Seminar Proceedings. His main research interests include Cryptography, Computer Network and Security, Cellular Automata, VLSI Design.

José Antonio Corral Marfil is a faculty member of the University of Vic - Central University of Catalonia. He holds a Ph.D. from the Autonomous University of Barcelona. Dr. Corral specializes in case analysis of innovative and technological industries.

Marina Duarte is associate professor in the Mechanical Engineering Department, School of Engineering, Polytechnic of Porto. She holds M.Sc. degrees in Mechanical Engineering and in Educational Sciences, and she is a member of the Centre for Research and Development in Mechanical Engineering (CIDEM) of the Polytechnic of Porto as well as a collaborator in the Centre for Research and Intervention in Education (CIIE) of the University of Porto.

İlkim Ecem Emre is research assistant at Marmara University, Business Informatics Department. She earned a BA in Business Informatics from Marmara University in 2015. In 2017, she received her MSc from İstanbul University, Informatics Department. She has studied on data mining techniques for her thesis. Currently, she is PhD student at İstanbul University, Informatics Department. Her research areas are data mining, machine learning, knowledge society, and knowledge management.

Xavier Ferràs Hernández is the Dean of the Faculty of Business and Communication at the University of Vic - Central University of Catalonia. He has a Ph.D. from the University of Barcelona. Dr. Ferràs' research interests include technology ecosystems and disruptive technologies.

Francisco José García-Peñalvo did his undergraduate studies in Computing at the University of Salamanca and University of Valladolid and his Ph.D. at the University of Salamanca. Dr. García-Peñalvo is the head of the research group GRIAL (Research Group Interaction and eLearning). His main research interests focus on eLearning, Computers & Education, Adaptive Systems, Web Engineering, Semantic Web and Software Reuse. He has led and participated in over 50 research and innovation projects. He was Vice Chancellor for Innovation at the University of Salamanca between March 2007 and December 2009. He has published more than 300 articles in international journals and conferences. He has been guest editor of several special issues of international journals (Online Information Review, Computers in Human Behaviour, Interactive Learning Environments...). He is also a member of the program committee of several international conferences and reviewer for several international journals. Now, he is the Editor-in-Chief of the International Journal of Information Technology Research and the Education in the Knowledge Society Journal. Besides he is the coordinator of the multidisciplinary PhD Programme on Education in the Knowledge Society.

Daya Sagar Gupta is pursuing his Ph.D. in Computer Science and Engineering from Indian Institute of Technology (ISM), Dhanbad, India. He received his M. Tech in Computer Application with distinction from Indian Institute of Technology (ISM), Dhanbad, India and B. Tech in Computer Science and Engineering from UPTU Lucknow (UP), India. His research interest includes Cryptography and Information Security.

Elif Kartal is research assistant in İstanbul University Informatics Department. She graduated from İstanbul University Faculty of Science, Department of Mathematics in 2008, She received her MSc from İstanbul University, Informatics Department in 2011 with her thesis entitled Software Project Cost Estimation with Artificial Neural Networks. She has been in Ball State University (USA) as a visiting scholar for two months in 2013. In 2015, she received her PhD from İstanbul University, Informatics Department with her thesis entitled Machine Learning Techniques based on Classification and a Study on Cardiac Risk Assessment. Her research areas are artificial intelligence, machine learning, data mining, web design, programming languages, and e-learning.

Arsla Khan is a research associate at School of Business and Economics, University of Management and Technology (UMT), Pakistan. Her research work is accepted for publications by various International Indexed Journals.

Ignacio López-Forniés, PhD in Engineering Design and Manufacturing program, BEng in Industrial Organization Engineering, BEng in Industrial Technical Engineer from the University of Zaragoza (Spain), BA Hons in Consumer Product Design by Coventry University of Coventry (UK). Product and Graphic Designer, currently Coordinator and teacher of the Degree in Engineering in Industrial Design and Product Development at the University of Zaragoza (Spain). Consultor to several different companies and/or institutions, currently developing various research related to Design Methodologies, and more specifically to Prospective Design and Biomimetics, Ecodesign and Creativity applied to product development process, as well as innovative methodologies for learning Product Design.

Eduardo Manchado, PhD in Engineering Design and Manufacturing program from the University of Zaragoza (Spain), BA in Fine Arts from the Complutense University of Madrid (Spain), BA Hons in 3D Product Design by De Monfort University of Leicester (UK). Product and Graphic Designer, currently Coordinator of the Master Studies in Engineering Product Design and teacher at the Degree in Engineering in Industrial Design and Product Development at the University of Zaragoza (Spain). Consultor to different companies and/or institutions, currently investigating different aspects related to Product Design, and more specifically the fields of Emotional Design, Product Communication, and Product Service System Design, as well as innovative methodologies for learning Product Design.

Petra A. Nylund is an assistant professor at the University of Vic. She has a Ph.D. in Management from IESE and an M.Sc. in Engineering and Business Management from KTH, Stockholm. Her research and teaching focus on different aspects of innovation and strategy. She is currently enthusiastic about the development of innovation ecosystems. She is also an expert in the econometric analysis of panel data. Her work has been published in journals such as The Leadership Quarterly, Technovation, and Journal of Knowledge Management. She has been a business consultant and corporate development manager in the telecoms industry of Africa, Latin America, and Europe.

Zeki Özen is research assistant at İstanbul University, Informatics Department. He graduated from Sakarya University Computer Programming Department in 2002 and İstanbul University, Faculty of Science, Department of Physics in 2008. He received his MSc in 2012 and PhD in 2016 from İstanbul University, Informatics Department. His research areas are source code comparison and similarity, informatics law, relational database systems, machine learning, keystroke dynamics, and e-Learning.

Vasileios Paliktzoglou has several years of experience in universities such as the University of Pisa, the Aegean University, Robert Gordon University, Mediterranean University College, University of Wales and the University of Eastern Finland having an active role as: Researcher, Tutor, Supervisor, Lecturer and Programme Manager. His working know-how varies in the field of research, industry and the military. He has also been involved as IT Teaching Consultant in some of the leading corporate educational institution in Greece, in the Hellenic Army as IT Specialist, and keeping the balance between academia and industry, he has worked as a Project Manager and IT Consultant in UK, Italy and Greece. He is active in research as editor and reviewer in academic journals. He is an author with papers presented at international conferences, and published at academic journals. His research interests are in the fields of innovation and entrepreneurship, social media, web 2.0 and e-learning in which he is actively involved in several international research projects. He has a special interest and involvement in innovation and entrepreneurship, mentoring students with innovative entrepreneurial projects and as an active member of the Entrepreneurship Committee at Bahrain Polytechnic. Currently has an active role in the development of Bahrain Polytechnic as a faculty member.

Gustavo Pinto is a senior technician in the Mechanical Engineering Department, School of Engineering, Polytechnic of Porto. He holds a Bachelor's degree in mechanical engineering and a master's degree in automation, instrumentation and control. He is a member of the Centre for Research and Development in Mechanical Engineering (CIDEM) of the Polytechnic of Porto.

Ana María Pinto Llorente holds PhD in Education and Technology, Bachelor in English as a second language (University of Salamanca) and Bachelor in Education (Pontifical University of Salamanca). She is currently working as a professor in the Department of Education, Organization and Research Methods at the University of Salamanca. Her areas of work are related to technology, innovation and didactics. She has been the Coordinator of the Postgraduate Programme in Bilingual Education and the Director of the Special Course in Bilingual Education in Infant, Primary and Secondary education. Her research interests focus on Second Language, Blended Learning, E-Learning, and Computers & Education. She has an extensive experience of teaching Technology, Linguistics, Phonetics, English, and Technology in face-to-face and Blended Learning Modality. She has published different articles in national and international journals, and presented several papers in national and international conferences.

Abdul Rafay is a Professor of Finance & Accounting and an ex-chairperson Department of Finance at the University of Management & Technology (UMT), Pakistan. His subjects of interest are Financial Technology (FinTech), Corporate Finance & Banking, International Financial Reporting Standards, Financial Derivatives, Financial Restructuring and Corporate Tax Management. He is a Fellow member of the Institute of Chartered Accountants of Pakistan and possesses 21 years experience as a freelance advisor, consultant & trainer to a wide variety of public & private sector national and multinational companies. He has published more than 20 articles in academic and professional journals of repute in addition to contributing a Chapter "Structural Mix of Credit Portfolios in Islamic Banking System: Evidence from a South Asian Economy (2017)" to edited Book "Advances in Islamic Finance, Marketing, and Management: An Asian Perspective" published by Emerald Group Publishing Limited, UK.

Mª Cruz Sánchez-Gómez, Doctor and Bachelor in Education Sciences at the University of Salamanca, and a Master's degree in Speech Pathology at the Pontifical University of Salamanca. She is a Professor in the Department of Education, Organization and Research Methods at the University of Salamanca with a profile in research and diagnostic methods in education. Her areas of work are related to new technologies (she is a researcher of an excellence research group (GRIAL and

IBSAL), groups at risk of social exclusion (coordinator of the researches related to women and disability of INICO, university institute of integration in the community), and educational evaluation (with members of MIDE). In 2008 she received the First National Award of Research of Caja Madrid Social Work and in 2014 the Awards of Innovation and Education Research Perfecta Corselas. She run the Department of Continuing Education at the University of Salamanca from March 2007 until October 2010, and was in charge of all non-formal education at the University of Salamanca: courses, continuing education, summer courses, conferences, seminars and degrees of Salamanca University (specialist, expert, and master). At the moment, Mª Cruz is the head of the Department of Education, Organization and Research Methods at the University of Salamanca.

Antonio Sarasa-Cabezuelo obtained an MSc. degree in Mathematics from theUniversidad Complutense de Madrid (UCM), a BSc. degree in Computer Science from the Universidad Nacional de Educación a Distancia (UNED), MSc. degree in Computer Science from the Universitat Oberta de Catalunya, and PhD. in Computer Science from the Software Engineering and Artificial Intelligence Department (Facultad de Informática, UCM). Currently, he is a Associate Professor at the UCM's Computing Systems and Computation Department. His research is focused on the language-oriented development of XML-processing applications, and on the development of applications in the fields of digital humanities and e-Learning. He is a member of a Research Group on Implementation of Language-Driven Software and Applications (ILSA) at the Universidad Complutense de Madrid. The research results of Prof. Sarasa have been published in over 50 research papers in conferences and book chapters. He collaborated with RED.ES in the construction of AGREGA, the first federated repository of learning objects built in Spain. Currently, he is a member of the AENOR CTN36 committee. Linkedin: https://www.linkedin.com/in/antonio-sarasa-2b86141/?ppe=1.

Index

U

W

Purchase Print, E-Book, or Print + E-Book

IGI Global books can now be purchased from three unique pricing formats: Print Only, E-Book Only, or Print + E-Book. Shipping fees apply.

www.igi-global.com

Recommended Reference Books

ISBN: 978-1-5225-0267-8
© 2016; 477 pp.
List Price: $175

ISBN: 978-1-4666-9519-1
© 2016; 622 pp.
List Price: $235

ISBN: 978-1-5225-0164-0
© 2016; 1,663 pp.
List Price: $1,850

ISBN: 978-1-5225-0034-6
© 2016; 2,429 pp.
List Price: $1,925

ISBN: 978-1-4666-9634-1
© 2016; 866 pp.
List Price: $400

ISBN: 978-1-4666-9680-8
© 2016; 450 pp.
List Price: $300

Looking for free content, product updates, news, and special offers?

Join IGI Global's mailing list today and start enjoying exclusive perks sent only to IGI Global members. Add your name to the list at **www.igi-global.com/newsletters.**

Publishing Information Science and Technology Research Since 1988 — IGI Global DISSEMINATOR OF KNOWLEDGE

www.igi-global.com | Sign up at www.igi-global.com/newsletters | facebook.com/igiglobal | twitter.com/igiglobal

www.ingramcontent.com/pod-product-compliance
Lightning Source LLC
LaVergne TN
LVHW080604300926
849162LV00005B/1039